The Handbook of Environmental Chemistry

Volume 1 Part A

Edited by O. Hutzinger

The Natural Environment and the Biogeochemical Cycles

With Contributions by
P. J. Craig, J. Emsley, D. J. Faulkner, P. M. Huang,
E. A. Paul, M. Schidlowski, W. Stumm, J. C. G. Walker,
P. J. Wangersky, J. Westall, A. J. B. Zehnder, S. H. Zinder

With 54 Figures

Springer-Verlag Berlin Heidelberg GmbH 1980

Professor Dr. Otto Hutzinger
Laboratory of Environmental and Toxicological Chemistry
University of Amsterdam, Nieuwe Achtergracht 166
Amsterdam, The Netherlands

ISBN 978-3-662-13541-9 ISBN 978-3-540-38516-5 (eBook)

DOI 10.1007/978-3-540-38516-5

Library of Congress Cataloging in Publication Data
Main entry under title: The Natural environment and the biogeochemical cycles. (The Handbook of environmental chemistry; v. 1, pt. A–). Includes bibliographies and index. 1. Biogeochemical cycles. 2. Environmental chemistry. I. Craig, Peter, 1944–. II. Series: Handbook of environmental chemistry; v. 1, pt. A–.
QD31.H335 vol. 1, pt. A, etc. [QH344] 574.5'222s
[574.5'222] 80-16608

Originally published by Springer-Verlag Berlin Heidelberg New York in 1980.

Softcover reprint of the hardcover 1st edition 1980

2152/3140-543210

Preface

Environmental Chemistry is a relatively young science. Interest in this subject, however, is growing very rapidly and, although no agreement has been reached as yet about the exact content and limits of this interdisciplinary discipline, there appears to be increasing interest in seeing environmental topics which are based on chemistry embodied in this subject. One of the first objectives of Environmental Chemistry must be the study of the environment and of natural chemical processes which occur in the environment. A major purpose of this series on Environmental Chemistry, therefore, is to present a reasonably uniform view of various aspects of the chemistry of the environment and chemical reactions occurring in the environment.

The industrial activities of man have given a new dimension to Environmental Chemistry. We have now synthesized and described over five million chemical compounds and chemical industry produces about hundred and fifty million tons of synthetic chemicals annually. We ship billions of tons of oil per year and through mining operations and other geophysical modifications, large quantities of inorganic and organic materials are released from their natural deposits. Cities and metropolitan areas of up to 15 million inhabitants produce large quantities of waste in relatively small and confined areas. Much of the chemical products and waste products of modern society are released into the environment either during production, storage, transport, use or ultimate disposal. These released materials participate in natural cycles and reactions and frequently lead to interference and disturbance of natural systems.

Environmental Chemistry is concerned with reactions in the environment. It is about distribution and equilibria between environmental compartments. It is about reactions, pathways, thermodynamics and kinetics. An important purpose of this Handbook is to aid understanding of the basic distribution and chemical reaction processes which occur in the environment.

Laws regulating toxic substances in various contries are designed to assess and control risk of chemicals to man and his environment. Science can contribute in two areas to this assessment; firstly in the area of toxicology and secondly in the area of chemical exposure. The available concentration ("environmental exposure concentration") depends on the fate of chemical compounds in the environment and thus their distribution and reaction behaviour in the environment. One very important contribution of Environmental

Chemistry to the above mentioned toxic substances laws is to develop laboratory test methods, or mathematical correlations and models, that predict the environmental fate of new chemical compounds. The third purpose of this Handbook is to help in the basic understanding and development of such test methods and models.

The last explicit purpose of the Handbook is to present, in concise form, the most important properties relating to environmental chemistry and hazard assessment for the most important series of chemical compounds.

At the moment three volumes of the Handbook are planned. Volume 1 deals with the natural environment and the biogeochemical cycles therein, including some background information such as energetics and ecology. Volume 2 is concerned with reactions and processess in the environment and deals with physical factors such as transport and adsorption, and chemical, photochemical and biochemical reactions in the environment, as well as some aspects of pharmacokinetics and metabolism within organisms. Volume 3 deals with anthropogenic compounds, their chemical backgrounds, production methods and information about their use, their environmental behaviour, analytical methodology and some important aspects of their toxic effects. The material for volume 1, 2 and 3 was each more than could easily be fitted into a single volume, and for this reason, as well as for the purpose of rapid publication of available manuscripts, all three volumes were divided in the parts A and B. Part A of all three volumes is now being published and the second part of each of these volumes should appear about six months thereafter. Publisher and editor hope to keep materials of the volumes one to three up to date and to extend coverage in the subject areas by publishing further parts in the future. Plans also exist for volumes dealing with different subject matter such as analysis, chemical technology and toxicology, and readers are encouraged to offer suggestions and advice as to future editions of "The Handbook of Environmental Chemistry".

Most chapters in the Handbook are written to a fairly advanced level and should be of interest to the graduate student and practising scientist. I also hope that the subject matter treated will be of interest to people outside chemistry and to scientists in industry as well as government and regulatory bodies. It would be very satisfying for me to see the books used as a basis for developing graduate courses in Environmental Chemistry.

Due to the breadth of the subject matter, it was not easy to edit this Handbook. Specialists had to be found in quite different areas of science who were willing to contribute a chapter within the prescribed schedule. It is with great satisfaction that I thank all 52 authors from 8 contries for their understanding and for devoting their time to this effort. Special thanks are due to Dr. F. Boschke of Springer for his advice and discussions throughout all stages of preparation of the Handbook. Mrs. A. Heinrich of Springer has significantly contributed to the technical development of the book through her conscientious and efficient work. Finally I like to thank my family, students and colleagues for being so patient with me during several critical phases of preparation for the Handbook, and to some colleagues and the secretaries for technical help.

I consider it a privilege to see my chosen subject grow. My interest in Environmental Chemistry dates back to my early college days in Vienna. I received significant impulses during my postdoctoral period at the University of California and my interest slowly developed during my time with the National Research Council of Canada, before I could devote my full time to Environmental Chemistry, here in Amsterdam. I hope this Handbook may help deepen the interest of other scientists in this subject.

Amsterdam, May 1980 O. Hutzinger

Contents

The Atmosphere

M. Schidlowski

The Hydrosphere

J. Westall and W. Stumm

Chemical Oceanography

P. J. Wangersky

Chemical Aspects of Soil

E. A. Paul and P. M. Huang

The Oxygen Cycle

J. C. G. Walker

The Sulfur Cycle

A. J. B. Zehnder and S. H. Zinder

The Phosphorus Cycle

J. Emsley

Metal Cycles and Biological Methylation

P. J. Craig

Natural Organohalogen Compounds

D. J. Faulkner

Volume 2, Part A: **Reactions and Processes**

Transport and Transformation of Chemicals: A Perspective.
G. L. Baughman and L. A. Burns
Transport Processes in Air. *J. W. Winchester*
Solubility, Partition Coefficients, Volatility and Evaporation Rates.
D. Mackay
Adsorption Processes in Soil. *P. M. Huang*
Sedimentation Processes in the Sea. *K. Kranck*
Chemical and Photo Oxidation. *T. Mill*
Atmospheric Photochemistry. *T. E. Graedel*
Photochemistry at Surfaces and Interphases. *H. Parlar*
Microbial Metabolism. *D. T. Gibson*
Plant Uptake, Transport and Metabolism. *I. N. Morrison and A. S. Cohen*
Metabolism and Distribution by Aquatic Animals. *V. Zitko*
Laboratory Microecosystems. *A. R. Isensee*
Reaction Types in the Environment. *C. M. Menzie*
Subject Index

Volume 3, Part A: **Anthropogenic Compounds**

Mercury. *G. Kaiser and G. Tölg*
Cadmium. *U. Förstner*
Polycyclic Aromatic and Heteroaromatic Hydrocarbons. *M. Zander*
Fluorocarbons. *J. Russow*
Chlorinated Paraffins. *V. Zitko*
Chloroaromatic Compounds Containing Oxygen. *C. Rappe*
Organic Dyes and Pigments. *E. A. Clarke and R. Anliker*
Inorganic Pigments. *W. Funke*
Radioactive Substances. *G. C. Butler and C. Hyslop*
Subject Index

List of Contributors

Dr. P. J. Craig
School of Chemistry
Leicester Polytechnic
Leicester, LE1 9BH, U. K.

Dr. J. Emsley
Dept. of Chemistry
King's College, Strand
London WC2R 2LS, U. K.

Dr. D. J. Faulkner
Scripps Institution of Oceanography
La Jolla, CA 92093, USA

Dr. P. M. Huang
Dept. of Soil Science
University of Saskatchewan
Saskatoon, S7N OWO Canada

Dr. E. A. Paul
Dept. of Soil Science
University of Saskatchewan
Saskatoon, S7N OWO Canada

Prof. M. Schidlowski
Max-Planck-Institut für Chemie
(Otto-Hahn-Institut)
D-6500 Mainz
Federal Republic of Germany

Prof. W. Stumm
EAWAG
CH-8600 Dübendorf, Switzerland

Dr. J. C.G. Walker
College of Engineering
University of Michigan
Ann Arbor, MI 48109, USA

Dr. P. J. Wangersky
Dept. of Oceanography
Dalhousie University
Halifax, N. S., B3H 4J1 Canada

Dr. J. Westall
EAWAG
CH-8600 Dübendorf, Switzerland

Dr. A. J. B. Zehnder
EAWAG
CH-8600 Dübendorf, Switzerland

Dr. S. H. Zinder
Div. of Environment and Nutritional
Sciences
School of Public Health
University of California
Los Angeles, CA 90024, USA

The Atmosphere

M. Schidlowski
Max-Planck-Institut für Chemie (Otto-Hahn-Institut)
D-6500 Mainz, Federal Republic of Germany

The atmosphere forms a gaseous envelope surrounding the Earth with decreasing density up to a height of some 500 km where it passes gradually into space. It is composed of a mixture of gases called air constituting a compressible fluid tied to the planet by gravitational attraction. As a volatile intermediate between the solid surface and outer space, the atmosphere exercises crucial controls on surface conditions, notably
(1) by absorbing the bulk of ultraviolet solar radiation and
(2) by its pronounced greenhouse properties,
both effects ultimately making planet Earth hospitable to life. Apart from water, air is the principal medium flowing around and sustaining, as part of the natural environment, the terrestrial biosphere. With its constituents (including water vapor) kept in permanent circulation by the exposure to solar heat, the atmosphere is also responsible for weathering and erosion processes which have shaped the face of the Earth since the oldest geological past.

The atmospheric gas mixture exercises a pressure on the ground which is equal to the weight of the vertical gas column per unit cross-sectional area. Hence, the average atmospheric pressure at sea level of $1{,}012 \times 10^6$ dyn cm^{-2} ($= 1{,}012$ mb) corresponds to the weight of the vertical column of air overlying an area of 1 cm^2 of the Earth's surface. Dividing this pressure by the acceleration due to gravity (980 cm s^{-2}), we obtain a value of 1030 g for the mass of the air in this column. With the total surface area of the Earth being 5.1×10^{18} cm^2, we arrive at a total mass of 5.27×10^{21} g for the terrestrial atmosphere.

Chemical Composition

As a result of a complex interplay of mixing and diffusion processes (along with chemical reactions which affect the life times of individual gases), the densities of most atmospheric constituents vary with altitude. Up to about 100 km the atmosphere is reasonably well mixed and homogeneous ("homosphere"); above this altitude the lighter gases become progressively more abundant as their density profiles are determined by diffusion rather than mixing processes ("heterosphere"). For most practical purposes, however, the mixture of gases prevailing in the 10–15 km thick atmospheric bottom layer, the *troposphere,* provides a reasonably accurate sample of average air. With about 80% of the mass of the atmosphere concentrated in the troposphere and the latter well mixed, the composition of ground level air comes very close to the overall composition of the atmosphere. When dealing with the atmosphere in environmental terms, i.e., as a medium for sustaining life in the widest sense, it is actually the composition of the troposphere which is of concern.

With the exception of a few trace constituents, tropospheric air consists of a virtually constant mixture of gases. Table 1 gives a compilation of the main components and the principal trace constituents. It is obvious from this summary that the three most abundant non-variable or "permanent" gases of the atmosphere (nitrogen, oxygen, argon) account for more than 99.9% of total atmospheric composition.

The most conspicuous variable in the lower atmosphere is *water vapor* whose concentration is highest in tropical air and at a minimum in cold continental air masses of the circumpolar regions. The presently observed *carbon dioxide* content of ~325 ppm has undergone an increase from some 290 ppm since the turn of the century as a result of fossil fuel burning (see, inter alia [22, 23]). With a rate of increase between 0.2 and 0.7% per year, this rise

Table 1. Composition of the atmosphere [20, 46, 48]

Constituent	Formula	Abundance by volume (percent, ppm, ppb)
Nitrogen	N_2	78.084 ± 0.004%
Oxygen	O_2	20.948 ± 0.002%
Argon	Ar	0.934 ± 0.001%
Water vapor	H_2O	Variable (%–ppm)
Carbon dioxide	CO_2	325 ppm
Neon	Ne	18 ppm
Helium	He	5 ppm
Krypton	Kr	1 ppm
Xenon	Xe	0.08 ppm
Methane	CH_4	2 ppm
Hydrogen	H_2	0.5 ppm
Nitrous oxide	N_2O	0.3 ppm
Carbon monoxide	CO	0.05–0.2 ppm
Ozone	O_3	Variable (0.02–10 ppm)
Ammonia	NH_3	4 ppb
Nitrogen dioxide	NO_2	1 ppb
Sulfur dioxide	SO_2	1 ppb
Hydrogen sulfide	H_2S	0.05 ppb

Sources: Junge, 1972; US Standard Atmosphere, 1976; Walker, 1977 may entail the hazard of inadvertent climate modifications as CO_2 is an important infrared absorber [2, 25, 42, 50, and others]. Apart from ozone and a conspicuous group of inert gases other than argon, the trace constituents of the ppm-range are, to a large extent, of biologic origin having often a strong anthropogenic component [e.g., carbon monoxide (automobile exhausts) or nitrous oxide (nitrate fertilizers)]. The same applies to the trace gases of the ppb-range. Here, notably a major contribution to the SO_2 budget is clearly anthropogenic, stemming from the combustion of coal and oil.

Because of its environmental significance as principal UV-absorber, *ozone* is a crucial component among the atmospheric trace gases. It is formed in the stratosphere by dissociation of molecular oxygen

$$O_2 \xrightarrow{h\nu} O + O$$

and subsequent recombination of the atomic oxygen with remaining O_2, this giving rise to triatomic ozone

$$O_2 + O \rightarrow O_3.$$

At the same time, ozone is being destroyed by photodissociation

$$O_3 \xrightarrow{h\nu} O_2^+ + O$$

and by reaction with both atomic oxygen

$$O_3 + O \rightarrow O_2 + O_2.$$

and stratospheric NO_x-components, e.g.,

$$O_3 + NO \rightarrow NO_2 + O_2.$$

These processes result in a photochemical equilibrium which leads to the buildup of a stationary ozone concentration with a maximum in an altitude between 22 and 23 km. From the sites of production in the stratosphere, part of the ozone gets into the troposphere and is finally destroyed near the ground. The overall mixing ratio of ozone in the atmosphere varies between <1 ppmv to about 10 ppmv.

It has been pointed out that the intricate photochemical equilibrium responsible for maintaining the stationary O_3-level of the atmosphere may be seriously disturbed by several anthropogenic trace gases, notably

(1) Nitrogen oxides released by supersonic transport and decomposition of nitrate fertilizers, and

(2) industrially produced chlorofluoromethanes (such as CF_2Cl_2 or $CFCl_2$).

These gases are destroyed photochemically in the upper stratosphere, with their breakdown products (NO_x and Cl-radicals) increasing natural destruction rates of ozone

$$O_3 + NO \rightarrow NO_2 + O_2$$

$$O_3 + Cl \rightarrow ClO + O_2,$$

thus tending to fix photostationary O_3-concentrations at considerably lower levels than the present one [17, 18, 30, 49, and others]. Increasing pollution of the environment is, therefore, apt to affect severely both the radiation balance and the chemistry of the atmosphere [9, 11].

With the exception of the noble gases, all atmospheric constituents undergo cycles which are often biologically mediated. Their virtually constant concentrations or "mixing ratios" are sustained by a delicate balance between sources and sinks. The cycles of nitrogen, oxygen and carbon dioxide as well as of a number of minor constituents (CH_4, H_2, N_2O, CO) are, to a large extent, biological or microbiological [19, 20]. It is obvious from a quantitative evaluation of "mixed" cycles that turnover rates in the biological branches are considerably larger than in the inorganic ones. In the case of several trace gases, sources are mainly (micro) biological while sinks are physico-chemical (e.g., H_2S, NH_3, CH_4). Atmospheric constituents with almost pure inorganic cycles are water vapor and ozone. With the impact of life on the gaseous shell of the Earth thus being profound, the bulk of the present atmosphere may be aptly regarded as an integral part of the biosphere.

Only the lightest atmospheric gases, hydrogen and helium, have a chance to leave the Earth's gravity field ("Jeans escape"). In the case of hydrogen, the escape rate is limited by its upward diffusion which can be estimated with fair accuracy. As this upward transport is independent of atmospheric background parameters, the present escape rate of 10^8 atoms cm^{-2} s^{-1} or some 2.5×10^{10} g hydrogen per year [16, 48] should also hold for the geological past. The calculated total loss to space of 1.1×10^{20} g hydrogen over the 4.5×10^9 yr of the Earth's history has probably accounted for the only change of the overall oxidation state of the atmosphere-hydrosphere-crust system since the formation of the planet.

Origin of the Atmosphere

During the last two decades numerous space probes have supplied a wealth of detailed information on the composition and structure of the atmospheres of our neighbor planets. A comparison of these planetary atmospheres with the gaseous shell of the Earth shows that the terrestrial atmosphere is indeed exceptional. Whereas the atmospheres of Mars and Venus contain between 93 and 98% carbon dioxide, the Earth's atmosphere is basically made up of nitrogen and oxygen accounting together for some 99% of its total composition.

Since there is reason to believe that all terrestrial planets started off with gaseous envelopes of about identical composition, we may infer that the Earth's original atmosphere did not differ too much from the reducing gas shells of the other planets. Accordingly, our modern oxygenic atmosphere must be the ultimate result of an evolution which had covered major parts of the Earth's history. It is well established by now that the totality of terrestrial life ("biosphere") has played a decisive role in this evolution, notably in the buildup of an atmospheric oxygen reservoir.

The Primitive Atmosphere

Apart from the evidence furnished by comparative planetology in favor of a primary anoxygenic atmosphere, the very fact that life originated on Earth constitutes conclusive proof of reducing conditions on the juvenile planet. Such conditions were shown to be a necessary prerequisite for abiotic synthesis of organic compounds and early chemical evolution [27, 29, 35]. With these requirements placing strong limits on the ancient environment, there has long been a broad consensus on an oxygen-free atmosphere at the dawn of the Earth's history.

On the other hand, however, there have been differences of opinion regarding the definite oxidation state of the oldest atmosphere, centering principally on the question whether highly reduced gases such as methane (CH_4) and ammonia (NH_3), or "pre-oxidized" species like carbon dioxide (CO_2) and nitrogen (N_2), were the dominant constituents. As hydrogen is the most abundant element in the cosmos, it was just logical to assume that chemical equilibria during the condensation of solar matter to planets should have favored the formation of hydrogen compounds like CH_4 and NH_3 [28, 45]. This conclusion was necessarily based on the premise that the Earth had formed from largely *undifferentiated* solar matter.

A most important source of information on the primitive terrestrial atmosphere are the noble gases, notably the primordial noble gases produced during the cosmic process of nucleosynthesis (in contrast to the radiogenic species formed *within* the Earth as a result of radioactive decay of elements such as U and K). It is known that primordial rare gases are depleted on Earth by factors between 10^{-7} and 10^{-11} as compared to their cosmic abundances [5, 43]. While, for instance, helium, neon and argon figure among the ten most abundant elements of the cosmos, the group as a whole is so scarce on Earth as to have escaped detection until the end of the last century.

This conspicuous depletion of noble gases has often been explained in terms of a removal of the primordial gas fraction from the newly-formed planet as a result of some "solar gale sweeping" during the passage of the sun through a T Tauri phase. There is increasing evidence, however, that major element fractionations had occurred already in the circumsolar region *prior* to the accretion of planetary bodies, i.e., *the noble gas depletion as presently observed on Earth was probably established already in those parts of the cosmic gas cloud from which the protoplanet grew* [1]. Primordial solid condensates within this chemically differentiated cloud of cosmic dust have probably served as carriers of occluded gases which were the ultimate source of the terrestrial volatiles now mainly concentrated in the Earth's oceans and atmosphere. During gravitational accretion and impacting as well as by subsequent volcanic degassing the bulk of these gases were transferred to, and assembled at, the planetary surface.

According to this model, the scarcity of terrestrial rare gases is primarily due to pre-accretional element fractionations in the circumsolar plasma resulting in a substantial depletion of these gases in the parent dust cloud of the proto-Earth. There is reason to believe that other volatile elements were depleted in this cloud in corresponding proportions, with the Earth thus accreting from material generally low in volatiles *inclusive of hydrogen.* This would place limits on the abundance of hydrogen on the primitive Earth and thereby on the oxidation state of the early terrestrial degassing products in general.

It has been argued that a highly reduced gas phase (with ample CH_4 and NH_3) could have been discharged only from a highly reduced mantle which, in such case, should have contained metallic iron [48]. The Fe^{3+}/Fe^{2+} ratio observed in mantle rocks is, however, high enough as to exclude the existence of a metallic iron phase; moreover, there is no evidence that the overall oxidation state of the mantle has markedly changed since the establishment of the Earth's outer silicate shell. Hence, the volatiles released at the very start of the degassing process should not have been more reduced than present-day volcanic emanations equilibrated with "primary" (i.e., basaltic) magmas [26]. Accordingly, the gaseous shell of the juvenile Earth should have come close in composition to modern volcanic gases, a conclusion already arrived at by Rubey [33, 34] from geochemical balance considerations.

If we accept that the oldest terrestrial atmosphere has reflected the average composition of volcanic emanations, then water vapor and CO_2 should have made up the bulk of the latter, followed by a suite of minor (and partially more reduced) species like N_2, H_2, H_2S, SO_2, CO, CH_4, NH_3, HCl, and HF. Altogether, these minor constituents should not have accounted for more than perhaps a few volume percent. The reducing character of this mixture of gases was mainly determined by its hydrogen content which averages about 0.5% in modern volcanic emanations.

Water Vapor and Runaway Greenhouse Effect

There is no doubt that the water vapor degassed was just a transient component of the ancient atmosphere (as it is today) as a result of being subjected to rapid condensation. We may exclude with fair accuracy that the Earth was ever veiled in a thick steam atmosphere as this would have given rise to "runaway" greenhouse conditions (Fig. 1). With water vapor strongly absorbing in the infrared, its continuous release *without* concomitant condensation would have steadily increased surface temperatures. In case the curve of increasing water vapor pressure had not intersected the saturation line, more and more water vapor would have accumulated in the atmosphere, making the temperature rise irreversible. The ultimate result of such "runaway greenhouse" are thermal regimes as presently prevailing on Venus with temperatures around 750°K.

Hence, temperatures on the primitive Earth must have been in the range favorable for a condensation of water vapor and a subsequent formation of oceans. There is hardly any doubt that the principal requirement for fixing initial surface temperatures within this range was an optimum distance from the sun. With Venus closer to the sun, surface temperatures apparently were so high here that the track of rising water vapor pressure failed to intersect the saturation line (cf. Fig. 1). Furthermore, the subsequent removal from the

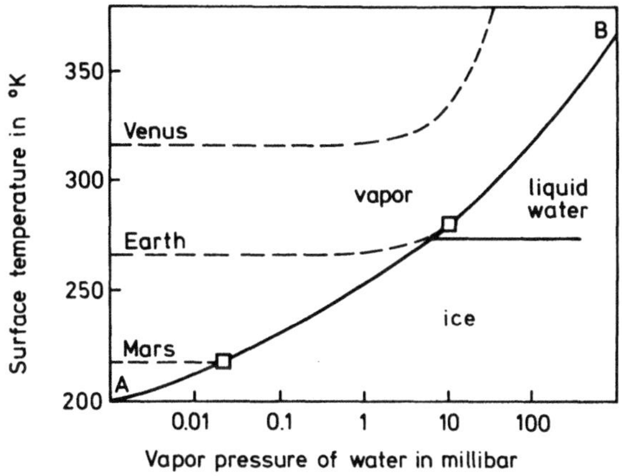

Fig. 1. The runaway greenhouse effect after Goody and Walker [14]. Dashed curves show rise of initial surface temperatures on Mars, Earth and Venus as a function of growing water vapor pressure by which the infrared opacity of the planetary atmospheres is progressively increased. The temperature rise is halted when the curves of water vapor pressure intersect the saturation line (solid line *A–B*) and vapor is removed as either liquid water (Earth) or ice (Mars). Because of the small distance of Venus from the sun, initial surface temperatures were too high here as to allow an intersection of the two curves. Hence, the temperature of the atmosphere steadily increased with increasing water vapor pressure. (Published with permission of Prentice-Hall, Inc., Englewood Cliffs, N.J.).

terrestrial atmosphere of carbon dioxide as another strong infrared absorber was contingent upon a previous formation of a hydrosphere. With CO_2 soluble in water and consequently forming bicarbonate and carbonate ions susceptible to precipitation as sedimentary carbonate, the system

$$CO_{2\,(gas)} \rightleftharpoons CO_{2\,(aq)} \rightleftharpoons HCO^-_{3\,(aq)} \rightleftharpoons CO^{2-}_{3\,(aq)} \rightleftharpoons CaCO_{3\,(solid)} \tag{1}$$

acts as a "pump" transferring atmospheric CO_2 to the crust. In this way, about 80% of the carbon dioxide originally degassed from the mantle has come to be stored in the sedimentary shell, the balance being tied up as fossil organic carbon in the same reservoir, with just roughly one per mil of the total residing in the atmosphere-ocean system in a steady state. As a result of the limits imposed by both the runaway greenhouse and the buffer capacity of (1), we may assume that the water vapor and carbon dioxide contents of the primitive atmosphere were approximately equal to their present values.

It is obvious, therefore, that the initial stages of atmospheric evolution were governed by a cosmic chance parameter, i.e., the distance of the Earth from the sun. A smaller distance resulting in higher initial surface temperatures on the planet would have precluded the condensation of water vapor and a subsequent transfer of the bulk of atmospheric carbon dioxide to the crust. It can be shown that the terrestrial atmosphere would possess a CO_2 content on the order observed on Venus (5.3×10^{23} g) if the totality of carbon presently stored in the sedimentary shell (about 5.5×10^{22} g) were released to the atmosphere again. In this case, we would obtain an atmospheric reservoir of 2.0×10^{23} g CO_2, while the present CO_2-burden of the atmosphere-ocean system is on the order of 10^{20} g only (with just 2.6×10^{18} g stored in the atmosphere). With all sedimentary carbon converted to CO_2, the present nitrogen content of the atmosphere of 78% (3.9×10^{21} g) would account for about 1% of total atmospheric composition only, thus approaching the mixing ratios of N_2 reported for the atmospheres of Mars and Venus.

Origin of Atmospheric Oxygen

A comparison of the main constituents of the Earth's gaseous shell (Table 1) with the probable composition of the primitive atmosphere shows that the basic difference is the presence of free oxygen. Nitrogen as the most abundant gas can be expected to have accumulated over the whole of the Earth's history as have the rare gases (notably argon). As pointed out before, it is fairly certain that the carbon dioxide content was kept approximately constant as from the very start of the terrestrial degassing process due to the buffering capacity of the carbon dioxide-bicarbonate-carbonate system, with marine silicate equilibria most probably responsible for sustaining the slightly alkaline character of seawater favorable for carbonate precipitation [15, 41]. The crucial problem of atmospheric evolution is, therefore, the buildup of modern partial pressures of molecular oxygen.

Since equilibrium concentrations of molecular oxygen in the volatile fraction of basaltic magmas are on the order of 10^{-7} to 10^{-8} atm only [26], free oxygen cannot stem from volcanic exhalations. Hence, we have to look for a non-geologic source, supplying this gas by splitting gaseous oxides such as water vapor and carbon dioxide in a thermodynamic "uphill" reaction. *With solar radiation providing the most important energy source for such process, the presence of O_2 in the terrestrial atmosphere must be ultimately due to a photochemical effect.* Here, we may distinguish between an inorganic and an organic effect.

An *inorganic photochemical effect* is provided by photolysis of water vapor and carbon dioxide in the upper atmosphere by energy-rich solar radiation ($\lambda < 2000$ Å) resulting, inter alia, in the formation of oxygen and hydrogen as breakdown products. Since the light hydrogen may escape from the Earth's gravity field, this process will lead to a net accumulation of oxygen in the atmosphere. The oxygen equivalent of the present loss rate of hydrogen to space amounts to some 2×10^{11} g O_2 per year which is negligible as compared to turnover rates of oxygen in the biological cycle which are on the order of 10^{17} g yr^{-1}. An integration of this inorganic oxygen source over the 4.5×10^9 yr of the Earth's history gives 9×10^{20} g O_2 which is less than 3% of the present oxygen budget of photosynthetic origin (cf. budget column in Fig. 2).

Although quantitatively unimportant, this process was certainly the oldest terrestrial source of free oxygen. According to Berkner and Marshall [4], photolytic water splitting may have sustained oxygen pressures in the prebiological atmosphere of about 10^{-3} of the present level. As this figure is based, however, on an equality of the rates of photolysis and hydrogen escape (which is not justified because of an extensive recombination of the disrupted water molecules), it is likely to be too high by several orders of magnitude. It has been proposed that the oxidation state of the primitive atmosphere was most probably the result of an interplay between the photolytic oxygen source and the rate of release of reduced volcanic gases, notably hydrogen [48].

Because of the insufficiency of this inorganic process, the buildup of a free oxygen reservoir must have hinged upon the appearance of an *organic photochemical process* which was subsequently provided by the biologically mediated reduction of CO_2 to carbohydrates. This reduction is carried out most expediently by green plants and certain blue-green algae (cyanophytes). From CO_2 and water these forms are able to synthesize organic matter, with molecular oxygen released as a metabolic by-product:

$$2\,H_2O^* + CO_2 \xrightarrow{h\nu} CH_2O + H_2O + O_2^* \qquad \Delta G_o' = +112.5\ \text{kcal} \tag{2}$$

Quantitatively, this process of green plant photosynthesis (in which water acts as electron donor for the reduction of CO_2) is the most important biochemical reaction taking place on Earth. It is, therefore, reasonable to assume that the oxygen content of the terrestrial atmosphere is ultimately due to photosynthetic O_2-production. As will be detailed below, there is reason to believe that this form of photosynthesis had commenced as early as 3.7×10^9 yr ago, the oldest oxygen producers having been cyanophytes [8, 21, 39].

Evolution of the Photosynthetic Oxygen Budget

A working model for the quantitative evolution of the terrestrial oxygen budget of photosynthetic origin may be obtained by combining Eq. (2) with a terrestrial carbon isotope mass balance and a plausible concept for the accumulation of the stationary sedimentary mass through geologic time [36, 38]. The basic premises underlying this approach can be summarized as follows:

(a) *Stoichiometry of the Photosynthesis Reaction.* Assuming that terrestrial free oxygen stems from either cyanophytic or green plant photosynthesis, the stoichiometry of Eq. (2) must necessarily reflect quantitative relationships between the amount of carbon fixed as organic matter and the corresponding oxygen equivalent released. It is obvious from (2) that for each carbon atom incorporated in organic matter one O_2-molecule is being released to the environment. As is known from a census of the terrestrial carbon cycle, the largest depository of organic carbon (C_{org}) is within the Earth's sedimentary shell containing, on average, 0.5% C_{org} [32]. With the sedimentary mass totalling about 2.4×10^{24} g [13], we obtain a sedimentary C_{org} reservoir of 1.2×10^{22} g, exceeding the totality of carbon contained in the living biomass ($\sim 10^{18}$ g) by about 4 orders of magnitude. If we could trace, therefore, through geologic time the reservoir of sedimentary organic carbon, we might readily calculate the photosynthetic oxygen equivalent. To achieve this goal we have to resort to the following approaches (b) and (c).

(b) *Terrestrial Carbon Isotope Mass Balance.* Carbon is stored in the sedimentary shell in two different forms, i.e.,

(1) as *organic carbon* (mostly finely dispersed "kerogen" constituents in shales; rarely coal or oil) and
(2) as *carbonate carbon* (in limestone and dolomite).

A ^{13}C mass balance can be utilized to determine the partitioning of total sedimentary carbon between the C_{org} and C_{carb} reservoirs. From the isotopic composition of deep-seated carbon (diamonds, carbonatites) it is commonly inferred that primordial mantle carbon has an average $\delta^{13}C$ value close to –5‰ [PDB][1]. Since there is no reason to assume that an isotope fractionation has occurred during the transfer of this mantle carbon (in the form of CO_2) to the crust, the value of –5‰ should also hold for average crustal (and thus sedimentary) carbon. As a result of a kinetic fractionation effect governing biological carbon fixation, the sedimentary carbon reservoir as a whole has, however, been subjected to an isotopic disproportionation, resulting in the formation of a partial reservoir which is isotopically lighter (C_{org} with $\delta^{13}C = -25 \pm 5$‰) and another partial reservoir which is isotopically heavier (C_{carb} with $\delta^{13}C = 0.0 \pm 2.5$‰) than average crustal carbon.

It has been shown that the isotopic fractionation between the sedimentary C_{carb} and C_{org} reservoirs was always close to 25‰ as from the start of the sedimentary record 3.7×10^9 yr ago [10, 39]. With this fractionation about fixed, and an average $\delta^{13}C$ value of -5‰ for total sedimentary carbon inherited from the mantle, the relative proportions of the C_{org} and C_{carb} reservoirs must be coupled with their respective $\delta^{13}C$-averages by the following mass balance equation:

$$\bar{\delta}^{13}C_{prim} = R\bar{\delta}^{13}C_{org} + (1-R)\bar{\delta}^{13}C_{carb}. \quad (3)$$

Here, C_{prim} denotes primordial mantle carbon, R the ratio of organic carbon to total sedimentary carbon $[C_{org}/(C_{org}+C_{carb})]$ and $\bar{\delta}$ average δ-values of the respective carbon reservoirs. Within the constraints of Eq. (3), the isotopic compositions of the sedimentary C_{org} and C_{carb} reservoirs reflect a definite value of R, that is, the relative proportion of C_{org} within the total sedimentary carbon reservoir. With $\bar{\delta}^{13}C_{prim} = -5$‰, $\bar{\delta}^{13}C_{org} = -25$‰, $\bar{\delta}^{13}C_{carb} = 0$‰ and C_{prim} set $=1$, we get $R = 0.2$, indicating that C_{org} makes up roughly one-fifth (or 20%) of the total sedimentary carbon reservoir. As $\delta^{13}C_{org}$ and $\delta^{13}C_{carb}$ values show relatively little variation as a function of geologic time [38, 47], organic carbon should have always accounted for about 20% of total sedimentary carbon as from the very start of the sedimentary record some 3.7×10^9 yr ago.

(c) *Evolution of Sedimentary Mass Through Geologic Time*. To convert this basically constant ratio of C_{org}/C_{carb} into an absolute figure for the sedimentary C_{org} reservoir, we have to introduce this ratio into a model for the growth of the stationary sedimentary mass as a function of time [24]. A

1 $\delta^{13}C$ values are a conventional means of expressing $^{13}C/^{12}C$ ratios, indicating differences in ^{13}C content (in permil) relative to a standard. The standard mostly used is Peedee belemnite carbonate (PDB)

reasonable approximation for the stationary sedimentary mass M existing at time t is given by

$$M_t \cong M_{tp}(1-e^{-\lambda t}) \quad [g], \qquad (4)$$

where $M_{tp} = 2.4 \times 10^{24}$ g is the present sedimentary mass (with $tp = 4.5 \times 10^9$ yr being the age of the Earth), $\lambda = 1.16 \times 10^{-9}$ yr^{-1} a plausible value for the time-averaged terrestrial degassing constant, and t the time elapsed since the Earth's formation (i.e., the start of the terrestrial degassing process). According to the "principle of geochemical uniformitarianism" following from the constraints of the overall weathering balance [12], M_t should have always contained about 3% *total* carbon ($\Sigma C_{org} + C_{carb}$) as does the present sedimentary mass [32]. With the carbon isotope record indicating that about 20% of this total carbon were always organic carbon, we may calculate the stationary reservoir of photosynthetic oxygen at time t [$(O_2)_t$] according to Eqs. (2) and (4), getting

$$(O_2)_t \cong 0.53\, \varepsilon M_{tp}(1-e^{-\lambda t}) \quad [g], \qquad (5)$$

with $\varepsilon = 0.03$ being the fraction of total carbon in the sedimentary shell and 0.53 a stoichiometric conversion factor. Fig. 2 gives a graphic representation of the increase of the photosynthetic oxygen budget through time according to this function, with $(O_2)_t$ expressed as fraction of the present oxygen reservoir $(O_2)_{tp} = 0.53\varepsilon M_{tp}$ which is set = 1.

Implications of the Oxygen Model

According to the curve of oxygen evolution presented in Fig. 2, a photosynthetic oxygen reservoir of perhaps 50–60% of the present one should have existed already at the start of the sedimentary record, rising to 95% and more at about 2×10^9 yr ago and later. If the partitioning of this total oxygen between the "bound" and the "free" oxygen reservoirs had been the same as today (with ~4% of the total stored as free oxygen in the atmosphere, cf. budget column in Fig. 2), then the O_2-content of the ancient atmosphere should have increased proportionately, coming close to present levels already some 3×10^9 yr ago. Such inferences are, however, incompatible with the paleontological and geological evidence compiled in the lower part of Fig. 2 which indicates a virtually anoxic environment prior to 2×10^9 yr ago. Hence, we must necessarily assume that *the rise of free oxygen in the atmosphere was not primarily related to the process of photosynthetic oxygen production as such (which had apparently started long before the buildup of an atmospheric O_2-reservoir), but was rather a problem of the partitioning of oxygen between the principal geochemical reservoirs.* If the "model" curve of oxygen evolution shown in Fig. 2 is basically correct, then an almost modern reservoir of total photosynthetic oxygen must have been coupled with negligible O_2-pressures in the atmosphere during the early history of the Earth.

This would imply that the ancient oxygen cycle was practically short-circuited, with very effective O_2-consuming reactions responsible for an instantaneous sequestration by reducing constituents of the crust of all photo-

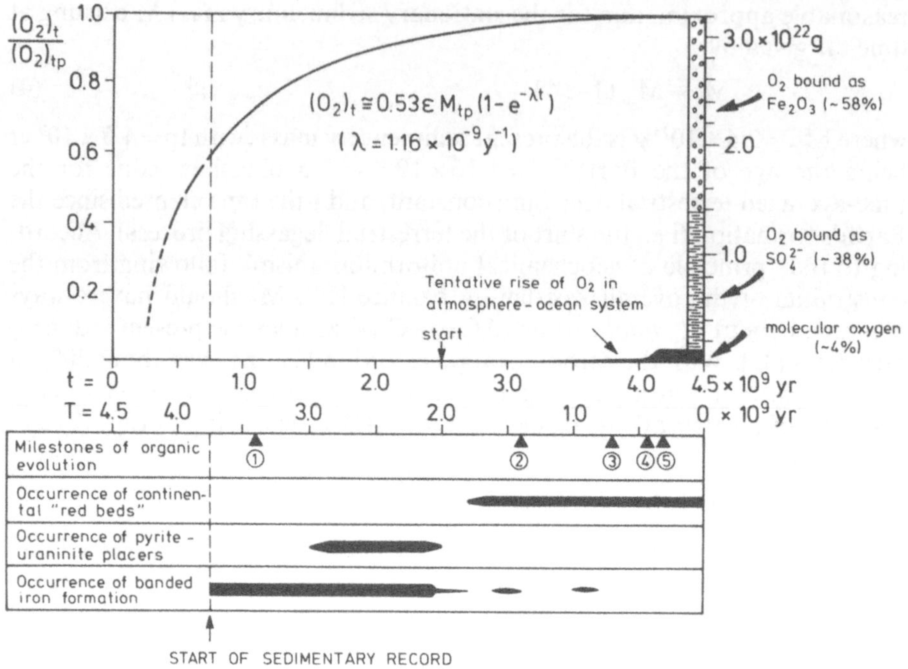

Fig. 2. Evolution of the stationary reservoir of photosynthetic oxygen according to Eq. (5), with reservoirs existing at times t expressed as fractions of the present one $[(O_2)_t/(O_2)_{tp}]$. Time scales t and T differ by the reversal of time's arrow (T being the conventional geological time scale). The column at the right side shows the partitioning of total oxygen between the "bound" and the "free" reservoirs in the present budget, the 3.2×10^{22} g of the latter being the stoichiometric equivalent of 1.2×10^{22} g organic carbon tied up in the Earth's sedimentary shell. Note that the 1.19×10^{21} g of free oxygen contained in the contemporary atmosphere account for about 4% of the total budget only, the bulk of photosynthetic oxygen having reacted with reducing constituents of the crust where it is now stored as either sulfate (SO_4^{2-}) or oxides of trivalent iron (e.g., Fe_2O_3). While the reservoirs of molecular and sulfate-bound oxygen have been determined independently, the iron-bound depository was calculated by difference and may be smaller than indicated if a sizable amount of oxygen has been consumed by oxidation of reduced gases of the primitive atmosphere.

The small curve below shows the probable accumulation history of free oxygen as inferred from paleontological and geological data listed in lower part of graph. Milestones of organic evolution indicated are: *1* appearance of oldest fossil bioherms ("stromatolites") of potential cyanophytic affinities; *2* rise of eukaryotes; *3* appearance of multicellular eukaryotes (eumetazoan faunas of Nama-Ediacara type); *4* conquest of continents by life (Upper Silurian); *5* appearance of exuberant continental floras (Upper Carboniferous)

synthetic oxygen produced. It is reasonable to assume that the process of banded ironstone formation characteristic of Early and Middle Precambrian times (Fig. 2) had provided an extremely effective oxygen sink keeping atmospheric P_{O_2} at negligible levels prior to 2×10^9 yr ago [7].

It should be noted that the proposed scheme of oxygen evolution requires that the bulk of the organics stored in the oldest sediments owe their origin to

O_2-releasing (cyanophytic) photosynthesis, and that contributions to the ancient C_{org}-budget by bacterial photosynthesis (yielding oxidation products other than free oxygen) be very small. Whereas this certainly holds for the major part of the Earth's history, it cannot be excluded that prior to 3×10^9 yr ago bacterial photosynthesis may have contributed a sizable share to the sedimentary C_{org}-reservoir. Hence, a revision of the oldest part of the curve may be called for in due course. On the other hand, the occurrence already in Early Archaean ($>3 \times 10^9$ yr) sediments of "stromatolites" (i.e., biosedimentary structures resulting from the matter behavior of microorganisms, including blue-green algae) is likely to support a very early start of cyanophytic photosynthesis as do several other lines of reasoning [36].

The Rise of Atmospheric Oxygen

Atmospheric oxygen is the strongest oxidizing agent able to persist at the Earth's surface. Stronger oxidants cannot exist as these would decompose water, the standard potential of the reaction

$$H_2O \rightleftharpoons 0.5\,O_2 + 2\,H^+ + 2\,e^- \qquad E^0 = +1.23 \text{ volts} \qquad (6)$$

thus giving an upper limit for the oxidation potential of all terrestrial near-surface environments. With (6) pH-dependent and E^0 decreasing with decreasing acidity, an upper boundary of $E^0 \sim +1.0$ (for pH = 4) would be still more realistic since pH-values in nature seldom drop below 4 [3]. Hence, the history of atmospheric oxygen is an integral part of the evolution of the terrestrial environment as a whole.

With our growth curve of "total" photosynthetic oxygen yielding no information on the accumulation of a free oxygen reservoir, the history of the latter is as yet based on geological and paleontological evidence alone which is summarized in the lower part of Fig. 2. The principal lines of argument are as follows:

(a) The abundance of *banded iron formation* (BIF) in marine sedimentary series of Early and Middle Precambrian age ($>2 \times 10^9$ yr) may be interpreted as reflecting the geochemical behavior of iron in an anoxygenic weathering cycle [6, 7]. Under such conditions, iron would have been washed from the continents as soluble Fe^{2+}-compounds, accumulating in the oceans and subsequently precipitating as iron-rich chemical sediments which were the precursors of present-day BIF-deposits. (In the contemporary exogenic cycle, Fe^{2+} is oxidized to Fe^{3+} which, because of its poor solubility, is retained in continental weathering crusts, the iron content of modern seawater thus being on the order of ppb only.)

(b) The occurrence of *detrital sulfides and uraninite* in ancient conglomerates covering the time span $2\text{–}3 \times 10^9$ yr [31, 37] would be in line with (a). In an oxygenic weathering cycle, sulfides as pyrite (FeS_2) are readily oxidized to oxides of trivalent iron and sulfate; likewise, tetravalent uranium present in uraninite (UO_2) will be oxidized to U^{6+}, giving rise to the soluble uranyl complex $[UO_2]^{2+}$. At present, these minerals do not survive weathering processes except in restricted low-temperature environments where chemical weathering is heavily retarded.

(c) The appearance of the oldest *continental red beds* (deriving their conspicuous coloration from abundant trivalent iron) has been interpreted as heralding the start of an oxygenic weathering cycle on the continents and thus the incipient buildup of an atmospheric oxygen reservoir [6]. With BIF and pyrite-uraninite placers disappearing from the record some 2×10^9 yr ago and continental red beds taking over at about the same time (cf. Fig. 2), the interval $2.0 \pm 0.2 \times 10^9$ yr has come to be widely accepted as the most probable transition period from an anoxic to an oxygenic atmosphere.

(d) With all eukaryotic life forms contingent on oxidative metabolism, the appearance of the *oldest eukaryotes* $1.4 \pm 0.1 \times 10^9$ yr ago [8, 40], is likely to indicate that the so-called "Pasteur level" of 10^{-2} of the present atmospheric oxygen pressure had been reached by then. Increased energy demands by the first eumetazoans evolving from primitive eukaryotes some 0.7×10^9 yr ago were probably such as to necessitate considerably higher oxygen concentrations in the environment. Fairly modern oxygen pressures should have prevailed as from Paleozoic times, notably the Carboniferous. The continuity of mammalian life as from the Triassic would exclude major excursions of atmospheric P_{O_2} during the last 200 million years.

Since most of these geological and paleontological data provide at best qualitative oxygen barometers, many details of the evolution of atmospheric oxygen are still open to discussion. The absence of a free oxygen reservoir prior to 2×10^9 yr ago as suggested by these findings is most probably related to the abundance in the ancient seas of hydrated ferrous ions which must have constituted ideal oxygen acceptors when the oldest cyanophytes started to release oxygen to their environment. With the biological cycle largely submerged below the ocean surface during the Precambrian, the dissolved Fe^{2+}-ions must have provided a very effective oxygen buffer keeping stationary O_2-concentration at extremely low levels. In principle, the 4% of the total oxygen budget presently contained in the atmosphere (cf. Fig. 2, budget column at right side) could be conveniently stored in the bound reservoir as well since the reducing capacity of the crust is by no means exhausted at the present level of sedimentary O_2-fixation. Only after the ancient seas had been swept of bivalent iron (which came to be precipitated as ferric iron in BIF), free oxygen could have accumulated in the oceans and, consequently, in the atmosphere.

With rising P_{O_2} in the atmosphere, oxidation of bivalent iron occurred in the continental weathering crusts, this further drying up Fe^{2+}-supplies to the oceans. Because of the sluggishness of oxidation weathering on the continents (where reducing constituents like Fe^{2+} do not occur in the form of a "buffering solution", but are unterred successively by denudation processes in small quantities), a new steady state must have been finally established for the free oxygen reservoir at a markedly elevated level at which the new sink function ultimately matched the rate of oxygen production. Hence, the oxygen burden of the present atmosphere is obviously due to the failure of thermodynamics to exercise instantaneous control over the redox balance of the atmosphere-hydrosphere-solid Earth system during operation of the present weathering cycle, with equilibrium between source and sink being attained only at a considerable P_{O_2} in the free reservoir.

References

1. Alfvén, H., Arrhenius, G.: Evolution of the Solar System. NASA: Washington, D.C., 1976, p. 485
2. Augustsson, T., Ramanathan, F.: J. Atm. Sci. *34,* 448 (1977)
3. Baas-Becking, L.G.M., Kaplan, I.R., Moore, D.: J. Geol. *68,* 243 (1960)
4. Berkner, L.V., Marshall, L.C.: Adv. Geophys. *12,* 309 (1967)
5. Brown, H.: The Atmospheres of the Earth and Planets, G.P. Kuiper, ed.; Univ. of Chicago Press, Chicago 1949
6. Cloud, P.E.: Science *160,* 729 (1968)
7. Cloud, P.E.: Econ. Geol. *68,* 1135 (1973)
8. Cloud, P.E.: Paleobiology *2,* 351 (1976)
9. Crutzen, P.J., Isaksen, I.S.A., McAfee, J.R.: J. Geophys. Res. *83,* 345 (1978)
10. Eichmann, R., Schidlowski, M.: Geochim. Cosmochim Acta *39,* 585 (1975)
11. Fishman, J., Crutzen, P.J.: J. Geophys. Res. *82,* 5897 (1977)
12. Garrels, R.M., Mackenzie, F.T.: Evolution of Sedimentary Rocks. Norton, New York 1971, p. 274
13. Garrels, R.M., Lerman, A.: Global Chemical Cycles and Their Alterations by Man, W. Stumm, ed.; 23, Abakon, Berlin 1977
14. Goody, R.M., Walker, J.C.G.: Atmospheres. Prentice-Hall, Englewood Cliffs, N.J. 1972
15. Holland, H.D.: The Chemistry of the Atmosphere and Oceans. Wiley, New York 1978
16. Hunten, D.M., Strobel, D.F.: J. Atm. Sci. *31,* 305 (1973)
17. Johnston, H.S.: Science *173,* 517 (1971)
18. Johnston, H.S.: J. Geophys. Res. *82,* 1767 (1977)
19. Junge, C.E.: Jb. Max-Planck-Ges. *1971,* 150
20. Junge, C.E.: Quart. J. Roy. Met. Soc. *98,* 711 (1972)
21. Junge, C.E., Schidlowski, M., Eichmann, R., Pietrek, H.: J. Geophys. Res. *80,* 4542 (1975)
22. Keeling, C.D.: Tellus *25,* 174 (1973)
23. Keeling, C.D., Bacastow, R.B., Bainbridge, A.E., Ekdahl, C.A., Guenther, P.R., Waterman, L.S., Chin, J.F.S.: Tellus *28,* 538 (1976)
24. Li, Y.H.: Amer. J. Sci. *272,* 119 (1972)
25. Manabe, S., Wetherald, R.T.: J. Atm. Sci. *32,* 3 (1975)
26. Matsuo, S.: Origin of Life, H. Noda, ed.; 21, Center Acad. Publ. Japan, Tokyo 1978
27. Miller, S.L.: J. Amer. Chem. Soc. *77,* 2351 (1955)
28. Miller, S.L., Urey, H.C.: Science *130,* 245 (1959)
29. Miller, S.L., Urey, H.C., Oró, J.: J. Mol. Evol. *9,* 59 (1976)
30. Molina, M. J., Rowland, F.S.: Nature *249,* 810 (1974)
31. Ramdohr, P.: Abh. dt. Akad. Wiss. Berlin, Kl. Chem. Geol. Biol. 1958, *3*
32. Ronov, A.B.: Sedimentology *10,* 25 (1968)
33. Rubey, W.W.: Bull. Geol. Soc. Amer. *62,* 1111 (1951)
34. Rubey, W.W.: Geol. Soc. Amer. Spec. Paper *62,* 631 (1955)
35. Rutten, M.G.: The Origin of Life by Natural Causes. Elsevier, Amsterdam 1971
36. Schidlowski, M.: Origin of Life, H. Noda, ed.; 3, Center Acad. Publ. Japan, Tokyo 1978
37. Schidlowski, M.: U.S. Geol. Surv. Prof. Paper 1980 (in press)
38. Schidlowski, M., Eichmann, R., Junge, C.E.: Precambrian Res. *2,* 1 (1975)
39. Schidlowski, M., Appel, P.W.U., Eichmann, R., Junge, C.E.: Geochim. Cosmochim. Acta *43,* 189 (1979)
40. Schopf, J.W., Oehler, D.Z.: Science *193,* 47 (1976)
41. Sillén, L.G.: Amer. Ass. Adv. Sci. Publ. *67,* 549 (1961)

42. SMIC-Report "Inadvertent Climate Modification"; MIT-Press, Cambridge, Mass. 1971
43. Suess, H.E.: J. Geol. *57,* 600 (1949)
44. Urey, H.C.: The Planets. Yale Univ. Press, New Haven, Conn. 1952
45. Urey, H.C.: Handbuch der Physik, S. Flügge, ed., *5,* 363. Springer, Berlin 1959
46. U.S. Standard Atmosphere, 1976; NOAA/NASA/U.S. Air Force: Washington, D.C., 1976
47. Veizer, J., Hoefs, J.: Geochim. Cosmochim. Acta *40,* 1387 (1976)
48. Walker, J.C.G.: Evolution of the Atmosphere. Macmillan, New York 1977, p. 124 and 222
49. Warneck, P.: Promet *5,* 11 (1975)
50. Zimen, K.E., Altenhein, F.K.: Naturwiss. *60,* 198 (1973)

The Hydrosphere

J. Westall, W. Stumm
EAWAG, Swiss Federal Institute of Technology
CH–8600 Dübendorf, Switzerland

Structure and Properties of Water

The Water Molecule

The electronic configuration of atomic oxygen is $1s^2\ 2s^2\ 2p^4$ and of atomic hydrogen is $1s^1$. In the water molecule the three nuclei are surrounded by ten electrons; the two 1s electrons of oxygen are confined to the vicinity of the oxygen nucleus, and the other eight electrons are in four approximately sp^3 hybrid orbitals which point to the corners of a distorted tetrahedron (Fig. 1). Two of the orbitals are bonding (contain bonding-pair electrons) and are directed along the O–H bond axes, and the two of the orbitals are non-bonding (contain lone-pair electrons) and are directed above and below the H–O–H molecular plane. The planar H–O–H bond angle in the isolated water

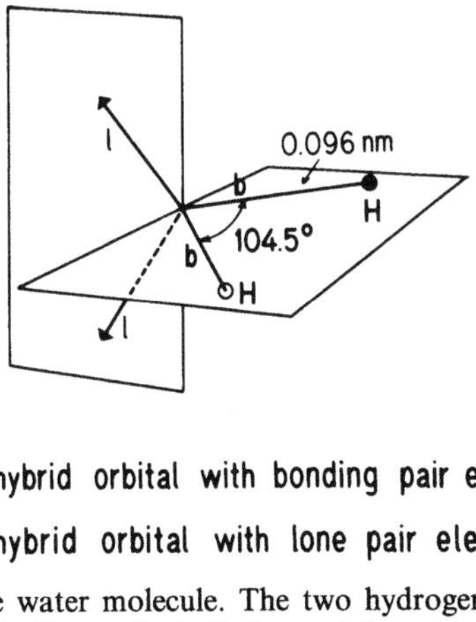

Fig. 1. The structure of the water molecule. The two hydrogen atoms and the two electron lone-pairs form the apices of a distorted tetrahedron, at the center of which is oxygen. (After [1])

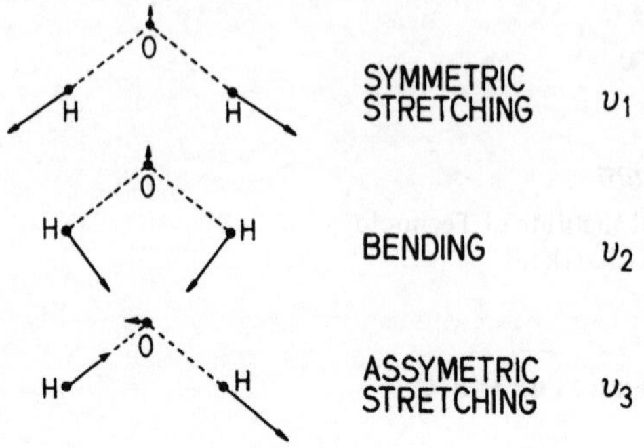

Fig. 2. The normal modes of vibration of the isolated water molecule. The principle IR frequencies associated with these vibrations are given in Table 1.

molecule is 104.5 °, compared to 109.5 ° for a regular tetrahedron; the O–H bond length in the isolated water molecule is 0.096 nm. The lone-pair electrons probably contribute an insignificant amount to the permanent dipole moment, but a significant amount to the induced dipole moment.

The isolated water molecule has C_{2v} symmetry; it has a two fold axis of rotational symmetry, C_2, the line bisecting the H–O–H angle, and a plane of reflection v passing through the axis and normal to the plane of the molecule. The water molecule has three normal modes of vibration (Fig. 2), the symmetric stretch, bending, and the assymmetric stretch. The prominent infrared frequencies associated with these normal modes are given in Table 1 for the isolated water molecule.

Table 1. Vibrational frequencies of an isolated H_2O molecule [1]

Transition between ground state and upper state with quantum numbers [a]			Absorption frequency (cm^{-1})
v_1	v_2	v_3	
0	1	0	1,594.59
1	0	0	3,656.65
0	0	1	3,755.79
0	2	0	3,151.4
0	1	1	5,332.0
0	2	1	6,874
1	0	1	7,251.6
1	1	1	8,807.05
2	0	1	10,613.12
0	0	3	11,032.36

[a] v_1: symmetric streching
v_2: bending
v_3: assymmetric streching

Hydrogen Bonding

Hydrogen bonding in water is the specific association of the hydrogen atom of one molecule with the lone pair electrons for another. While there is ample evidence of hydrogen bonding in ice and liquid water, little direct evidence of hydrogen bonding in water vapor exists. The energy associated with hydrogen bonds may be defined and calculated in different ways; values range from 5.4 to 18.8 $kJ \cdot mol^{-1}$ H-bond in liquid water and from 17.8 to 32.2 $kJ \cdot mol^{-1}$ H-bond in ice [1]. The O–H bond in the water molecule itself is weakened by participation of the molecule in hydrogen bonding. Hydrogen bonding is responsible for many of the extraordinary physical properties of water and their dependence on temperature and pressure.

Structure of Ice

The relationship between the physical properties of ice and its structure is relatively well understood. Although there are many polymorphs of ice, only one, ordinary hexagonal ice (ice I), is stable at normal low ($\leq$ 1 bar) pressures. The structure of ice I is well-established (Fig. 3): Each water molecule is bound

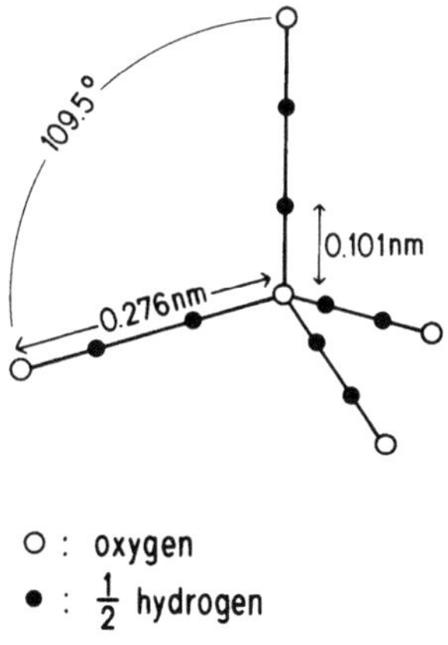

Fig. 3. The water molecule in ice. Although the oxygen atoms in the ice lattice are not arranged in perfect tetrahedra, the approximate dimensions are given. The hydrogen atom may occupy one of two positions (represented by shaded circles) between two oxygen atoms. (After [1])

to four other water molecules by intermolecular hydrogen bonds: the two hydrogen atoms of one water molecule are associated with electron lone-pairs of other water molecules, and the two electron lone-pairs are associated with hydrogen atoms of other water molecules. The tetrahedral coordination of water molecules yields an open lattice of oxygen nuclei arranged in layers of puckered hexagonal rings, similar to the "chair" form of cyclohexane. Each layer of rings is a mirror image of the next layer (Fig. 4). This arrangement of oxygen atoms is isomorphous to tridymite, a crystalline form of silica, and

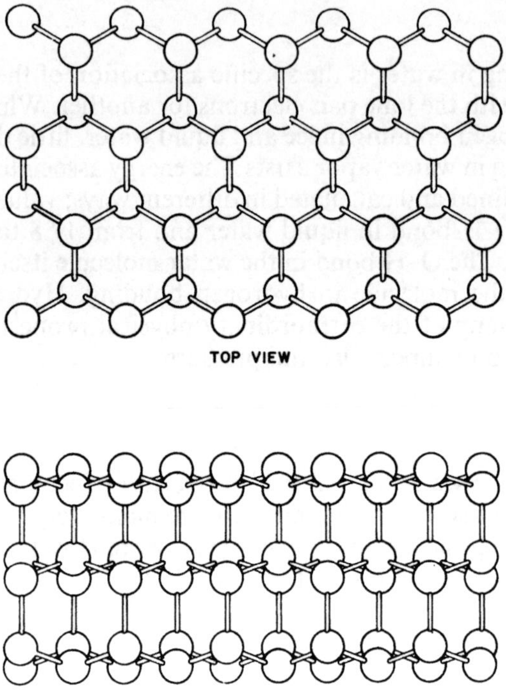

Fig. 4. The structure of ice I. The large open "shafts" in the crystal lattice are responsible for the low density of ice I. (From [2], reprinted with permission of American Institute of Physics)

wurtzite, a crystalline form of zinc sulfide. This open crystal structure of ice is the reason for the low density of ice (0.9168 g cm^{-3} at 0 °C) compared to that of liquid water (0.99987 g cm^{-3} at 0 °C). The distance between oxygen atoms in the ice lattice is 0.276 nm, and the O–O–O angle is 109.5 °. The OH bond length (0.101 nm) and HOH bond angle (104.5 °) of the water molecule in ice are not much different from those in an isolated water molecule. As shown in Fig. 3, a hydrogen atom can occupy one of two positions along the O–O axis, 0.101 nm from either of the two oxygen atoms. Early theoretical work by Pauling based on the residual entropy of ice at 0 °C predicted that over a sufficiently long time interval, the hydrogen atom occupies each of these positions half of the time. This was later confirmed by neutron diffraction studies.

Structure of Liquid Water

The time-averaged structure of liquid water, as interpreted from x-ray diffraction data, may be described as follows [1]: i) one water molecule does not approach another more closely than 0.25 nm, and the order imposed by one molecule on other molecules does not extend more than 0.80 nm; ii) about a central water molecule, there is a high concentration of other molecules at 0.29

nm, 0.45–0.53 nm, and 0.64–0.78 nm; this distribution is consistent with tetrahedral coordination; iii) each water molecule has approximately 4.4 neighbors in the first coordination shell. Thus liquid water is a highly structured liquid in which the tetrahedral coordination observed in ice is still evident.

Thermodynamics and Physical Properties

The isopiestic (1 bar) heat capacity of water, and the thermodynamic functions derived from the heat capacity, are given in Fig. 5. The heat capacity of liquid water is almost twice that of ice at 0 °C or steam at 100 °C. Whereas the heat capacity of ice and steam is primarily due to vibrational energy, there is in liquid water an additional configurational heat capacity related to the energy involved in alterations in the structure of water.The thermodynamic constants for the phase changes of water are given in Table 2. The relatively low heat of fusion of water is evidence that the hydrogen bonding in the ice structure remains intact to some degree during fusion; the extremely high heat of vaporization reflects the disappearance of hydrogen bonding upon vaporization.

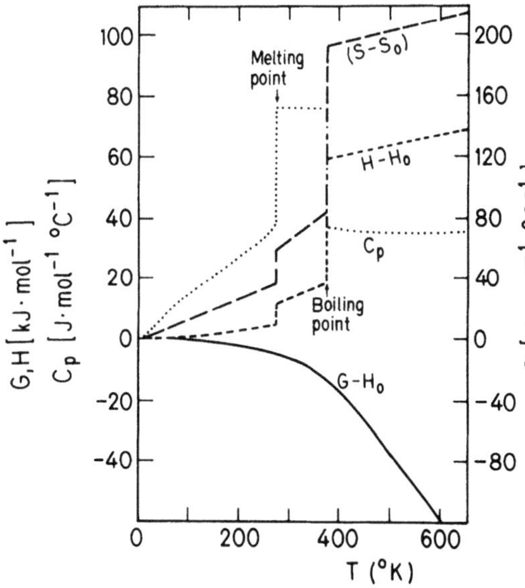

Fig. 5. Enthalphy, entropy, free energy, and isopiestic heat capacity of H_2O at 1 atmosphere pressure, calculated from heat capacity measurements:

$$H_T - H_0 = \int_0^T C_p \, dT + \Delta H_{pc} \qquad S_T - S_0 = \int_0^T \frac{C_p}{T} \, dT + \Delta S_{pc} \qquad G_T - G_0 = H_T - H_0 - TS_T$$

The subscript "pc" indicates phase change. Thermodynamic constants for phase change are given in Table 2. (Figure after [1], data from [3])

Table 2. Thermodynamic constants for phase changes of H_2O [1]

	Fusion [a]	Vaporization [a]	Sublimation [b]
Temperature °K	273.15	373.15	273.16
ΔC_p (J mol^{-1} °C^{-1})	37.28	-41.93	
ΔH (kJ mol^{-1})	6.01	40.66	51.06
ΔS (J mol^{-1} °C^{-1})	22.00	108.95	186.92
ΔV (cm^3 mol^{-1})	-1.621	3.01×10^4	
ΔE (kJ mol^{-1})	6.01	37.61	48.79

[a] At 1 atm
[b] At ice I-liquid-vapor triple point

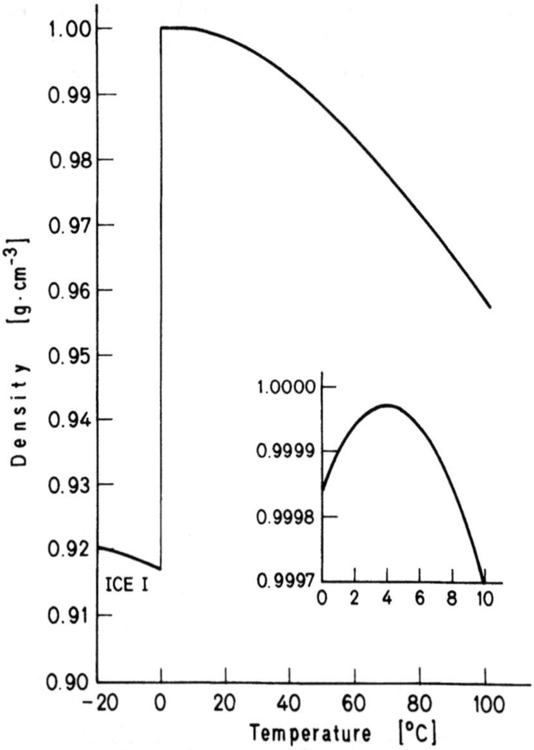

Fig. 6. The density (at 1 atm) of ice and liquid water as a function of temperature. The inset shows the density in the domain of the its maximum. (Data for ice from [1], data for water from [4])

P-V-T Relationships. The density of water reaches a maximum at 4 °C above the melting point (Fig. 6). This extraordinary maximum may be explained as the result of two counteracting phenomena: as the temperature of liquid water is increased, i) the open four-coordinated structure of water is further broken down, decreasing the volume; ii) the amplitude of anharmonic intermolecular vibrations increases, increasing the volume. Below 4 °C the first effect predominates.

The pressure-temperature curve for water is given in Fig. 7. The triple point, at which ice, liquid water, and water vapor coexist at equilibrium, is a convenient reference state: T = 0.01 °C, P = 6.11 mbar.

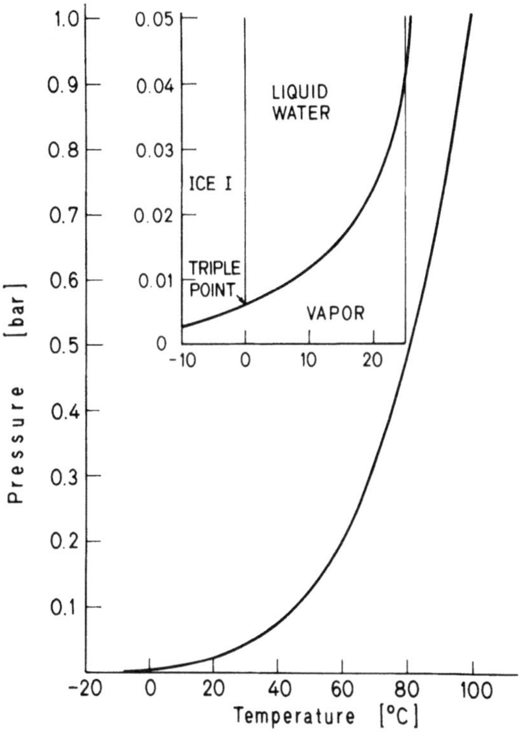

Fig. 7. The vapor pressure of ice and liquid water as a function of temperature. The inset shows the vincinity of the triple point. (Data from [1])

Dielectric Constant. The dielectric properties of water result from i) electronic and atomic polarization of molecules (formation of induced dipoles) and ii) the orientation of permanent dipoles. The high value of the dielectric constant of water is chiefly the result of the orientation of permanent dipoles, whereby not only the orientation of individual molecules, but also the mutual orientation of neighboring molecules, is important. For comparison, the dielectric constant of ice I is 91 at 0 °C, and that of liquid water is 88 at the same temperature. As the temperature is increased, the degree of orientation of the molecules, and the dielectric constant, decreases (Fig. 8).

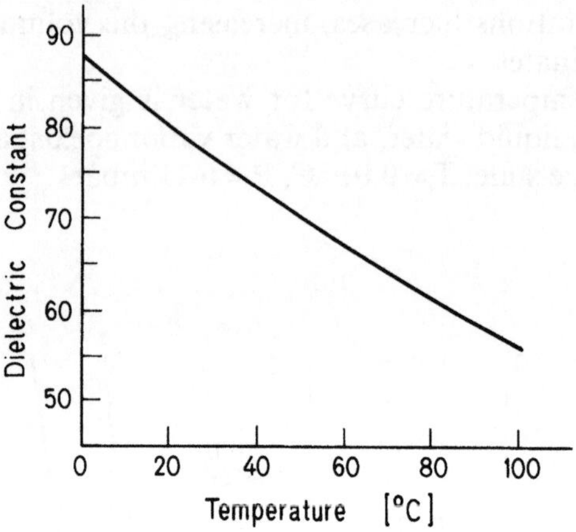

Fig. 8. The dielectric constant of water as a function of temperature. (Data from [5])

The preceding discussion dealt with the dielectric properties of water in a static electric field. In a high frequency alternating electric field, the permanent dipoles can no longer reorient with the variable electric field; then the observed dielectric constant is due solely to the induced dipoles. For liquid water the value of this high frequency dielectric constant, ε_∞ is between 4.5 and 6.0.

Viscosity. The shear viscosity of water can be defined by the equation

$$\tau_{zx} = -\mu \frac{dv_x}{dz}$$

where τ is the shear stress (force in the x-direction per unit area normal to z) to maintain the velocity gradient dv_x/dz in the z-direction, in the fluid of viscosity μ. The viscosity of water approximately doubles in the range between 25 °C and 0 °C (Fig. 9).

Models of Liquid Water

The structure of liquid water, unlike that of ice, remains a great mystery. The ultimate model for the structure of liquid water must include an explanation for the extraordinary properties of water described above [7]:

1) density maximum at 4 °C above the melting point
2) specific heat of the liquid approximately twice that of the solid
3) extraordinarily large heat of vaporization.

In addition, the model should be consistent with IR, Raman, X-ray, NMR and dielectric measurements made on water. There are many models which can reproduce some part of the experimental observations, but the mere fact that the model and the experiment agree does not confirm that the model is indeed a valid description of the physical situation. An entirely satisfactory model for liquid water has not yet been developed.

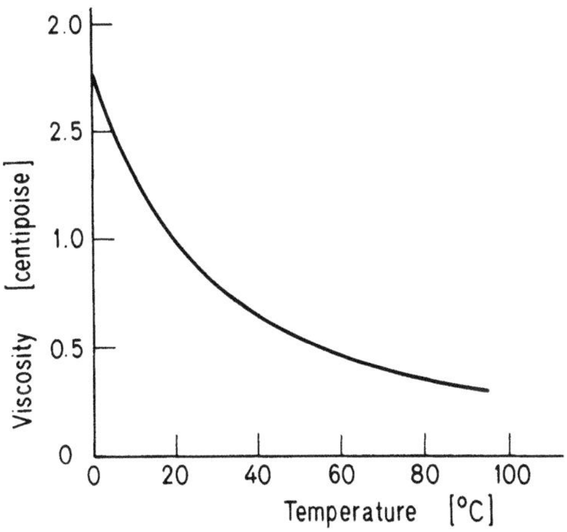

Fig. 9. The shear viscosity of water as a function of temperature. (1 centipoise = 10^{-3} N s m^{-2}) (Data from [6])

The different models for liquid water may be divided into two classes: the continuum models and the mixture models. Models in both classes achieve some success in explaining the properties of water [1, 7–11].

Continuum Models. The continuum theories ascribe essentially complete hydrogen bonding to water, at least at low temperature. Associated with the hydrogen bonds is a distribution of angles, distance and energies. The average bond energy varies with temperature, due to the changes in the distributions of bond lengths and bond angle distortions.

Pople's classic model [8] postulates that when ice melts the hydrogen bonds become more flexible and disorder is introduced into the structure by the bending and stretching of bonds rather than by the breaking of them. Thus the water molecules in the liquid phase, like those in the solid phase, retain a coordination number of four. All the molecules retain essentially the same environments, thereby giving rise to the continuum.

Mixture Models. The mixture models are based on the existence of a small number of distinctly different species in liquid water. Of historical interest are the small aggregate models based on monomers, dimers, and trimers of molecular water. The interstitial models use a hydrogen-bonded lattice within which non-hydrogen bonded water reside. Other models are based on equilibria between a "bulky" H_2O species and a "dense" H_2O species. The flickering cluster model treats the formation of hydrogen bonds as a cooperative phenomenon, in which the formation of one hydrogen bond promotes the formation of other hydrogen bonds in the vicinity. This cooperative making and breaking of hydrogen bonds leads to the idea of "flickering clusters."

Chemistry of Natural Waters

Kinetics and Chemical Equilibrium

The chemistry of natural waters involves reactions of numerous minerals, dissolved species, and gases, as well as reactions involving the biosphere. In addition to these reactions occurring within the system, there is transport of materials and energy into and out of the system, so that not only the speciation of materials within the system, but also the amount of materials in the system is constantly changing. This degree of complexity is seldom encountered in the laboratory.

In order to understand the dominant processes in this complex mélange of reaction and transport rates, it is necessary to consider some limiting cases. In principle, chemical reactions may be divided into two classes according to their rates: those which are much faster than the transport rates, and those which are much slower. The fast reactions may be described by thermodynamic equilibrium concepts, and the slow reactions may be treated as if they did not occur at all. The characteristic transport rate of a material in a natural system may be expressed by its residence time τ (valid at steady state):

$$\tau = \frac{\text{total mass of material in system}}{\text{rate of input and output of material}} .$$

The residence time of a material may be equal to the hydraulic residence time, or it may be much greater or much less, depending on the reactions that occur.

In a multi-component system there may be one chemical reaction rate, or one transport rate that is slow, and all others are fast by comparison; such a system could be described with chemical equilibrium among the many components as the amount of the rate-limiting component varies with time.

Thus equilibrium concepts are used as the basic vehicle for discussing the chemistry of natural waters. Discrepancies between equilibrium calculations and field data for natural systems give insight into those cases where chemical reactions are not sufficiently understood, where non-equilibrium conditions prevail, or where analytical data are not sufficiently accurate or precise [12–14].

The problem of determining the chemical equilibrium speciation in multicomponent systems including aqueous, gas, and solid phases may be expressed very simply as a linear algebraic problem solved by iterative techniques [15] (Table 3).

Carbonate Equilibria

The atmosphere contains CO_2 gas at a constant partial pressure, and the species CO_2 (aq), H_2CO_3, HCO_3^-, and CO_3^{2-} are found in relatively high concentrations in almost all natural waters; on account of these facts and the role of CO_2 in photosynthesis/respiration, the carbonate system is extremely important for the regulation of pH in natural waters [14, 16–18].

Table 3. Description of chemical equilibrium problem

Scalar[a]	Description	Matrix or vector	Description
a_{ij}	Stoichiometric coefficient of component j in species i	A	Matrix of a_{ij}
C_j	Free concentration of species i	C	Vector of C_i
		C*	Vector of log C_i
K_i	Stability constant of species i	K*	Vector of log K_i
T_j	Total analytical concentration of component j	T	Vector of T_j
X_j	Free concentration of component j	X	Vector of X_j
		X*	Vector of log X_j
Y_j	Residual in material balance equation for component j	Y	Vector of Y_j
Z_{jk}	Partial derivative $(\delta Y_j/\delta X_k)$	Z	Jacobian of Y with respect to X
	Mass law equations $\log C_i = \log K_i + \sum_j a_{ij} \log X_j$	$C^* = K^* + AX^*$	
	Mass balance equations $Y_j = \sum_i a_{ij} C_i - T_j$	$Y = {}^tAC - T$	
	Iteration procedure (Newton-Raphson) $Z_{jk} = \sum_i (a_{ij} a_{ik} C_i/X_k)$	$Z \cdot \Delta X = Y$; $\Delta X = X_{original} - X_{improved}$	

[a] *Indices:* "i" is used to denote any *species;* "j" and "k" are used to denote any *component*

Table 4. The carbonate equilibria

Species: CO_2 (g), $H_2CO_3^*$ [a], HCO_3^-, CO_3^{2-}

Equilibria:

$$K_H = \frac{[H_2CO_3^*]}{p_{CO_2}}$$

$$K_1 = \frac{[H^+][HCO_3^-]}{[H_2CO_3^*]}$$

$$K_2 = \frac{[H^+][CO_3^{2-}]}{[HCO_3^{2-}]}$$

Conservative quantities:

$$C_T = [CO_3^{2-}] + HCO_3^-] + [H_2CO_3^*]$$

$$[Alk] = [HCO_3^-] + 2\,[CO_3^-] + [OH^-] - [H^+]$$

$$[Acy] = 2\,[H_2CO_3^*] + [HCO_3^-] + [H+] - [OH^-]$$

[a] CO_2 gas will dissolve in water to form a dissolved gas species CO_2 (aq) and a hydrated species H_2CO_3. The symbol $[H_2CO_3^*]$ is used to represent the analytical sum of the two species $[H_2CO_3^*] = [CO_2(aq)] + [H_2CO_3]$. Since at 25 °C $[CO_2(aq)] \gg [H_2CO_3]$, operational acidity constants for $H_2CO_3^*$ can be defined and equilibrium calculations thereby simplified

Carbonate species and the equilibrium expressions for the reactions between the species are summarized in Table 4. Values for the equilibrium constants are found in Table 5.

Table 5. Values of carbonate equilibrium constants [a]

Temperature °C	$-\log K_H$ [b]	$-\log K_1$ [b]	$-\log K_2$ [c]
0	1.11	6.58	10.63
5	1.19	6.52	10.56
10	1.27	6.46	10.49
15	1.34	6.42	10.43
20	1.41	6.38	10.38
25	1.46	6.35	10.33
30	1.52	6.33	10.29
40	1.62	6.30	10.22
50	1.71	6.29	10.17

[a] Extrapolated to infinite dilution
[b] Data from Harned and Davies, J. Amer. Chem. Soc. *65*, 2030 (1943)
[c] Data from Harned and Scholes, J. Amer. Chem. Soc. *63*, 1706 (1941)

Conservative Quantities. The alkalinity, acidity, and total inorganic carbon, C_T are extensive (capacity) properties of a system; they are conservative. (Alkalinity, acidity, and C_T are expressed in units of concentration, e. g. molarity, molality, equivalents per liter or as mg/l $CaCO_3$.) The use of these conservative parameters facilitates the calculation of the effects of the addition or removal of acids, bases, carbon dioxide, and carbonates to aqueous systems.

Alkalinity and acidity are determined operationally by acidimetric and alkalimetric titrations to the appropriate pH end points (inflection points on the titration curve in Fig. 10. Alkalinity and acidity may be defined mathematically by the equations:

$$\mathrm{Alk} = [HCO_3^-] + 2[CO_3^{2-}] + [OH^-] - [H^+]$$
$$\mathrm{Acy} = 2[H_2CO_3^*] + [HCO_3^-] + [H^+] - [OH^-].$$

These equations express alkalinity as the net proton deficiency relative to a given reference level (H_2CO_3 and H_2O), and acidity as the proton excess relative to another reference level (CO_3^{2-} and H_2O). These concepts van readily be extended to account for non-carbonate buffering species. For example, in the presence of borate and ammonia, the alkalinity becomes:

$$\mathrm{Alk} = [HCO_3^-] + 2[CO_3^{2-}] + [OH^-] - [H^+] + [B(OH)_4^-] + [NH_3].$$

The total inorganic carbon is defined by:

$$C_T = [H_2CO_3] + [HCO_3^-] + [CO_3^{2-}].$$

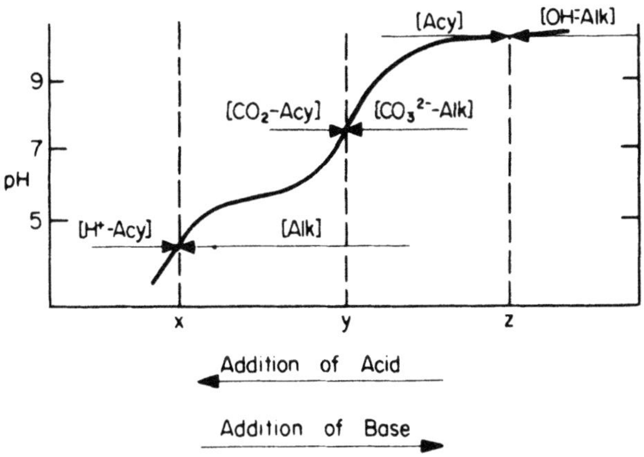

Term	Equivalence point	Definition	
		I Acid Neutralizing Capacity (ANC)	
Caustic alkalinity	z	$[OH^- - Alk] = [OH^-] - [HCO_3^-] - 2[H_2CO_3^*] - [H^+]$	(1)
Carbonate alkalinity	y	$[CO_3^{2-} - Alk] = [OH^-] + [CO_3^{2-}] - [H_2CO_3^*] - [H^+]$	(2)
Alkalinity	x	$[Alk] = [HCO_3^-] + 2[CO_3^{2-}] + [OH^-] - [H^+]$	(3)
		II Base Neutralizing Capacity (BNC)	
Mineral acidity	x	$[H^+ - Acy] = [H^+] - [HCO_3^-] - 2[CO_3^{2-}] - [OH^-]$	(4)
CO_2-acidity	y	$[CO_2 - Acy] = [H_2CO_3^*] + [H^+] - [CO_3^{2-}] - [OH^-]$	(5)
Acidity	z	$[Acy] = 2[H_2CO_3^*] + [HCO_3^-] + [H^+] - [OH^-]$	(6)

III Combinations

$[H_2CO_3^*] + [HCO_3^-] + [CO_3^{2-}] = C_T$	(7)	$[Alk] + [CO_2 - Acy] = C_T$	(11)
$[Alk] + [H^+ - Acy] = 0$	(8)	$[Alk] + [Acy] = C_T$	(12)
$[Acy] + [OH^- - Alk] = 0$	(9)	$[Alk] - [CO_3^{2-} - Alk] = C_T$	(13)
$[CO_3^{2-} - Alk] + [CO_2 - Acy] = 0$	(10)	$[CO_2 - Acy] - [H^+ - Acy] = C_T$	(14)

Fig. 10. Alkalinity and acidity titration curve for the aqueous carbonate system. The conservative quantities alkalinity and acidity refer to the acid neutralizing and base neutralizing capacities of a given aqueous system. These parameters can be determined by titration to appropriate equivalence points with strong acid and strong base. The equations given below define the various capacity factors rigorously

As shown in Fig. 10 certain conservative quantities remain constant for particular changes in chemical composition. The case of addition and removal of CO_2 is of special interest: respiration by aquatic organisms contributes CO_2 while photosynthesis consumes CO_2. An increase in CO_2 increases the acidity and the C_T but does not change the alkalinity. Alterately, the addition or removal of $CaCO_3$ does not affect the acidity. Addition of strong acid or strong base does not affect the C_T. The pH changes associated with these perturbations can be read from Fig. 11.

Dissolved Carbonate Equilibria. Two systems may be considered: i) a system closed to the atmosphere, and ii) a system in equilibrium with the atmosphere.

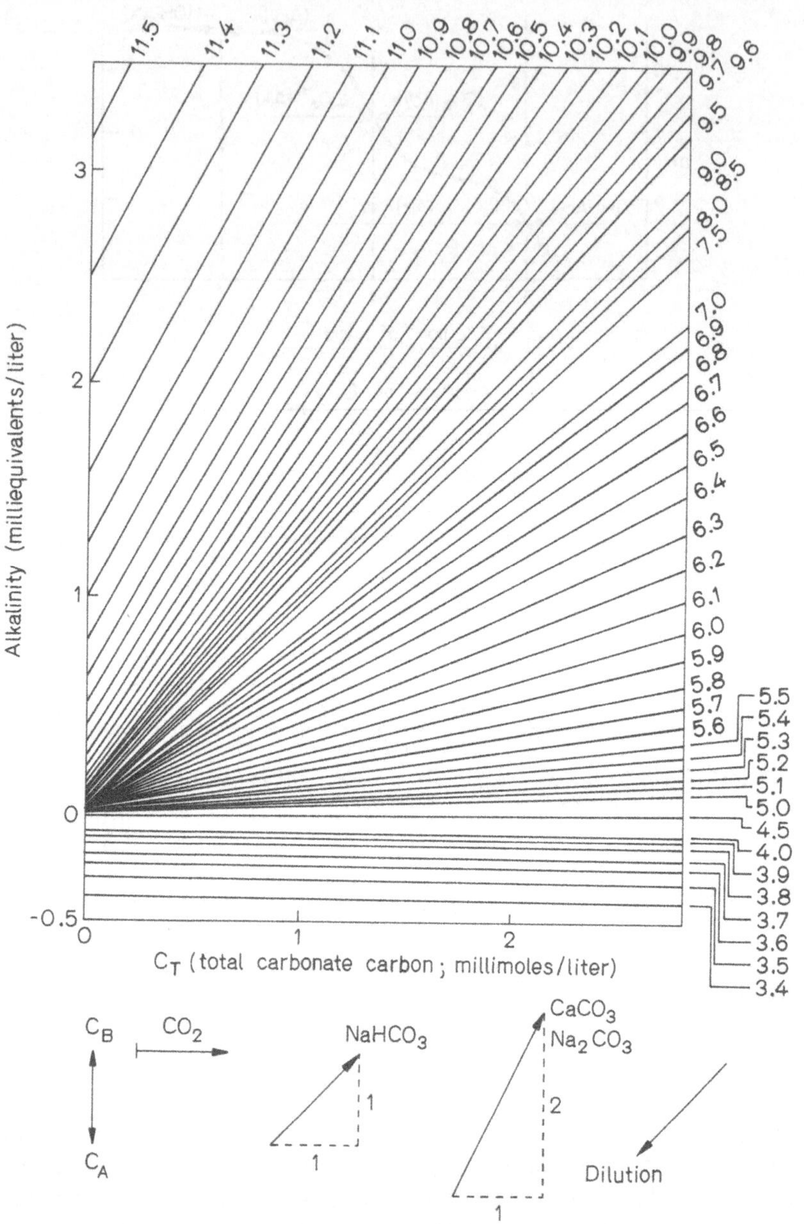

Fig. 11. Closed system capacity diagram: pH contours for alkalinity versus C_T (total carbonate carbon). The point defining the solution composition moves as a vector in the diagram as a result of addition (or removal) of CO_2, $NaHCO_3$, and $CaCO_3(Na_2CO_3)$ or C_B (strong base) or C_A (strong acid). (After [17])

Closed System. In this case $H_2CO_3^*$ is considered a non-volatile acid. The species $H_2CO_3^*$, HCO_3^-, and CO_3^{2-} are related by the equilibria

$$[H^+][HCO_3^-] \,/\, [H_2CO_3^*] = K_1$$
$$[H^+][CO_3^{2-}] \,/\, [HCO_3^-] = K_2.$$

The ionization fractions, whose sum equals unity, are defined:

$$\alpha_0 = [H_2CO_3^*] \,/C_T$$
$$\alpha_1 = [HCO_3^-] \,/C_T$$
$$\alpha_2 = [CO_3^{2-}] \,/C_T.$$

These equations can be rearranged and the ionization fractions expressed in terms of H^+:

$$\alpha_0 = \left[1 + \frac{K_1}{[H^+]} + \frac{K_1K_2}{[H^+]^2}\right]^{-1}$$

$$\alpha_1 = \left[\frac{[H^+]}{K_1} + 1 + \frac{K_2}{[H^+]}\right]$$

$$\alpha_2 = \left[\frac{[H^+]^2}{K_1K_2} + \frac{[H^+]}{K_2} + 1\right]^{-1}$$

The use of α_0, α_1, α_2, in computing the composition of a closed carbonate system is demonstrated in Fig. 12.

Systems Open to the Atmosphere. A very elementary system showing some of the characteristics of the carbonate system in natural waters is obtained by equilibrating pure water with a gas phase (e.g. the atmosphere) containing CO_2 at a constant partial pressure. Such a system will remain in equilibrium with p_{CO_2} despite any variation of pH by addition of strong acid or base. This simple model has its counterpart in nature when CO_2 reacts with bases of rocks, e.g., clays and silicates.

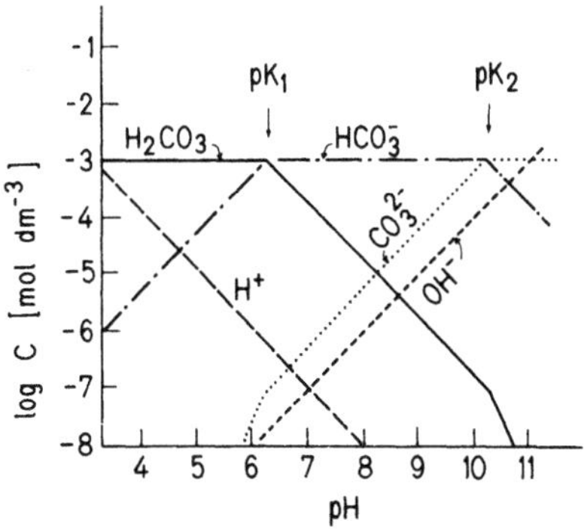

Fig. 12. Speciation in a solution of $C_T = 10^{-3}$ M as a function of pH. The concentration of each species is given by:
$[H_2CO_3^*] = \alpha_0 C_T$, $[HCO_3^-] = \alpha_1 C_T$, $[CO_3^{2-}] = \alpha_2 C_T$

(After [14])

The distribution of solute species for the CO_2/H_2O model is shown in Fig. 13. The equilibrium concentrations of all species may be expressed as a function of p_{CO_2} and H^+.

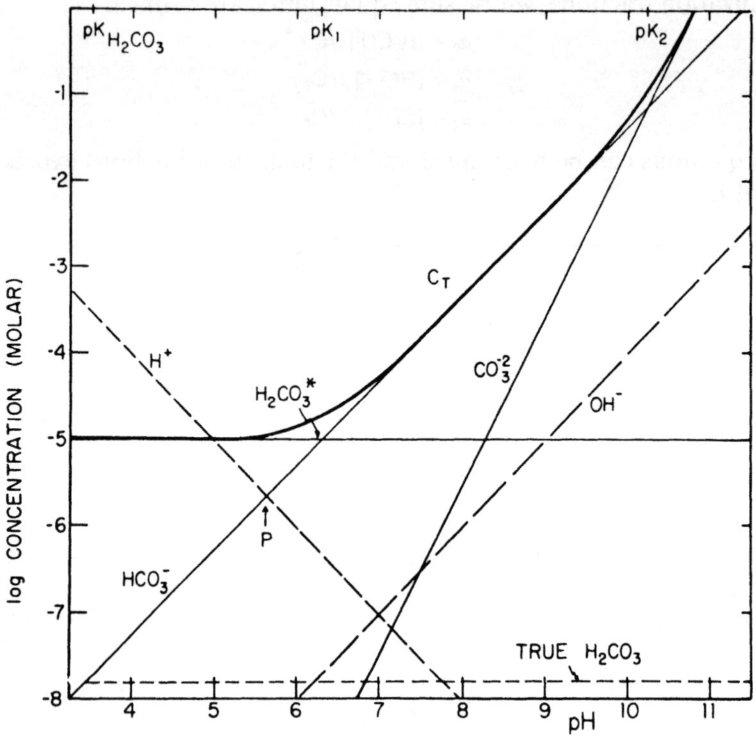

Fig. 13. Logarithmic concentration – pH equilibrium diagram for the aqueous carbonate system open to the atmosphere. Water is equilibrated with the atmosphere ($p_{CO_2} = 10^{-3.5}$ atm) and the pH is adjusted with strong base or strong acid. The constants (25 °C) $pK_H = 1.5$, $pK_1 = 6.3$, $pK_2 = 10.25$, pK (hydration of CO_2) $= -2.8$ have been used. The pure CO_2 solution is characterized by the proton condition $[H^+] = [HCO_3^-] + 2[CO_3^{2-}] + [OH^-]$ (see point *P*) and the equilibrium concentrations $-\log\ [H^+] = -\log\ [HCO_3^-] = 5.65$; $-\log[CO_2aq] = -\log[H_2CO_3^*] = 5.0$; $-\log[H_2CO_3] \sim 7.8$; $-\log[CO_3^{2-}] = 8.5$ (From [14], reprinted by permission of John Wiley and Sons, Inc.)

$$[H_2CO_3^*] = p_{CO_2} K_H$$

$$[HCO_3^-] = p_{CO_2} K_H \frac{K_1}{[H^+]}$$

$$[CO_3^{2-}] = p_{CO_2} K_H \frac{K_1 K_2}{[H^+]^2}$$

$$C_T = \frac{p_{CO_2} K_H}{\alpha_0}$$

The equilibrium pH of the system pure water and CO_2 at atmospheric pressure is ~ 5.65, as depicted by point P in Fig. 13.

Complexation Chemistry

Two principle types of metal-ligand bonds can be distinguished: i) ion pairs, in which metal and ligand retain their hydration spheres, and bonding is through one or more water molecules and principally electrostatic; ii) coordination complexes, in which the ligand is immediately adjacent to the metal and there is some degree of electron donation from the ligand to the metal [19, 20]. Certain organic ligands have more than one functional group capable of

Table 6. Formulation of stability constants[a]. (From [14], reprinted by permission of John Wiley and Sons, Inc.)

I. Mononuclear complexes

(a) Addition of ligand

$$M \xrightarrow[K_1]{L} ML \xrightarrow[K_2]{L} ML_2 \ldots \xrightarrow[K_i]{L} ML_i \ldots \xrightarrow[K_n]{L} ML_n$$

$$\text{——} \beta_2 \text{——}\rightarrow$$

$$\text{——} \beta_i \text{————————}\rightarrow$$

$$\text{——} \beta_n \text{——————————————}\rightarrow$$

$$K_i = \frac{[ML_i]}{[ML_{(i-1)}][L]} \qquad (1)$$

$$\beta_i = \frac{[ML_i]}{[M][L]^i} \qquad (2)$$

(b) Addition of protonated ligands

$$M \xrightarrow[{}^*K_1]{HL} ML \xrightarrow[{}^*K_2]{HL} ML_2 \ldots \xrightarrow[{}^*K_i]{HL} ML_i \ldots \xrightarrow[{}^*K_n]{HL} ML_n$$

$$\text{——} {}^*\beta_2 \text{——}\rightarrow$$

$$\text{——} {}^*\beta_i \text{————————}\rightarrow$$

$$\text{——} {}^*\beta_n \text{——————————————}\rightarrow$$

$${}^*K_i = \frac{[ML_i][H^+]}{[ML_{(i-1)}][HL]} \qquad (3)$$

$${}^*\beta_i = \frac{[ML_i][H^+]^i}{[M][HL]^i} \qquad (4)$$

II. Polynuclear complexes

In β_{nm} and ${}^*\beta_{nm}$ the subscripts *n* and *m* denote the composition of the complex M_mL_n formed. [If $m = 1$, the second subscript (= 1) is omitted.]

$$\beta_{nm} = \frac{[M_mL_n]}{[M]^m[L]^n} \qquad (5)$$

$${}^*\beta_{nm} = \frac{[M_mL_n][H^+]^n}{[M]^m[HL]^n} \qquad (6)$$

[a] The same notation as that used in L. G. Sillén and A. E. Martell: *Stability Constants of Metal-Ion Complexes*. Special Publ., No. 17, The Chemical Society, London, 1964, is used

forming coordinate bonds; complexes of metals with these "multidentate" ligands are referred to as chelates. Chelate complexes are generally much more stable than simple complexes or ion pairs. A standard notation for the expression of equilibrium constants for metal-ligand reactions is given in

Table 7. Classification of metal ions and other Lewis acids [a]

"A" and "B" metal ions [b]

A-Metal cations	Transition metal cations	B-metal cations
Electron configuration of inert gas. Low polarizability. "*Hard* Spheres"	1–9 outer shell electrons; spherically not symmetrical	Electron number corresponds to Ni^0, Pd^0 and Pt^0 (10 or 12 outer shell electrons). Low electronegativity, high polarizability. "*Soft* Spheres"
(H^+), Li^+, Na^+, K^+, Be^{2+}, Mg^{2+}, Ca^{2+}, Sr^{2+}, Al^{3+}, Sc^{3+}, La^{3+}, Si^{4+}, Ti^{4+}, Zr^{4+}, Th^{4+}	V^2, Cr^2, Mn^{2+}, Fe^{2+}, Co^{2+}, Ni^{2+}, Cu^{2+}, Ti^{3+}, V^{3+}, Cr^{3+}, Mn^{3+}, Fe^{3+}, Co^{3+}	Cu^+, Ag^+, Au^+, Te^+, Ga^+, Zn^{2+}, Cd^{2+}, Hg^{2+}, Pb^{2+}, Sn^{2+}, Tl^{3+}, Au^{3+}, In^{3+}, Bi^{+3}

According to Pearson's [c] *Hard and soft acids*

Hard acids	Borderline	Soft acids
All A-metal cations plus Cr^{3+}, Mn^{3+}, Fe^{3+}, Co^{3+}, UO^{2+}, VO^{2+}	All bivalent transition metal cations plus Zn^{2+}, Pb^{2+}, Bi^{3+}	All B-metal cations minus Zn^{2+}, Pb^{2+}, Bi^{3+}
As well as species like BF_3, BCl_3, SO_3, RSO_2^+, RPO_2^+, CO_2, RCO^+, R_3C^+	SO_2, NO^+, $B(CH_3)_3$	All metal atoms, bulk metals I_2, Br_2, ICN, I^+, Br^+
Preference for ligand atom		
N ≫ P		P ≫ N
O ≫ S		S ≫ O
F ≫ Cl		I ≫ F
Qualitative generalizations on stability sequence		
Cations	Cations	
$\text{Stability} \simeq \text{prop.} \frac{\text{Charge}}{\text{Radius}}$	Irving-Williams order [d] $Mn^{2+} < Fe^{2+} < Co^{2+} < Ni^{2+} < Cu^{2+} > Zn^{2+}$	
Ligands		Ligands
F > O > N = Cl > Br > I > S		S > I > Br > Cl = N > 0 > F
$OH^- > RO^- > RCO_2^-$		
$CO_3^2 \gg NO_3^-$		
$PO_4^{3-} \gg SO_4^{2-} \gg ClO_4^-$		

[a] After Stumm and Morgan, Aquatic Chemistry, Wiley Interscience, New York, 1970, p. 259
[b] S. Ahrland, S. J. Clatt and W. R. Davies, Quart. Rev. (London) *12*, 265 (1968)
[c] R. G. Pearson, J. Amer. Chem. Soc. *85*, 3533 (1963)
[d] H. Irving and R. J. P. Williams, J. Chem. Soc. *1953*, 3192

Table 8. Equilibrium model [a] for fresh water: Speciation of metal ions [b]

Metal	Total concentration	Inorganic fresh water[c]			Inorganic fresh water[c] + Organic material[d]						
		Free metal	Major inorganic complex		Free metal	Organic complexes[e]					
						Acetate	Citrate	Tartrate	Glycinate	Gutamate	Phthalate
Ca(II)	2.7	2.7	4.6	$CaHCO_3^+$	2.7	7.0	5.2	5.6	9.1	8.6	5.7
Mg(II)	3.7	3.7	5.1	$MgSO_4$	3.7	8.0	7.0	7.1	8.0	9.1	-
Fe(III)	Saturated[f]	17.7	8.7	$Fe(OH)_2^+$	17.7	19.0	7.2	-	15.1	-	-
Mn(II)	7.0	7.0	8.5	$MnSO_4$	7.0	11.3	9.7	-	11.5	11.1	-
Cu(II)	7.0	7.5	7.2	$CuCO_3$	9.9	13.1	7.0	11.3	9.4	9.4	11.4
Zn(II)	6.7	6.7	8.2	$ZnSO_4$	6.7	10.3	10.5	8.9	9.6	8.6	9.1
Cd(II)	7.7	7.7	9.2	$CdSO_4$	7.8	11.5	9.2	9.6	11.5	10.3	9.7
Pb(II)	7.0	8.0	7.1	$PbCO_3$	8.0	11.0	8.9	8.8	10.1	-	9.2
Ag(I)	9.0	9.2	9.5	AgCl	9.2	13.8	17.5	-	13.3	-	-

[a] Computation made with the program MINEQL [15]

[b] All concentrations given as -log (mol/liter)

[c] pH = 7.0, T = 25 °C; Free Ligand Concentration: $pSO_4 = 3.4$; $pHCO_3 = 3.1$; $pCO_3 = 6.1$; pCl = 3.3

[d] 7×10^{-6} mol/liter of each of the organic ligands; this corresponds to 2.3 mg/liter soluble organic carbon of the approximate composition $C_{13}H_{17}O_{12}N$

[e] Concentrations refer to sum of all complexes, e. g., CuCit, CuHCit, $CuCit_2$

[f] Solution in equilibrium with $Fe(OH)_3(s)$

Table 6. Complexation reactions are generally fast and equilibrium concepts are applicable for natural waters.

The concept of hard und soft acids and bases has been established to assess the tendency of metals and ligands to form ion-pair or coordinative complexes, and the relative stabilities of metals with groups of ligands (Table 7). In natural waters the inorganic ligands capable of influencing metal speciation are OH^-, CO_3^{2-}, SO_4^{2-}, and Cl^-. In addition, organic matter with O and N functional groups may be present at concentrations high enough to influence metal speciation [21]. However, an equilibrium computation for the speciation of major and minor metal ions with organic functional groups of the type and concentration expected in natural waters show that the organics may have little influence (Table 8).

Collections of stability constants are available [22, 23].

Solubility Equilibria

Dissolution and precipitation reactions are responsible for the concentrations of major cations and anions in natural waters. It is difficult to generalize about rates of precipitation and dissolution other than to recognize that they are usually slower than reactions between dissolved species. Kinetic data for geochemically important solid-solution reactions are generally lacking. Frequently the solid phase formed initially is metastable with respect to another more stable phase. Examples of such metastability include the formation of aragonite under certain conditions instead of calcite, the more stable form of calcium carbonate, and the oversaturation of quarz in most natural waters. The solubilities of most inorganic salts increase with temperature. However, some compounds of geochemical interest ($CaCO_3$, $CaSO_4$) decrease in solubility with increasing temperature. The dependence of solubility with pressure is slight, but critical in some cases (e.g. $CaCO_3$ in the ocean).

Surface Equilibria

Significant for the transport of many trace substances in natural waters are the surfaces of suspended particulate matter. The surfaces may be those of clay minerals, alumina or silica particles, iron or manganese oxide coatings, organic particles or organic coatings. In many cases, the surfaces may be represented as having surface hydroxyl groups $\equiv$ROH which undergo hydrolysis, complexation, and ligand exchange reactions analogous to those of the same functional groups on dissolved molecules [24]. This "surface complexation" model has been used, for example, to describe the adsorption of metals on silica and organic acids on alumina (Fig. 14).

Photosynthesis and Respiration

Photosynthesis may be viewed conceptually as an energy consuming process leading to products of high oxidation intensity (organic matter). Respiration is the reverse reaction, the combination of the products of photosynthesis with

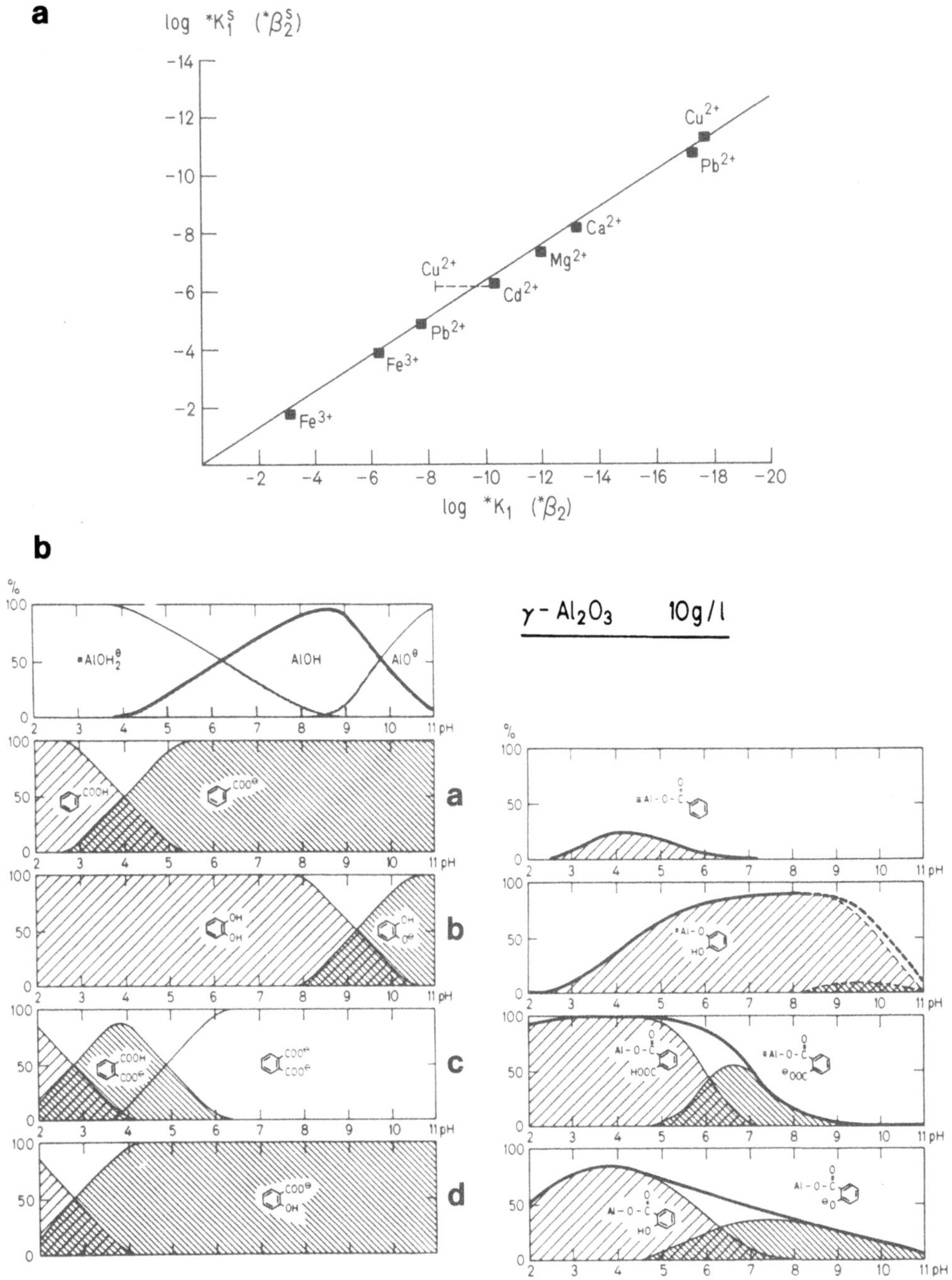

Fig. 14 a, b. Adsorption on oxide surfaces [24, 25]. **a** Metal Ions: Correlation between the stability constants for the surface complexes ($^*K_1^s$ $^*\beta_2^s$) and the corresponding hydroxo-complexes ($^*K_1$, $^*\beta_2$). **b** Organic Acids: Predominance range of the individual species in solution and on the surface. Left side: Above: γ–Al_2O_3 surface, *a* benzoic acid, *b* pyrocathecol, *c* phthalic acid, *d* salicylic acid. Right side: The calculated adsorption curves γ·Al_2O_3 10 g/l; organic acids 0.5 mmol/l; 0.1 *M* $NaClO_4$)

the release of energy. The photosynthesis/respiration reaction may be written (simplified) as

$$CO_2 + H_2O \underset{\text{respiration}}{\overset{\text{photosynthesis}}{\rightleftarrows}} CH_2O + O_2.$$

In the aquatic environment the reaction may be written more precisely to include trace elements and to reflect the constitution of algal protoplasm.

$$106\,CO_2 + 16\,NO_3^- + HPO_4^{2-} + 122\,H_2O + 18\,H^+ \underset{R}{\overset{P}{\rightleftarrows}} C_{106}\,H_{263}\,O_{110}\,N_{16}\,P_1 + 138\,O_2.$$

The relationship between photonsynthesis/respiration and the nutrient cycle is seen in Fig. 15.

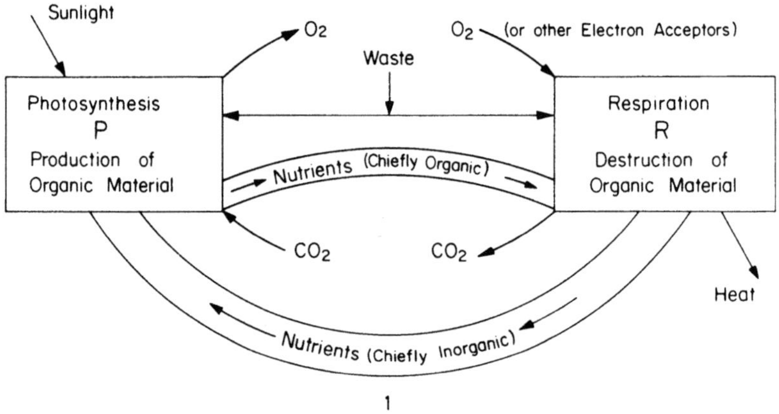

Fig. 15. Photosynthesis and the nutrient cycle. Solar energy and inorganic nutrients are converted by photosynthesis into chemical energy stored within the organic material of algae and plants. Heterotrophic consumers mobilize this energy by degrading organic matter during respiration and other oxidative processes

Redox Chemistry

Only a few elements are major participants in redox processes in natural waters: C, N, O, S, Fe, Mn. The redox reactions in which these elements participate are often biologically mediated; some of these processes are given in Table 9.

Theoretically, the redox intensity of a system can be defined and measured in a way similar to pH. The theoretical pε is related to the redox potential, E_H, of the system by the Nernst equation:

$$p\varepsilon = \frac{E_H}{2.3\,RT/F} = \frac{E^0}{2.3\,RT/F} + \frac{1}{n}\log\frac{\pi\,[ox_i]^{n_i}}{\pi\,[red_i]^{n_i}}.$$

Table 9. Sequence of microbially mediated redox reactions [a]

In presence of excess organic matter [b]

Aerobic respiration
$1/4\ CH_2O + 1/4\ O_2 = 1/4\ CO_2 + 1/4\ H_2O$

Denitrification
$1/4\ CH_2O + 1/5\ NO_3^- + 1/5\ H^+ = 1/4\ CO_2 + 1/10\ N_2 + 7/20\ H_2O$

Nitrate reduction
$1/4\ CH_2O + 1/8\ NO_3^- + 1/4\ H^+ = 1/4\ CO_2 + 1/8\ NH_4^+ + 1/8\ H_2O$

Reduction of Mn-oxides
$1/4\ CH_2O + 1/2\ MnO_2(s) + H^+ = 1/4\ CO_2 + 1/2\ Mn^{2+} + 3/4\ H_2O$

Fermentation reactions
$3/4\ CH_2O + 1/4\ H_2O = 1/4\ CO_2 + 1/2\ CH_3OH$

Reduction of Fe-oxides
$1/4\ CH_2O + FeOOH(s) + 2\ H^+ = 1/4\ CO_2 + Fe^{2+} + 7/4\ H_2O$

Sulfate reduction
$1/4\ CH_2O + 1/8\ SO_4^{2-} + 1/8\ H^+ = 1/4\ CO_2 + 1/8\ HS^- + 1/4\ H_2O$

Methane formation
$1/4\ CH_2O = 1/8\ CO_2 + 1/8\ CH_4$

In presence of excess oxygen [c]

Aerobic respiration
$1/4\ O_2 + 1/4\ CH_2O = 1/4\ CO_2 + 1/4\ H_2O$

Sulfide Oxidation
$1/4\ O_2 + 1/8\ HS^- = 1/8\ SO_4^{2-} + 1/8\ H^+$

Fe(II) oxidation
$1/4\ O_2 + Fe^{2+} + 3/2\ H_2O = FeOOH(s) + 2\ H^+$

Mn(II) oxidation
$1/4\ O_2 + 1/2\ Mn^{2+} + 1/2\ H_2O = 1/2\ MnO_2(s) + H^+$

Nitrification
$1/4\ O_2 + 1/8\ NH_4^+ + 1/8\ H_2O = 1/8\ NO_3^- + 1/4\ H^+$

[a] After [14], pp. 336–337
[b] Water contains incipiently O_2, NO_3^-, SO_4^{2-}, HCO_3^-. Examples: hypolimnetic layers of eutrophic lake, sediments, sewage treatment plant digester
[c] Water contains incipiently organic material, HS^-, NH_4^+, Fe(II), Mn(II). Examples: aerobic waste treatment, self purification in streams, epilimnetic waters

However since many redox reactions are slow to reach equilibrium, and some do not couple readily, there may be several apparent levels of pε in the same local environment. Additionally, the standard tool for measuring redox intensity, the platinum electrode, is subject to the same problems of non-equilibrium and non-coupling, and may yield still another value for redox intensity [26, 27]. Although redox reactions may be far from equilibrium, they tend to occur in the order of their energy yield. This may be seen in the framework of the biochemical redox cycle (Fig. 16). The oxidation of organic matter produ-

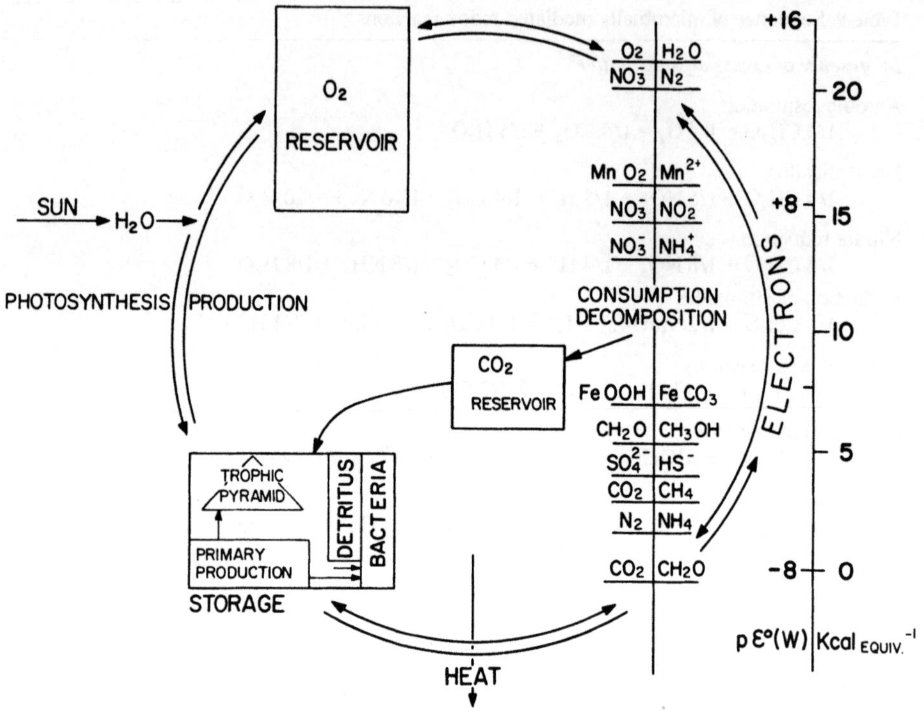

Fig. 16. The biochemical redox cycle. Heterotrophs utilize the oxidants which yield the greatest amount of energy, as long as these oxidants are available; then less energy yielding oxidants are utilized

ced in photosynthesis yields energy; the amount of energy depends on the nature of the oxidant, or electron acceptor. The energetically most favorable oxidant is oxygen; after oxygen is depleted there follows a succession of organisms capable of reducing NO_3^-, MnO_2, FeOOH, SO_4, CO_2, with each oxidant yielding successively less energy for the organism mediating the reaction.

Transformation of Organic Carbon

In fresh water the concentrations of organic carbon are typically a few milligrams per liter, but occasionally concentrations may be as high as 50 mg/l. In the oceans concentrations may range from 0.5 to 1.2 mg C per liter, the concentrations being higher near the surface, and practically constant in deep waters.

Organic carbon in natural waters is often expressed in collective parameters: dissolved organic carbon (DOC); total organic carbon (TOC); or in chemical or biological oxygen demand (COD or BOD). TOC measurements are generally made by combustion of the sample, after removal of inorganic carbon, and analysis for the CO_2 released. COD is the equivalent amount of oxygen consumed by an organic sample during permanganate or dichromate oxidation. BOD is the amount of oxygen consumed by organisms mediating the oxidation of the organic substrate.

Whereas these collective parameters are useful in gaining a rough idea of the organic load in natural waters, analyses for individual compounds and compound classes are necessary [28, 29]. Only a small fraction of the natural organic materials found in natural waters has been identified structurally, primarily amino acids, sugars and their biopolymers, fatty acids, alcohols and their esters, sterols, hydrocarbons, pigments, and vitamins. In addition there is a large amount of poorly defined, relatively high molecular weight material of partially aromatic character with phenolic and carboxylic acid functional groups known collectively as gelbstoffe, separated operationally into a higher molecular weight, less soluble *humic* fraction, and a more soluble *fulvic* fraction. Among the anthropogenic organic compounds, extensive work on the identification of mineral oil residues, pesticides, and chlorinated hydrocarbons has been done.

The Composition of Fresh Waters and the Hydrologic Cycle

Natural waters acquire their chemical characteristics by chemical reactions with solids, liquids, and gases with which they have come into contact during the various parts of the hydrologic cycle. Waters vary in their chemical composition, but these variations are at least partially understandable if the environmental history of the water and the chemical reactions of the rock-water-atmosphere systems are considered.

As a first approximation the composition of seawater may be interpreted as the result of a gigantic acid-base reaction: acids of volcanoes (SO_2, HCl, H_2S) have reacted with bases of rocks (oxides, carbonates, silicates) to yield the dissolved components of seawater [30–32].

Similarly the composition of fresh water may be interpreted as the result of the reaction of carbon dioxide from the atmosphere with bases of rocks [33, 34]. Water acts as a transport medium as well as a chemical reagent in the conversion of rock to dissolved matter, soil, sediments and sedimentary rock.

Rainwater

The dissolved and suspended materials in rainwater are transported to the atmosphere from three principle sources: marine spray, terrigenous dust, and anthropogenic by-products [35–37]. The exact composition of rainwater over land is highly variable and depends on local meteorological conditions, i.e., the movement of air masses over regions affected by various local inputs. Whereas it is often possible to track the course of individual storms and to relate the composition of the rainwater to the course of the storm, it is very difficult to reach generally applicable conclusions about such phenomena on account of the extreme variability of weather conditions. It is possible, however, to take yearly averages for the concentration of elements in rainwater und relate them to prevailing meteorological and geographical conditions.

In regions affected by marine air masses sea spray is the primary source of the elements Cl, Na, Ca, Mg, Sr in rainwater. Because Cl is a major component of seawater and under normal circumstances is relatively rare in terrigenous or anthropogenic atmospheric input, the relationship of

$$[X]_{rain}/[Cl]_{rain} \text{ to } [X]_{sea}/[Cl]_{sea}$$

can be used to gain an idea of the amount of an element X originating from non-marine sources. The elements Si, Fe, Al, As, and many trace metals are characteristic of terrigenous air masses. The concentration of Al or Si can be used as a normalizing factor for terrigenous air masses in the same way that Cl is used for marine air masses.

The primary atmospheric input of many trace metals, as well as N, S, and C in many regions is anthropogenic. The burning of fossile fuel releases great quantities of metals; the release of N- and S-oxide, and further oxidation results in the formation of strong acids and acid rain.

Table 10. Examples of typical weathering reactions [a]

I. Congruent dissolution reactions

$SiO_2(s) + 2\,H_2O = H_4SiO_4$
quartz

$CaCO_3(s) + H_2CO_3^* = Ca^{2+} + 2\,HCO_3^-$
calcite

$Al_2O_3 \cdot 3\,H_2O(s) + 2\,H_2O = 2\,Al(OH)_4^- + 2\,H^+$
gibbsite

$Ca_5(PO_4)_3(OH)(s) + 4\,H_2CO_3^* = 5\,Ca^{2+} + 3\,HPO_4^{2-} + 4\,HCO_3^- + H_2O$
apatite

II. Incongruent dissolution reactions

$NaAlSi_3O_8(s) + H_2CO_3^* + 9/2\,H_2O$
albite (Na-feldspar)

$= Na^+ + HCO_3^- + 2\,H_4SiO_4 + 1/2\,Al_2Si_2O_5(OH)_4(s)$
kaolinite

$7\,NaAlSi_3O_8(s) + 6\,H_2CO_3^* + 20\,H_2O$
albite (Na-feldspar)

$= 6\,Na^+ + 6\,HCO_3^- + 10\,H_4SiO_4 + 3\,Na_{0.33}Al_{2.33}Si_{3.67}O_{10}(OH)_2(s)$
Na-montmorillonite

$Al_2Si_2O_5(OH)_4(s) + 5\,H_2O = 2\,H_4SiO_4 + Al_2O_3 \cdot 3\,H_2O$
kaolinite; gibbsite

$CaMg(CO_3)_2(s) + Ca^{2+} = Mg^{2+} + 2\,CaCO_3(s)$
dolomite; calcite

[a] After [14], p. 390

Weathering Process

Dissolution reactions (chemical weathering) take place because many constituents of the earth's crust are thermodynamically unstable in the presence of water and the atmosphere [33, 38]. Most important among the weathering reactions is the incongruent (partial) dissolution of aluminum silicates, which may be represented schematically as:

$$\text{Me–Al-silicate} + H_2CO_3^* + H_2O = n\,HCO_3^- + H_4SiO_4 + Me^{n+} + \text{Al-Silicate}$$

where Me is a cation such as Na^+, K^+, Ca^{2+} Mg^{2+} (Table 10). In essence, primary minerals are converted to secondary minerals. The secondary minerals are frequently structurally ill-defined or x-ray amorphous. The structural breakdown of aluminum silicates is accompanied by a release of cations and usually of silicic acid.

Minerals of the kaolinite group are the main alteration products of feldspar weathering. In addition to kaolinite, montmorillonites and micas are often found as intermediates. The relatively slow chemically weathering process may be accelerated by physical weathering, the process by which rocks are mechanically broken up to expose a greater (relative) surface area.

The composition of fresh waters is generally dominated by these weathering reactions. While Ca^+, HCO_3^-, H^+, and Mg^{2+} may be controlled by dissolution of carbonate rock, the source of Na^+, K^+, H_4SiO_4, and possible also of Ca^{2+} and Mg^{2+}, is the silicate minerals that make up much of the rock in contact with underground waters and streams. The composition of natural

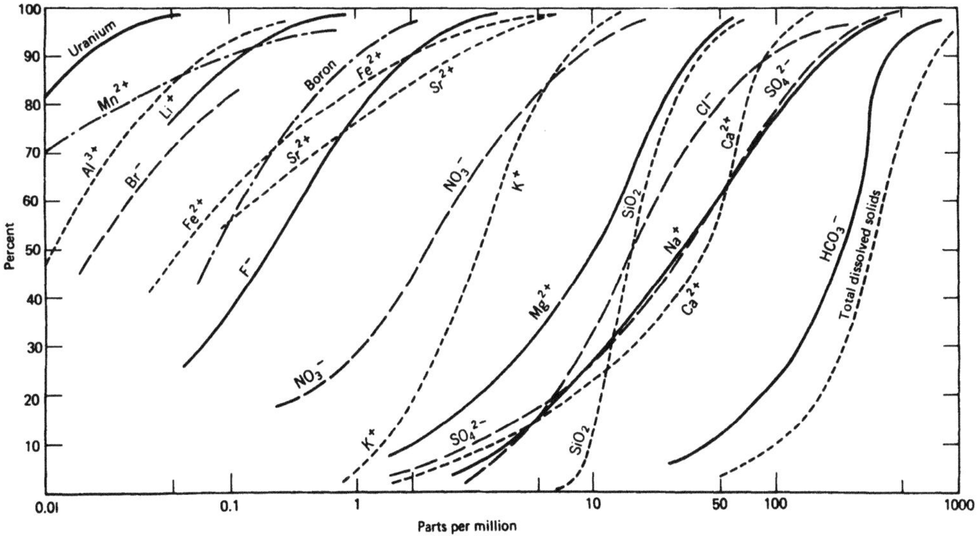

Fig. 17. Cumulative distribution of inorganic species terrestrial waters. The curves represent the percentage of waters with concentrations below that indicated on the horizontal axis. Data are mostly from the United States from various sources. (From [34], reprinted by permission of John Wiley and Sons, Inc.)

Table 11. Examples of natural waters. (From [14], reprinted by permission of John Wiley and Sons, Inc.)

	1	2	3	4	5	6	7	8	9	10	11
Type [a]	Stream	Stream	Stream	Lake Erie	Ground water	Ground water	Ground water	Ground water	Ground water	Ground water	Closed basin lake
Type of rocks being drained	Granite	Quartzite	Sandstone [b]		Granite	Gabbro plagioclase	Sandstone	Shale	Limestone	Dolomite	Soda lake
*p*H	7.0	6.6	8.0	7.7	7.0	6.8	8.0	7.3	7.0	7.9	9.6
*p*Na	4.0	4.6	4.3	3.4	3.4	3.0	3.3	2.6	3.0	3.5	0.0
*p*K	4.7	5.1	4.8	4.3	4.0	4.5	4.0	4.2	3.7	-	1.7
*p*Ca	4.0	4.3	3.1	3.0	3.5	3.1	3.0	2.5	2.7	2.8	4.5
*p*Mg	4.6	5.1	4.0	3.4	3.8	3.2	3.5	2.5	3.4	2.8	4.6
pH_4SiO_4	3.8	4.2	4.1	4.7	3.2	3.0	3.9	3.5	3.7	3.4	2.8
$pHCO_3$	3.6	4.0	2.9	2.7	2.9	2.5	2.6	2.1	2.3	2.2	0.4
*p*Cl	5.3	5.8	5.3	3.6	4.0	3.5	3.7	4.0	3.2	3.3	0.3
pSO_4	4.5	4.7	3.7	3.6	4.2	4.0	3.2	2.2	3.4	4.7	2.0
- log (ionic strength)	3.5	3.8	2.7	2.5	2.8	2.4	2.4	1.7	2.2	2.2	0.0

Note: $pX = -\log[X]$

[a] 1–3: "Small Streams in New Mexico", U.S. Geol. Surv. Bull., *1535 F* (1961). 4: J. Kramer, Geochim. Cosmochim. Acta, *29*, 921 (1965). Types 5–10 are from U.S. Geol. Surv. Bull., *440 F* (1963). 5: Granite McCormick Co. (Table 1). 6: Harrisburg (Table 2). 7: Home Wood (Table 3). 8: Cuyahoga (Table 5). 9: Edwards limestone (Table 6). 10: Precambrian dolomite (Table 7). 11: "Albert Summer Lake Basin, Oregon", North. Ohio Geol., J. L. Rau, Ed., 1966, p. 181

[b] With slate and limestone beds

waters from several geological environments is shown in Table 11. The variation in the composition of fresh waters is shown in Fig. 17.

Because H_2CO_3 is usually the acid in the attack of water on the primary silicates, HCO_3^-, is the predominant anion in most fresh waters. Because of respiration by organisms, the CO_2 concentration in soils can be a few hundred times that in the atmosphere. Correspondingly groundwaters contain higher concentrations of dissolved materials and are at lower pH.

Minor Elements

On a global scale, man's activity has had little effect on the flux of heavy metals in the hydrogeosphere [39]. However, in centers of civilization, man's input can easily exceed natural fluxes of heavy metals.

Table 12. Atmospheric vs fluvial transport of trace metals [35]

Element	Atmospheric rainout	Stream load	Rain/Stream %	
Al	33,000	17,000,000	0.2	LP
Ti	2,700	840,000	0.3	LP
Sm	3	890	0.3	LP
Fe	49,000	9,900,000	0.5	LP
Mn	3,000	160,000	1.9	LP
Co	62	3,500	1.8	LP
Cr	720	17,000	4.2	LP
V	1,900	24,000	7.9	LP
Ni	1,200	13,000	9.2	LP
Sn	-	2,900	-	
Cu	2,600	11,000	23.6	AP
Cd	510	1,200	42.5	AP
Zn	10,000	25,000	40.0	AP
As	2,900	3,000	96.7	AP
Se	200	180	111.1	AP
Sb	340	1,000	34.0	AP
Mo	310	700	44.3	AP
Ag	10	130	7.7	LP (?)
Hg	410	52	788.5	AP
Pb	5,700	4,700	121.3	AP

LP = lithophile: AP = atmophile
Rainout and stream load values are in units of 10^8 g/yr

Certain trace elements are transported primarily through the atmosphere, while others are transported primarily through streams (Table 12). A comparison of natural and anthropogenic contributions of trace metals in the atmosphere is given in Table 13.

Table 13. Natural and anthropogenic sources of atmospheric emissions. See [35] for assumptions and discussion of sources of data

Element	Continental dust flux	Volcanic dust flux	Volcanic gas flux	Industrial particulate emissions	Fossil fuel flux	Total emissions (industrial + fossil fuel)	Atmospheric interference factor (%)
Al	356,500	132,750	8.4	40,000	32,000	72,000	15
Ti	23,000	12,000	-	3,600	1,600	5,200	15
Sm	32	9	-	7	5	12	29
Fe	190,000	87,750	3.7	75,000	32,000	107,000	39
Mn	4,250	1,800	2.1	3,000	160	3,160	52
Co	40	30	0.04	24	20	44	63
Cr	500	84	0.005	650	290	940	161
V	500	150	0.05	1,000	1,100	2,100	323
Ni	200	83	0.0009	600	380	980	346
Sn	50	2.4	0.005	400	30	430	821
Cu	100	93	0.012	2,200	430	2,630	1,363
Cd	2.5	0.4	0.001	40	15	55	1,897
Zn	250	108	0.14	7,000	1,400	8,400	2,346
As	25	3	0.1	620	160	780	2,786
Se	3	1	0.13	50	90	140	3,390
Sb	9.5	0.3	0.013	200	180	380	3,878
Mo	10	1.4	0.02	100	410	510	4,474
Ag	0.5	0.1	0.0006	40	10	50	8,333
Hg	0.3	0.1	0.001	50	60	110	27,500
Pb	50	8.7.	0.012	16,000	4,300	20,300	34,583

All fluxes are in units of 10^8 g/yr
Interference factor = (total emissions/continental + volcanic fluxes) × 100

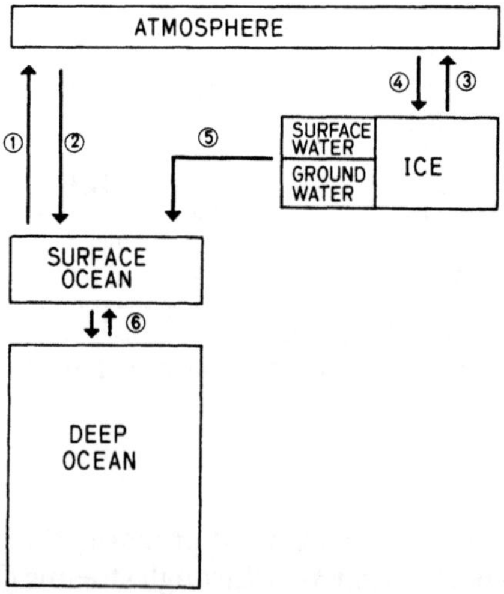

Fig. 18. Major reservoirs in the hydrologic cycle. See Table 14 for magnitudes of reservoirs and fluxes

The Hydrologic Cycle

The dynamic equilibrium or steady state existing between evaporation, precipitation, percolation, runoff, and circulation comprises the hydrologic cycle [36, 40]. The hydrologic cycle is driven by solar energy; not only does the sun provide energy for evaporation, but the uneven heating of the earth's surface causes circulation of air masses and ocean currents.

A schematic diagram of the reservoirs and fluxes in the hydrologic cycle is given in Fig. 18. The capacity of the reservoirs and magnitude of the fluxes are given in Table 14. The capacities and fluxes can be expressed in units of volume, or in units of depth distributed over the land surface, or ocean surface, of the earth. The residence time is a measure of how long a given water molecule may reside in a given reservoir; it is defined by

$$\tau = \frac{V}{Q} = \frac{\text{volume}}{\text{input/output}}.$$

Table 14. Reservoirs and fluxes in the hydrologic cycle

Reservoir	Volume $[10^{15}\ m^3]$	Equivalent depth [a] [m]	Residence time [y]	Total flux through reservoir [b] $[10^{15}\ m^3\ y^{-1}]$
Atmosphere	0.0015	0.003 T	0.0036	0.43
Ocean-Surface	57.	160. O	80	1.07
Ocean-Deep	1100.	3200. O	1600	0.71
Glaciers	26.	170. L		
Lakes and rivers	0.20	1.3 L		0.11
Groundwater	7.6	50. L		
Fluxes [b]				
Description	Flux $[10^{15}\ m^3\ y^{-1}]$	Equivalent depth [a] $[my^{-1}]$		
1 Surface ocean to atmosphere	0.36	1.00 O		
2 Atmosphere to surface ocean	0.32	0.90 O		
3 Land to atmosphere	0.072	0.47 L		
4 Atmosphere to land	0.11	0.70 L		
5 Land to surface ocean	0.036	0.23 L 0.10 O		
6 Surface ocean/ deep ocean	0.71	2.00 O		

[a] Equivalent depth based on the following surface areas:
T $0.510 \times 10^{15}\ m^3$, surface area of earth
O $0.7 \times T$, surface area of oceans
L $0.3 \times T$, surface area of land
[b] See Fig. 18 for identification of fluxes

The primary reservoir of water is the oceans, which hold 97% of the earth's water. About 5% of this amount is in the upper mixed layer, where biological activity occurs, and the remainder in the deep ocean, where the residence time is relatively long. The next largest reservoir is ice, containing about 3% of the earth's water; again this reservoir is relatively immobile. The other reservoir of significant size is groundwater; groundwater moves too slowly to be a significant transport mechanism in the hydrologic cycle. Surface waters and atmospheric water comprise a very small fraction of the earth's water; the fluxes into and out of these reservoirs are relatively high and the residence time short.

The mixing of deep ocean water with surface ocean water occurs at a rate almost twenty times greater than the inflow of river water. Thus a change in composition in surface ocean water induced by a change in river water composition would be slow. The ocean mixing is the hydrologic transport process of greatest magnitude. The next greatest transport process in the hydrologic cycle is evaporation from the ocean; however 90% of this vapor falls back to the ocean and 10% is transported over the land. Of the water that falls on land, about ⅔ evaporates again and about ⅓ runs off to the ocean. The world wide precipitation over land is unevenly distributed and may range from 0 to 10 m per year; the average is 0.7 meters per year. According to the annual amount of precipitation the following biotopes result: 0–0.25 m desert; 0.25–0.40 m grass lands or open wood land; 0.40–1.25 m dry forests; > 1.25 m wet forests.

In assessing the impact of a pollutant on a body of water, it is important to consider the residence time of the water and the pollutant. The pollutant may remain in the system for a longer time than water (e.g. phosphorous recycling in lakes) or for a shorter time than water (e.g. volatile organic compounds). In general, systems with longer hydrologic residence times are more difficult to perturb, but more difficult to correct, while those with shorter residence time react more rapidly.

References

1. Eisenberg, D., Kauzmann W.: The Structure and Properties of Water. Oxford University Press, London 1969
2. Davis, C.M., Litovitz, T.A.: J. Chem. Phys. *42*, 2563 (1965)
3. Dorsey, N.E.: Properties of Ordinary Water-Substance in all its Phases; Water-vapor, Water, and all the Ices. Chem. Soc. Monograph No. 81, Reinhold Publishing Corp., New York 1941
4. Kell, G.S.: J.Chem. Eng. Data *12*, 66 (1967)
5. Malmberg, C.M., Maryott, A.A.: J. Res. Natn. Bur. Stand. *56*, 1 (1956)
6. Stokes, R.H., Mills, R.: International Encyclopedia of Physical Chemistry and Chemical Physics. Pergamon Press, Oxford 1965, Vol. I
7. Krindel, P., Eliezer, I.: Coord. Chem. Rev. *6*, 217 (1971)
8. Pople, J.A.: Proc. R.Soc. A *205*, 163 (1951)
9. Horne, R.A. (Ed.): Water and Aqueous Solutions, Structure, Thermodynamics, and Transport Processes. Wiley, New York 1972
10. Horne, R.A.: Marine Chemistry. The Structure of Water and the Chemistry of the Hydrosphere. Wiley, New York 1969

11. Kavanau, J.L.: Water and Solute-Water Interactions. Holden-Day Inc., San Francisco 1964
12. Stumm, W. (Ed.): Equilibrium Concepts in Natural Water Systems, Adv. Chem. Series 67, Amer. Chem. Soc., Washington 1967
13. Lerman, A.: In: Non-Equilibrium Systems in Natural Water Chemistry. Adv. Chem. Series 106, Amer. Chem. Soc., Washington 1971
14. Stumm, W., Morgan, J.J.: Aquatic Chemistry. Wiley-Interscience, New York 1970
15. Westall, J., Zachary, J., Morel, F.: MINEQL, A Computer Program for the Calculation of Chemical Equilibrium Composition of Aqueous Systems. Technical Note No. 18, Ralph Parsons Laboratory, Massachusetts Institute of Technology, Cambridge 1976
16. Garrels, R.M., Christ, C.: Minerals, Solutions, and Equilibria. Harper and Row, New York 1965
17. Deffeyes, K.S.: Limnol. Oceanogr. *10*, 412 (1965)
18. Dryssen, D., Sillén, L.G.: Tellus *19*, 110 (1967)
19. Basolo, F., Pearson, R.G.: Mechanism of Inorganic Reactions, A Study of Metal Complexes in Solution. Wiley, New York 1967
20. Burgess, J.: Metal Ions in Solution. Wiley, New York 1978
21. Singer, P. (Ed.): Trace Metals and Metal Organic Interactions in Natural Waters. Ann Arbor Science, Ann Arbor 1973
22. Sillén, L.G., Martell, A.E.: Stability Constants of Metal-Ion Complexes. Special Publ. No. 17 and 25, Chem. Soc., London 1964 and 1971
23. Smith, R.M., Martell, A.E.: Critical Stability Constants, Vol. I–IV. Plenum Press, New York 1976
24. Schindler, P.W.: Thalassia Jugoslavica *11*, 101 (1975)
25. Kummert, R.: Ph.D. Thesis, Nr. 6371, Federal Institute of Technology, Zürich 1979
26. Whitfield, M.: Limnol. Oceanogr. *19*, 857 (1974)
27. Whitfield, M.: Limnol. Oceanogr. *17*, 383 (1972)
28. Concepts in Marine Organic Chemistry. Proc. NATO/ONR Symposium, Edinburgh, 6–10 Sept. 1977. Marine Chemistry *5*, 303 (1978)
29. Hutzinger, O., Tulp, M., Zitko, V.: In: Aquatic Pollutants: Transformation and Biological Effects, Hutzinger, O., van Lelyveld, L.H., Zoeteman, B.C.J. (Eds.). Pergamon Press, Oxford 19;
30. Sillén, L.G.: In: Oceanography Sears, M. (Ed.), Amer. Assn. for the Adv. Sci., Washington 1961
31. Goldberg, E.D.: In: Oceanography, Sears, M., (Ed.), Amer. Ass. for the Adv. Sci., Washington 1961
32. Kramer, J.R.: Geochim. Cosmochim. Acta *29*, 921 (1965)
33. Garrels, R.M., Mackenzie, F.: In: Equilibrium Concepts in Natural Water Systems. Adv. Chem. Series 67; Amer. Chem. Soc., Washington 1967
34. Davis, S.N., DeWiest, R.: Hydrogeology. John Wiley and Sons, New York 1966
35. Lantzy, R.J., Mackenzie, F.T.: Geochim. Cosmochim. Acta *43*, 511 (1979)
36. Holland, H.D.: The Chemistry of the Atmosphere and Oceans. John Wiley and Sons, New York 1978
37. Kitano, Y. (Ed.): Geochemistry of Water, Benchmark Papers in Geology, Vol. 16; Dowden, Hutchinson, Ross, Stroudsburg, Pa. 1975
38. Bricker, O.P., Garrels, R.M.: In: Principles and Applications of Water Chemistry, Faust, S.D., Hunter, J.V., (Eds.). Wiley, New York 1967
39. Goldberg, E.D.: The Health of the Oceans, UNESCO Press, Paris 1976
40. Broeker, W.S.: Chemical Oceanography, Harcourt, Brace, Janovich, Inc., New York 1974

Chemical Oceanography

P. J. Wangersky

Department of Oceanography, Dalhousie University
Halifax, N.S. B3H 4J1, Canada

Earlier sections in this Handbook have discussed the ways in which the properties of water affect the chemistry of fresh water and the hydrological cycle. In this section we will examine the ways in which sea water differs from fresh water, how these differences affect the reactions occurring in sea water, and how these reactions are further affected by the addition of compounds resulting from man's activities.

Properties of Sea Water

The properties of sea water are in general determined by the properties of water. The anomalous latent heats of fusion and vaporization, the high heat capacities, and the high boiling and freezing points, all properties resulting from the great degree of structure present in liquid water and which serve to make water a most unusual liquid, are also present in sea water.

What makes sea water different from either fresh water or the various non-marine salt waters, such as the Dead Sea or the Great Salt Lake, is the constancy of composition of its salt content. Sea water is a medium of moderately high (~0.7 *M*) ionic strength. The relative composition of its

Table 1. Major ions of sea water

Cations	g/kg at S = 35‰	Anions	g/kg at S = 35‰
Na^+	10.77	Cl^-	19.354
Mg^{2+}	1.29	SO_4^{2-}	2.712
Ca^{2+}	0.412	Br^-	0.067
K^+	0.399	F^-	0.0013
Sr^{2+}	0.0079		

major constituents (Table 1) is constant enough so that the determination of one major component allows us to calculate the amounts of the other components present in the sample. Because of the accuracy and reproducibility of the measurement, the quantity usually chosen for measurement has been the chlorinity, a measure of the halide content of the sample. The analytical method normally used was the Mohr titration with silver nitrate, using potassium chromate as the indicator. Chlorinity was defined in terms of the atomic weight of silver, as the mass in grams of silver needed to precipitate the halogens in 328.5233 g of sea water, and was commonly expressed in parts per thousand (‰). The salinity, the grams of sea salt present in 1 kg of sea water after drying according to a specified technique, could then be calculated from the chlorinity by a relationship determined by an International Commission in 1899.

This technique for measuring salinity is seldom used now; instead, the salinity is calculated from the conductivity of the sea water, a technique as accurate as the Mohr titration and considerably faster. However, since much of the older work was presented in terms of salinity and chlorinity, by definition the two units are related by the equation

$$S(‰) = 1.80655 \; Cl \; (‰).$$

The relationship between salinity and conductivity at any temperature to be found in the oceans can be derived from the International Oceanographic Tables [1]. The units in which salinity and chlorinity are expressed are currently under review, and within the next few years will be revised to conform to ISO usage.

The total range of salinity in the open ocean is small; except for regions near shore, the salinity stays between 32–37.5‰. Salinities below this range generally indicate an admixture of fresh water, and a possible distortion of the ionic ratios. An implication of the constancy of ionic composition is that the mixing time of the ocean is relatively short, compared to the time constants of the processes of addition and removal of materials.

Within the range of normal variation, the distribution of salinity in the oceans is dependent upon those processes driving the oceanic current systems. This dependence of composition upon physical factors, rather than upon local additions and removals, is one of the major differences between sea water and even large bodies of fresh water, and is a function of scale. Man may alter the ionic ratios in a lake, and perhaps in an estuary, by a program of dredging and dumping; it is difficult even to imagine the size of a program which might alter the ionic ratios in oceanic sea water.

The pH of fresh water may range from 1.5 to 11; the total range of sea water pH is between 7.5 and 8.3, with most values falling between 7.8–8.2. This constancy in pH helps to supply a constant environment for the development of marine life.

The concentration of particulate matter in sea water is similarly relatively constant. The particulate organic carbon, which normally makes up some 30–50% of the total particulate matter, varies only between 3–20 μg C/l in the deep water, and surpasses 50 μg C/l in the surface waters only during plankton blooms. Thus the total particle load at any point in the surface waters may vary by a factor of 4 over the year, while the deeper water shows little or no seasonal variation. The equivalent variation in streams and lakes may be a factor of 1000 or more over a distance of a few tens of meters or a time span of a few days.

The organic carbon content of sea water displays an even greater constancy than does the particulate matter. By far the largest part of the organic matter is present as dissolved material. While the total organic carbon is 50–100 times as great as the particulate organic carbon, the vertical variability is only a factor of 3, from 0.5 mg C/l in deep water to 1.5 mg C/l in surface water, and the horizontal variability is seldom as much as a factor of 2.

The overall impression of sea water, then, is of a medium of high and almost constant ionic strength and composition, almost constant pH, and low and relatively constant particle load and total organic content, altogether a different environment from the fresh waters, which can be highly variable both in space and in time. This constancy places limitations on the kinds of reactions which can proceed in the oceans.

Conservative and Non-Conservative Properties

Most of the major and many of the minor constituents of sea water vary directly with the salinity. Changes in the concentrations of these elements reflect dilutions, concentrations, and mixings of water masses. While these constituents may be involved in biological or chemical reactions, these are either too local or involve too small a proportion of the constituent to alter ionic ratios appreciably. The involvement of magnesium in photosynthesis, and therefore in phytoplankton growth, is a good example.

Constituents of sea water whose concentrations are thus a function only of salinity are called "conservative" materials. These constituents include those major ions, such as Na^+, K^+, Mg^{2+}, Cl^-, and SO^{2-}, whose concentrations are not altered locally by biological activity. There is no general rule for the minor and trace inorganic ions; the nature of the relationship to salinity must be determined in each case. The dissolved inorganic gases in the surface layer, other than O_2, N_2O, CO_2, and CO, are in equilibrium with air. Once this layer is out of contact with air, the concentrations of these gases become truly conservative. Those gases which can be formed by radioactive decay, such as Rn, ^{40}Ar, and ^3He, must be counted as non-conservative, although the amount of argon added to sea water by this mechanism is not great enough to make the total concentration of argon non-conservative. The distributions of conservative constituents can be completely explained in terms of water mass movement and mixing; the departures from conservative behavior are perhaps more interesting to the environmental chemist.

While conservative materials, like happy families, all have much the same history, each of the non-conservative constituents varies in its own particular fashion. Most of the non-conservative constituents are in some way involved with biological activity, although this cannot be accepted as a general rule. For example, the effect of pressure on chemical reactions is to push them in the direction of lower volume. The bulkiest material in aqueous solutions is water itself; any change which will abstract water molecules from the water structures will therefore be favored by increased pressure. This includes such changes as greater dissociation, decreased ion pairing, and increased solubility. Some particulate materials which are only sparingly soluble in surface waters may be considerably more soluble at depth.

However, by far the most important cause of non-conservative behavior is biological activity. In the case of the nutrients, NO_3^-, NO_2^-, NH_4^+, and PO_4^{2-}, the distributions are determined by their use in surface waters by phytoplankton and by their regeneration at depth, largely by bacteria. The annual cycle of nutrient accumulation and depletion found in temperate and boreal waters is essentially the response of the system to the imposition and relaxation of a further restraint, that of daylight. In most cases the cycle is reinforced by the introduction of deeper, nutrient-rich water by increased turbulence during those periods when the cooling of the surface layer makes the upper part of the water column isothermal, but even without such mixing, regeneration *in situ* during the winter would result in an accumulation of nutrients in the surface waters sufficient to produce at least a small spring bloom.

The distributions of Ca^{2+} and SiO_4^{2-} can be considered together, since certain features of their distributions are the result of similar processes. SiO_4^{2-} is undersaturated everywhere in the water column; it is bound into diatom tests in the surface water, and at the death of the diatom sinks out of the surface layer. As it sinks through the water column it begins to dissolve, until, below 1,000 m, solution is nearly complete. Ca^{2+}, on the other hand, is close to saturation as Mg-containing calcite in the surface waters. The degree of saturation decreases with depth beacuse of the effects of pressure and of the addition of CO_2 to the water from *in situ* respiration. The details of the manner in which this change occurs are still very much in dispute. The net effect, however, is the solution at depth of $CaCO_3$ sedimenting in the form of coccoliths and pteropod and foraminiferal tests. In the cases of both SiO_4^{2-} and Ca^{2+} the result is a transport into the deeper water; without subsequent upwelling or advection into the surface waters of this deeper water, the surface waters would ultimately become depleted in both substances to the point where populations of some organisms could not be maintained.

The non-conservative fixed gases may be divided into two classes, those which are involved in biological activities and those which are produced by radioactive decay. Those produced by radioactive decay reflect the distribution of the parent element as well as the processes leading to equilibration with the atmosphere, and may in many cases also reflect diffusion from the sediments. These gases are useful as tracers of physical processes, particularly in deep waters.

An especially useful subset of these gases is that of the gases produced by reactions in the atmosphere. Thus, naturally produced ^{14}C has been used to date sediments and to estimate the ages of water masses, while bomb-produced ^{14}C has given us information on rates of incorporation of CO_2 into sea water and sediments, and rates of water mass circulation. Bomb-produced tritium, incorporated into water molecules, has furnished rather precise rates of short-term mixing and circulation.

The fixed gases involved in biological activities follow no single fixed rule of distribution. Thus, O_2 is in equilibrium with the atmosphere in the very surface layer; below this layer, its distribution is dominated by biological activity. It is taken up in respiration and emitted in photosynthesis. In those depths where the rate of photosynthesis is greater than that of respiration, the net result is an increase in O_2, often to levels well above equilibrium with the atmosphere; below these depths, the O_2 will decrease, reaching a minimum at some intermediate depth. The increase following the oxygen minimum is attributed to diffusion and mixing with the O_2-rich deeper water, which was at equilibrium with the atmosphere at the time the water cooled and sank.

The CO_2 distribution is somewhat more complicated, since in addition to its use in photosynthesis and release in respiration, it undergoes dissociation according to the reaction

$$CO_2 + H_2O \rightleftharpoons H_2CO_3 \rightleftharpoons H^+ + HCO_3^- \rightleftharpoons 2H^+ + CO_3^{2-},$$

and is therefore involved in the maintenance of pH in sea water. Furthermore, it is coupled to the dissolution of $CaCO_3$ through the reaction

$$CaCO_3 \rightleftharpoons Ca^{2+} + CO_3^{2-}.$$

The ramifications of this cross-linkage have led to the growth of a sub-school of chemical oceanography and to the publication of a great many strongly-worded papers. An entry into this literature may be secured through the paper by Edmond and Gieskes [2].

Little is known of the biological reactions leading to the production of N_2O. That they are biological can be seen from the inverse relationship to be found between N_2O and O_2 in the surface waters [3]. Even less is known of the oceanic production of CO, although it has been shown that at least one pathway is through photolysis of dissolved organic matter [4]. Contrary to assumptions made in the past, it now appears that the ocean may be a source rather than a sink for both N_2O and CO.

The organic matter, both particulate and dissolved, in oceanic sea water is almost entirely biological and marine in origin. While there is still no general agreement as to its molecular composition, the bulk of the material is of a molecular weight large enough to suggest a polymeric structure. Certainly, monomeric sugars and amino acids are present only in trace quantities. A much larger portion of the material has been called "humic acids", in an analogy with the soil humates, although the compounds in question are no longer separated from sea water by the classical soil chemistry techniques. In current terms, any organic compound collected from sea water by adsorption on XAD-2 resin is considered a humic acid. The marine origin of these compounds has been demonstrated through their $^{13}C/^{12}C$ ratios.

In coastal waters, sea water is often green rather than blue, and emits a blue-white fluorescence. Both the color and fluorescence are due to the presence of compounds of high molecular weight, of still unknown composition, collectively called "Gelbstoffe". It is believed that these compounds are derived from the exudates and decomposition products of the fixed algae, and products resembling Gelbstoffe can be derived by the condensation of polyphenols and carbohydrates from these fixed algae [5]. Reactions leading to the destruction of these compounds are not known, although there is reason to believe they are subject to photodegradation [6].

The dissolved materials are differentiated from the particulate largely by convention, the dissolved fraction being those materials passing through a filter of 0.45 μm pore size, and the particulate being the material retained. Both fractions follow roughly the same distribution patterns. The distribution of the particulate fraction tends to be less continuous than that of the dissolved, with occasional small patches which are considerably (3–5 times) higher than the background value.

The trace inorganics seem to follow no general rule, with a few acting like conservative elements, others following biological patterns, and still others being partitioned between the ionic, organic, and particulate phases. Our understanding of the behavior of trace metals has not been helped by our difficulty in collecting and analyzing for them; as our ability to avoid contamination has improved, the accepted values for many of these materials have steadily decreased. Interlaboratory calibration programs suggest that not many trace metals are measured accurately even now.

Regions of the Oceans

The oceans can be subdivided into regions both vertically and horizontally. The vertical divisions occur at clearly marked natural boundaries, in some cases at actual phase changes; the horizontal divisions are usually water mass or current boundaries, are much less easily denoted, and often cannot be given a definite latitude and longitude position.

Horizontal Divisions

We can divide the oceans horizontally into regions of enrichment and regions of impoverishment. The regions of enrichment are regions where deeper water containing a supply of nutrients is brought into the surface layer. The nutrients serve to support phytoplankton growth at levels considerably higher than those which can be maintained simply by *in situ* regeneration. The result is a region where productivity at all trophic levels is increased, and the distribution of all of the sea water constituents involved in biological activities show the effects of this involvement most strongly. The major fisheries of the world are all contained in these regions; therefore, any interference by man which affects the productivity of these regions can result in problems at the global scale.

The most important of these regions are the coastal zones. These are areas where the oceans are still shallow enough so that the whole water column, including the upper layers of the sediment, is available for mixing at least part of the year. The bottom topography of these regions is usually rough, bringing about complex current patterns with considerable vertical advection. These regions include the Atlantic coastal shelf fisheries; Georges Bank and the Grand Banks in the western North Atlantic, the North Sea and the Baltic Sea in the eastern North Atlantic, and comparable areas along the other continental margins.

A special case of coastal zone enrichment is the area of upwelling. These occur where the prevailing winds along a coast drive the surface waters out to sea, away from the coast. To replace these surface waters, deeper water is brought to the surface. This water is generally drawn from mid-depths, somewhere less than 200 m, and therefore contains elevated levels of inorganic nutrients. Such upwelled water can support a considerable population of phytoplankton; the continual renewal of nutrients can lead to a perpetual plankton bloom, with a concommitant high level of productivity for zooplankton and fish. The best-known example of the upwelling situation is that off the Peruvian coast, where the anchoveta fishery, in good years, has made up some 10% of the world catch for all fish species. The dependence of the upwelling on the prevailing winds has been emphasized by the occasional year when changes in the local wind system have stopped the upwelling and have permitted the intrusion of warm nutrient-poor surface water from the north into this coastal zone. This condition, known as "El Niño", results in a sudden drop in the anchoveta population and a catastrophic mortality in the sea bird populations dependent upon the fish. While the Peruvian upwelling is perhaps the best known, similar situations exist along the western coasts of most of the continents, and in other areas with prevailing offshore winds.

Similar conditions of nutrient enrichment occur wherever deeper water is brought to the surface, either by local winds or by water circulation patterns. These conditions are found along shear zones between current systems, such as the slope water along the northern and western edge of the Gulf Stream, and in regions where current systems diverge. While these regions of enrichment are not as spectacular as the regions of upwelling, the great krill fishery in the Antarctic regions, which supported much of the world population of baleen whales, results from this mechanism of upward transport.

In the immediate area of upwelling, the inorganic nutrient and silicate levels will be high, and the dissolved organic carbon and nitrogen, along with the dissolved O_2, relatively low. The particle count will also be low. Often growth of phytoplankton does not begin as soon as the water parcel reaches the surface; a lag of one to several days may be required before maximum growth rates are reached. This lag has been interpreted as the time required for the resident phytoplankton to exude enough organic materials to chelate toxic trace metals, and thus render them physiologically unavailable [7]. This hypothesis is still far from proven.

The bloom of phytoplankton, characteristically diatoms, thus begins some little distance "downstream" from the point of upwelling. Once the phytoplankton growth has begun, the nutrient levels drop and the dissolved O_2, dissolved organic material, and particulate load all increase. Since upwelling tends to be intermittant, we actually find a series of batch cultures proceeding downcurrent, with heavy grazing pressure from zooplankton and herbiverous fish. The distributions of properties in each of the bodies of upwelled water then follow the time pattern to be expected of a bacterized batch culture. In a region where upwelling is sufficiently active, the transport of organic carbon to the deeper waters and the microbial utilization of this carbon may be great enough to reduce the dissolved oxygen at mid-depths almost to zero. On occasion, hydrogen sulfide has been found in the water column below the Peruvian upwelling system.

If the upwelling areas might be considered the market gardens of the ocean, the oceanic central gyres must surely be the oceanic deserts. These are regions, like the Sargasso Sea and the North Pacific Gyre, where the surface waters show a relatively stable temperature structure, no strong currents, and little exchange with other water masses. The nutrient content, particulate concentration, and productivity are all very low. The inorganic nutrient content, for example, is so low that Sargasso Sea water is sometimes used as a blank for nutrient determinations. If one wished to examine distributions of various components displaying a minimum of biological interference, these would be the regions to sample.

However, even in these regions one cannot afford to be unwary in sampling. In the past, investigators have found what seemed to be seasonal changes in such properties as the particulate organic carbon content of Sargasso Sea water, extending into deep water, and implying a rapid sinking rate for even the smaller particles. It now seems probable that the apparent spring increase in particle load was actually derived from a profile taken in a cold-core ring, an eddy detached from the Gulf Stream. These rings travel through the Sargasso Sea with lifetimes of the order of a few years, either dissipating in the Sea or, more frequently, rejoining the Gulf Stream at some point south and west of their point of detachment. While they exist, they retain the characteristics, including the flora and fauna, of the water mass from which they came; they are essentially small universes with only limited exchange with the surrounding water masses. Without a knowledge of the position and number of the rings, such as one can get from satellite surface temperature data, sampling in the oceanic gyres may result in mistaken interpretations of misleading data.

Vertical Divisions

The ocean is much more easily and naturally divided into zones in the vertical dimension. If we start from the air-sea interface, the first layer we encounter is the surface film. This film, a few molecular layers thick, contains hydrophobic and surface-active materials from both natural and anthropogenic sources. It is normally present everywhere on the sea surface, but is visible only where winds, currents, or Langmuir circulation pile the materials several molecular layers thick, usually in windrows.

The composition of this layer is a matter of some dispute. Most workers feel that it is composed of hydrocarbons, lipids, fatty acids, and those substances more soluble in these compounds than in sea water. However, Baier et al. [8], using internal reflectance infrared spectroscopy, concluded that the predominant compounds in the surface film were glycoproteins and proteoglycans. Since the methods normally used for the determination of the composition of the surface film, gas chromatography and coupled gas chromatography – mass spectrometry, are methods which would not measure these larger, less volatile materials, Baier's results cannot be discounted out of hand; this question can only be resolved by the maintenance of an organic carbon budget throughout the separation and determination procedures.

While this layer is composed of surface-active and hydrophobic materials, not all of these materials present in sea water are contained in this layer. If the layer is removed carefully, it will be reconstituted from material in the body of the water. This procedure can be repeated a number of times before the material suitable for the formation of a film has been exhausted from the water parcel. The rate of reconstitution can be increased by aeration, the surface-active material being carried to the interface on the surfaces of the rising bubbles [9]. In some manner an equilibrium is maintained between the material at the surface and the material in the body of the water. The mechanisms involved in the equilibrium have not been elucidated.

Although this surface film contains only a small proportion of the dissolved organic materials, it is important beyond its actual concentration because of its position and composition. As a surface film, the material is the first liquid with which airborne particulates and vapors come in contact. Hydrophobic compounds carried by aerial transport could be extracted into this layer, and concentrated to levels well beyond any they could achieve in the water column. The chemistry of this surface film has been reviewed in some detail by Wangersky [10].

The next natural division, the surface layer, extends from the surface film to the thermocline. In general, this is a layer low in inorganic nutrients, although in situations where light is limiting the growth of phytoplankton, the nutrients can accumulate to levels almost as high as those of the deep water. At the very surface all of the fixed gases are in equilibrium with the atmosphere. The depth to which this equilibrium persists depends upon the extent of biological use or production of the gas and the degree of turbulence in the water column.

The particulate load in this part of the water column is high, but drops quickly. In most parts of the ocean, by 100 m the POC values are approaching those of the deep water. Although we are inclined to think of particulate distributions as relatively uniform, there is reason to believe that in the upper layers of the ocean the particles are present in the form of clumps and thin layers superimposed upon a low background level. There have been reports of the accumulation of particles into layers at density discontinuities, but until proper devices for the continuous recording of particle densities become more plentiful, we will lack sufficient data to decide whether these layers are a common feature of the open ocean.

Aside from those few reactions favored by higher pressures, such as the solution of carbonates and silicates, most of the reactions of interest to the environmental chemist occur in this top layer of water. Bacterial activity is at its highest level here, unhampered by temperature and pressure effects; the primary productivity upon which life in the rest of the water column and in the sea floor depends all takes place in this upper layer; gas exchange with the atmosphere, the injection of gases into the water column by breaking waves, and the many reactions facilitated by the presence of bubbles, are all surface layer phenomena; and the reactions caused, either primarily or secondarily, by the penetration of light into the water, all take place above the permanent thermocline. Any material introduced into the oceans by atmospheric transport must pass through the surface layer. If these materials can react with sea water or with the organisms present, it is in this layer that the reactions will occur.

The third layer, the thermocline, is a transition region between the surface layers and the deep water, and a barrier to many kinds of activities, as well. This is the region of rapid density change; mixing goes on more easily within the water masses above and below the thermocline than across it. In a sense, this is a circular argument, since the depth of the thermocline depends upon the energy available for mixing in the surface layers. The depth of the permanent thermocline, which varies over a range of 100–1,000 m, is a measure of the energy input into the ocean surface from the action of wind on that surface.

Within the thermocline there is a rapid gradation of many of the properties of the water masses. We would expect that all of the conservative properties would follow the change in salinity, and that in this intermediate layer they would display a gradation in values between those of the water masses above and below.

The non-conservative properties of biological interest will follow quite different distributions. In each case, the distribution seen is a balance between utilization and renewal, but the processes involved are not always the same. Oxygen, for example, is utilized at all depths of the ocean, even though rates of utilization decrease greatly in the deeper water. The main features of the O_2 distribution, however, are brought about by the rapid decrease of O_2 in the layers where utilization exceeds production by photosynthesis, and the replenishment through mixing and diffusion from the deeper water, which had been in equilibrium with the atmosphere at the time of sinking. Thus an oxygen minimum would be expected in the region of the top of the thermocline. In some parts of the ocean there can be two minima. The intermediate minimum is associated with the intrusion of a layer of water in fairly recent contact with the surface.

The other non-conservative substances will show distributions which result from the interaction of their particular patterns of utilization and replenishment. In the case of N and P, regeneration is the dominant method of replenishment, while with silicate it is re-solution at depth. In all of these cases, the thermocline is also a region of transition from the high values typical of deep water to the low values to be found in the regions of utilization; however, with N, P, and Si, the dominance of replenishment over utilization may occur well above the thermocline.

The water below the thermocline has not seen the surface in some time.

The properties of this water are basically the properties it carried down from the surface, modified by biological and chemical reactions occurring *in situ,* by the addition of some materials through advection and diffusion, and of others by sedimentation from the surface layers. Nutrient levels are high, since regeneration and re-solution have largely proceeded to completion. Organic materials, both dissolved and particulate, are low and relatively uniform. Because of the high pressure and low temperatures, microbial activity is low. Oxygen consumption in the deep waters of the North Atlantic has been calculated as 0.00013 ml/l/yr, as compared to 0.21 ml/l/yr for surface waters [11]. Altogether, the deep water is a region where very little seems to happen, and what does, happens very slowly. Even here, we must insert a *caveat*; our methods of measurement in oceanography have all emphasized long-term averages and uniform gradients. We have chosen to minimize the importance of individual events, possibly because there has been no good way even to identify such events. We have studied the climate and ignored the weather. With the advent of continuous recording methods, we have begun to discover the importance of the short-term event in the distribution of physical properties. We are still a very long way from any similar understanding of short-term changes in chemical properties, particularly in deep water, where there are few instruments capable of continuous recording of chemical parameters.

The greatest part of the activity, both chemical and biological, occurring below the thermocline occurs in the water-sediment interface. It has been calculated [12] that some 95% of the POC reaching the sea floor is decomposed rather than incorporated into the sediment. Moreover, their values for POC did not include the large components with rapid sinking rates which Bishop, et al. [13] calculated to be a major fraction of the particulate matter delivered to the sediments. The scarcity of such larger organic particles in the deep sea sediments argues for a complete utilization in what would be a short period by geological standards.

Along with this activity, whether due to bacteria or to the efforts of invertebrate bottom dwellers, must go changes in the O_2 and CO_2 content of the interface. It is difficult to document such changes, since this interface is so difficult to sample, and so loosely packed that diffusion and exchange with the water column above are little hindered. *In situ* experiments, carried out in shallow waters by divers and in deep waters by manned submersibles, may be the only useful approach to the study of reactions in this layer.

Reactions at deeper levels within the sediment must largely be inferred from changes in distributions with depth. When we argue on these grounds, we must assume that the supply of material to the sea floor has been constant in both quantity and composition at least over the last half million years. Since this period of time encompasses a major portion of the last glacial period, the assumption is difficult to justify. However, since we have no other starting place, we must perforce begin there.

Certain kinds of reactions can safely be assumed from the evidence found in most deep sea sediments. There is some solution of silicates in all regions, and of carbonates in those regions and at those depths where carbonates reach the sea floor. At least in the regions of slow sedimentation, there can be a formation of authigenic silicates, in particular zeolites such as phillipsite. Most of the organic matter reaching the sea floor is oxidized to CO_2, presumably by bacterial activity. This is demonstrated both by the disappearance of organic carbon and the change in redox potential with depth in the sediment.

While the redox potential is difficult to measure directly, since the act of taking the sample from the sea floor, through the oxygenated water column, and into the air produces changes which make any reading taken on shipboard more than a little fictional, we can find evidence of *in situ* changes in the distribution of some of the components of the sediment. Thus, the accumulation of manganese as encrustations and micronodules at shallow depths within some oceanic sediments signals the depth in that sediment at which the redox potential becomes reducing enough to mobilize Mn, presumably as Mn^{2+}, into the interstitial water. At a somewhat lower depth, the redox potential required to mobilize Mo can sometimes be shown; this depth is not so obvious, because the amount of Mo present is too small to produce visible phenomena. In most oceanic sediments the amount of organic matter is too small to produce a redox potential capable of mobilizing the $Fe(OH)_3$ coated on clay particles. Limits can therefore be set on the possible range of redox potentials in a sediment through proper interpretation of the transition metal distributions within that sediment. If the organic content of the sediment is high enough, as in some coastal regions of high productivity, the sediment may be anoxic below a thin oxidized surface layer. Such sediments display a completely different set of biological and chemical reactions.

In the top layers of a normal sediment the activities of organisms may enable reaction products in the interstitial waters to escape into the water column; while the depth of homogenization of the sediment by this mechanism has arbitrarily been taken as 10 cm, a better limit on such mixing can be set by looking for sharply layered structures in the sediment. Individual burrows may extend downward into the sediment, passing through such layers, for several tens of centimeters, but sharp banding can usually be observed below 20 cm. Reactions occurring in the deeper layers of the sediment, below the zone in which the sediment is mixed by burrowing organisms, have little effect on the chemistry of the ocean.

Controls on Composition and Reactions

While there has been considerable dispute concerning the history of the oceans and the probable composition of sea water in the distant past, it is now generally accepted that the composition has remained much the same for a very long time. Since sea water is seemingly far from being a saturated solution for most of its major components, we must consider the reactions which keep it uniform both in time and in space.

Any chemical system which exists essentially unchanged over time or space must either be in equilibrium or in a steady state; it is not always easy to choose between these possibilities. In a steady state, rates of addition of the various components are balanced by the rates of removal of products. While chemical equations can be written to describe the mechanisms of removal, these equations say nothing about possible rates. In a system of low total volume, the adjustment of rates to achieve a steady state can be difficult. In a system whose total volume is so much greater than the volume of the additions, as in the oceans, the adjustment of rates can be crude without disturbing the steady state noticeably.

Control of concentrations in a system at or approaching equilibrium is a much more positive matter, and prediction of eventual conditions is more precise. However, the compositions of all of the phases involved must be known, as well as the relevant equilibrium constants. It is not possible to write equations involving only the components in solution in sea water and emerge with a system which adequately controls the composition of the solution. Sillén [14] has shown that such control might be accomplished if the solid phases present in the sediments are included in the equations. These equilibria suggest that the composition of sea water may be under a considerably greater degree of control than could be exerted by a steady state system.

The question of degree of control enters into the discussion of the maintenance of the pH, which normally varies from 7.8 to 8.2. The major contributor of H^+ to sea water is the CO_2 system, described in an earlier section (4.2.2). Minor contributions are present also from the borate system,

$$H_3BO_3 + H_2O \rightleftharpoons B(OH)_4^- + H^+,$$

and from the ionization of water,

$$H_2O \rightleftharpoons H^+ + OH^-.$$

The buffering capacity of the system so defined is relatively small; it has been calculated as approximately 0.25 m*M* [15]. If reactions with major ions involving the formation of ion pairs are considered, the calculated buffer capacity increases to 0.72 m*M*. This is a reasonable estimate of the instantaneous buffer capacity, that which we would measure by titration in a beaker of sea water. If we increase the size of our sample container and the time constant of our reactions to approximate oceanic conditions, we should include the sedimentary solid phases in our system; the addition of these increases the buffer capacity of our system by a factor of 400 over that calculated from dissolved components alone. Thus, the ultimate control of pH over geological time would seem to be based on the heterogeneous system which includes the sedimentary components.

Fixing the pH at values close to 8 sets definite limits on the kinds of reactions which can occur in sea water. For purposes of concentration and fractionation, we may be interested in reactions taking place at pH 3 or 11, but we must always bear in mind that if it doesn't happen near pH 8, it doesn't happen in sea water.

The redox potential of sea water is fixed by the presence of dissolved O_2 almost everywhere in the oceans. The exceptions are the deeper waters of a few seas, such as the Baltic and the Black Sea, a few coastal trenches, such as the Cariaco Trench, and occasionally the mid-depths in particularly productive areas, such as the Peruvian upwelling zone. While the chemistry of these areas is quite singular, it is far removed from normal sea water, and should be treated as a special topic. Everywhere else in the oceans, the redox reactions are dominated by the presence of molecular oxygen. All purely chemical reactions occurring in the greater part of the ocean must be those possible at the Eh and pH set by the natural controls.

A third and not quite so obvious control is set by the presence of organic materials. The total concentration of organic material present is small, but its effect is disproportionately important. Bishop et al. [13] have shown that a major portion of the particulate flux to the deep ocean occurs as large, rapidly sinking particles, and Fowler and Small [16] reported that many smaller particles are incorporated into copepod fecal pellets, protected from reaction with sea water by an organic skin. Bacterial attack breaks up this skin, but only after the pellet has descended below 1,000 m. Thus a mechanism exists for the rapid transport of surface materials to deeper water, permitting the effect of seasonal occurrences, such as the spring bloom, to be felt throughout the water column.

Since many of the organic compounds occurring naturally in sea water are surface-active, we would expect them to accumulate at phase interfaces. We have already discussed the effects of such accumulation at the air-sea interface. We would expect similar accumulations to occur on the surfaces of particles passing through the water column, and in fact such accumulations have been found [17–19]. Neihof and Loeb [20] have also shown that any particle entering sea water immediately collects an organic surface coating, and that the composition of this coating, rather than the bulk composition of the particle, then determines the surface charge on the particle. Thus all non-living particles entering the oceans, no matter what their composition, quickly assume the same surface charge, at least as long as enough surface-active material is present in solution.

The coating of organic material regulates more than just the surface charge; the surrounding solution is in contact only with the organic coating, and for purposes of reaction, the bulk composition of the particle is irrelevant. In considering solution-solid equilibria in sea water, we usually determine the composition of the solid in equilibrium with the solution by grinding up the solid and determining its bulk composition. Actually, it is only the outer few molecular layers whose composition is important; we determine bulk composition because the analysis of just the outer few molecular layers can be performed by only a few, relatively expensive techniques, such as ESCA. However, it should be evident that the bulk of a particle cannot react with sea water; if there is an impenetrable outer skin present, the most soluble particle will be as safe from solution as a sugar cube wrapped in cellophane. An outer layer of surface-active organic materials may thus prevent or delay reactions which would otherwise be favored.

This principle applies even in the absence of organic coatings. Cooke [21] has shown that the solubility of calcite in sea water depends upon the pressure and upon the magnesium content of the calcite; with increasing pressure, calcite must contain less magnesium in order to remain in equilibrium with the ambient sea water. Thus, a low-magnesium calcite will acquire an outer coating containing that amount of magnesium stable at the ambient pressure. As the particle sinks, the coating will dissolve and be replaced successively by coatings containing less magnesium, until the depth is reached at which the coating and the bulk of the particle have the same composition. Below this depth, the whole particle will go into solution. At every point in this journey, it is only the outer coating of the particle which is in equilibrium with the surrounding sea water.

The organic compounds in sea water are also involved in the distributions of many of the trace metals, through the twin "black box" activities of chelation and complexation. The roles of these reactions in maintaining the concentrations of trace metals is far from understood, and has become muddier than necessary by the use of the terms as explanations for any otherwise inexplicable phenomena. We are beginning to be able to measure the amount of the metals present in some form other than the ionic; the increase in the ionic form after oxidation of the organic materials certainly suggests some combination with these organic materials. What little we know of the structures of the larger organic molecules present in sea water certainly suggests that both chelation and complexation are likely. Thus the distribution of trace metals may depend upon the kinds and amounts of organic materials present.

The impingement of light on the surface layers of the ocean is an energy source seldom considered except by phytoplankton biologists. It is also a driving force for many chemical reactions, and while the energy contributed per unit area is not great, the total area of sea surface is very great indeed. Many of the organic materials present in sea water can take part in photochemical reactions, and can involve some of the inorganic constituents as well. The photochemistry of sea water will be discussed in detail in a later section of this Handbook, Vol. 2, Part A.

Organic materials are greatly involved in the productivity of the oceans; the total amount of inorganic nutrients is roughly constant in the oceans, and the continuance of primary productivity depends upon the regeneration of these nutrients into inorganic form from the organic material in which they are freed into the water column. Any interruption of this cycle, any considerable irreversible incorporation of these materials into intractable organic compounds, would result in the eventual impoverishment of the surface waters.

Specific organic compounds are also involved in many biological reactions, sometimes as pheromones, as stimuli for feeding, perhaps as determinants of the direction of planktonic succession. In these roles, the amounts of material necessary for effective action is small, and identification of the compounds involved must surely proceed from biological evidences of activity, rather than from direct isolation of the compounds from sea water.

Anthropogenic Materials

Man's contributions to the oceans can be of two kinds; he can add more of something which is already present, and he can add substances completely new to the system. When he adds "more of the same, only louder", he stands in danger of exceeding the ability of the oceans to cope with the material. The proper systems exist, but they are just overwhelmed by the magnitude of the problem.

A typical example might be the discharge of nutrients into coastal waters by an industrial process or a sewage treatment plant. A large increase in primary productivity brought about by added nutrients could cause the sediments and the deeper water to become anoxic during the summer, leading to a variety of unpleasant surprises during the period of mixing in the autumn. That this is not a purely local, small-scale problem was shown by the major fish kill which occurred from similar causes in the New York Bight in the mid-1970's. The exact cause or causes of the high planktonic productivity which led to anoxic conditions near the bottom are not known, so there can be no guarantee that the condition will not recur.

While this sort of problem can be severe, the processes involved are all understood and the consequences are predictable. The warning signs are usually posted well in advance, and it is only our reluctance to heed them which leads us into disaster. Much more subtle problems stem from the introduction into an ecosystem of materials altogether new to the system. Usually the materials are spread widely through the system before we understand the possible perils. An example might be the insecticide DDT. By the time we began to search for it in nature, it had spread throughout the globe.

It is possible to gain some insight as to the effects of a new material on an ecosystem through laboratory tests of its effect on representative members of that ecosystem. However, before we can decide which members will be the most susceptible, or which physical features of the system the most subject to change, we must be able to predict pathways of transport and regions of concentration. Many of the organic materials added to the marine environment, such as DDT and the PCB's, are preferentially adsorbed on the surfaces of particles and in the surface film. Therefore, while it is of some interest to know the effects of these chemicals on phytoplankton, it is much more important to understand their effects on particle feeders and on the bacteria living on particles and in the surface film.

Before we can decide on safe levels for a particular material in the environment, we must know whether it remains dispersed in this environment, or whether it tends to collect in discrete entities. A substance which tends to adsorb on particles may be present at a safe level of parts per billion as measured in the total sample, but at unsafe levels of parts per million in the particulate fraction, and perhaps at lethal levels of tenths of a percent on the particle surfaces. Similarly, hydrophobic gases, which might remain at low concentrations in the air in an air-water system, might accumulate to considerable local concentrations in the hydrophobic surface film.

It is not enough to study the effects of a possible pollutant on one or a few organisms, even if these organisms are the most important components of the ecosystem at risk. We must know a good deal about the chemistry of the material, and about its probable distribution in the ecosystem, before we can make any statement concerning its effects on the environment.

Acknowledgement

This article was written while the author was a Norges Teknisk-Naturvitenskapelige Forskningsrad Visiting Senior Scientist at the Institutt for Marin Biokjemi, Universitetet i Trondheim, Norway.

References

1. UNESCO.: International Oceanographic Tables. National Institute of Oceanography of Great Britain, and UNESCO. *1966,* 118 pp.
2. Edmond, J.M., Gieskes, J.M.T.M.: Geochim. Cosmochim. Acta *34,* 1261 (1970)
3. Yoshinari, T.: Nitrous oxide in the sea. Mar. Chem. *4,* 189 (1976)
4. Wilson, D.F., Swinnerton, J.W., Lamontagne, R.A.: Science *168,* 1577 (1970)
5. Sieburth, J.McN., Jensen, A.: J.exp.mar.Biol. Ecol. *3,* 275 (1969)
6. Zika, R.G.: Ph.D. thesis, Dalhousie University, 1977, 346 pp.
7. Barber, R.T., Ryther, J.H.: J.exp.mar.Biol.Ecol. *3,* 191 (1969)
8. Baier, R.E., Goupil, D.W., Perlmutter, S., King, R.: J.Rech.Atmos. *8,* 571 (1974)
9. Jarvis, N.L.: Limnol. Oceanogr. *12,* 213 (1967)
10. Wangersky, P.J.: Ann.Rev.Ecol.Syst. *7,* 161 (1976)
11. Riley, G.A.: Bull.Bingham Oceanog. Collection *13,* 1 (1951)
12. Eadie, B.J., Jeffrey, L.M.: Mar. Chem. *1,* 19 (1973)
13. Bishop, J.K.B., Edmond, J.M., Ketten, D.R., Bacon, M.P., Silker, W.B.: Deep-Sea Res. *24,* 511 (1977)
14. Sillén, L-G.: Oceanography Sears, M. (ed.). Washington: Amer.Ass.Adv.Sci.Publ. *67,* 1961, pp. 549–581
15. Morel, F., McDuff, R.E.M., Morgan, J.J.: Mar. Chem. *4,* 1 (1976)
16. Fowler, S.W., Small, L.F.: Limnol. Oceanogr. *17,* 293 (1972)
17. Chave, K.E.: Science *148,* 1723 (1965)
18. Suess, E.: Geochim. Cosmochim. Acta *34,* 157 (1970)
19. Suess, E.: Geochim. Cosmochim. Acta *37,* 2435 (1973)
20. Neihof, R., Loeb, G.I.: Limnol. Oceanogr. *17,* 7 (1972)
21. Cooke, R.C.: Mar.Chem. *5,* 75 (1977)

Chemical Aspects of Soil

E.A. Paul, P.M. Huang
Department of Soil Science, University of Saskatchewan
Saskatoon, Canada

Introduction

The lithosphere or the solid outer mantle of the earth comprising the outer portion of the earth's crust is exposed to the atmosphere over one-quarter of the area of the globe and covered by the hydrosphere over the remainder. Where the earth's rocks are exposed they, however, usually are in close contact with water and living organisms (the biosphere). The physically and chemically weathered components of the exposed lithosphere form the unconsolidated surface usually known by geologists as the regolith. Further weathering and upward or downward translocation of soluble components in the presence of living organisms and their detritus often result in the formation of an orderly sequence of horizons (Birkeland [4]). Soil is characterized by the ability to support plant and microbial life and the formation of distinct recognizable pedological features across most of the earth's non-aqueous surface.

The interaction of the components plays as great if not a greater role than the chemical characteristics of the individual parts in determining the role of the earth's surface in environmental geochemistry. An understanding of this role can best be obtained from knowledge of the interrelated structure and composition. This chapter includes some indication of the geochemical classification and distribution of the elements, a discussion of the major components of soils, and information on the mechanisms of weathering reactions and soil formation.

Composition of the Earth's Crust

Geochemical Classification and Distribution of the Elements

The elements can be classified into five main groups according to their geochemical character (Rankama and Sahama [34]). *Lithophile elements* ion-

Table 1. The average amounts of the elements in crustal rocks (Berry and Mason [3])

Element[a]	Geochemical classification[b]	$\mu g\ g^{-1}$	Element	Geochemical classification	$\mu g\ g^{-1}$
O	At, Bi, Li	466,000	Hf	Li	5
Si	Li	277,200	Dy	Li	5
Al	Li	81,300	Sn	Si	3
Fe	Ch, Si	50,000	B	Li	3
Ca	Li	36,300	Yb	Li	3
Na	Li	28,300	Er	Li	3
K	Li	25,900	Br	Li	3
Mg	Li	20,900	Ge	Si	2
Ti	Li	4,400	Be	Li	2
H	At, Bi, Li	1,400	As	Ch	2
P	Bi, Li, Si	1,180	U	Li	2
Mn	Li	1,000	Ta	Li	2
F	Li	700	W	Li	1
S	Ch	520	Mo	Si	1
Sr	Li	450	Cs	Li	1
Ba	Li	400	Ho	Li	1
C	At, Bi, Li, Si	320	Eu	Li	1
Cl	Li	200	Tl	Ch	1
Cr	Li	200	Tb	Li	0.9
Zr	Li	160	Lu	Li	0.8
Rb	Li	120	Hg	At, Ch	0.5
V	Li	110	I	At, Li	0.3
Ni	Si	80	Sb	Ch	0.2
Zn	Ch	65	Bi	Ch	0.2
N	At, Bi	46	Tm	Li	0.2
Ce	Li	46	Cd	Ch	0.2
Cu	Ch	45	Ag	Ch	0.1
Y	Li	40	In	Ch	0.1
Li	Li	30	Se	Ch	0.09
Nd	Li	24	A	At	0.04
Nb	Li	24	Pd	Si	0.01
Co	Si	23	Pt	Si	0.005
La	Li	18	Au	Si	0.005
Pb	Ch	16	He	At	0.003
Ga	Ch, Li	15	Te	Ch	0.002
Th	Li	7	Rh	Si	0.001
Sm	Li	7	Re	Si	0.001
Gd	Li	6	Ir	Si	0.001
Pr	Li	6	Os	Si	0.001
Sc	Li	5	Ru	Si	0.001

[a] Omitting those present in less than 0.001 $\mu g\ g^{-1}$: Ne, Kr, Xe and the short-lived radioactive elements
[b] At = Atmophile; Bi = Biophile; Ch = Chalcophile; Li = Lithophile; Si = Siderophile

ize readily or form stable oxyanions and occur mainly in oxygen compounds. *Chalcophile elements* ionize less readily and tend to form sulphides and covalent compounds with Se and Te. *Siderophile elements* do not readily form compounds with O and S and occur mainly as native elements. *Atmophile elements* are present mainly in atmospheric gases. *Biophile elements* tend to be associated with organisms and thus accumulate in the horizons most affected by organisms in soils

The abundance of elements in the earth's crust 10 miles thick decreases in the order: O, Si, Al, Fe, Ca, Na, K, and Mg (Clarke [8], Ahrens [1]). Only these eight elements are present in amounts higher than 1 % (Table 1) and make up nearly 99 % of the earth's crust. The elements Ti, H, P, and Mn, are present in amounts between 0.1 and 1 %. The remaining elements together make up less than 0.5 % of the earth's crust. Nevertheless, many of these trace elements are of concern to plant growth, animal nutrition and health and the well being of mankind (National Academy of Sciences [31, 32]).

Igneous, Metamorphic and Sedimentary Rocks

The earth's crust consists mainly of the lithophile elements which make up the majority of the rocks and are common in silicates and soils. About 80 elements are distributed in approximately two thousand compounds or minerals. Only a few dozen make up the bulk of surficial rocks. The earth's crust consists of 95 % igneous and metamorphic rocks. The remaining 5 % are sedimentary rocks (Clarke and Washington [9]).

The cooling of magma or silicate melts of the lithophile elements has resulted in the formation of *igneous rocks*. They can be classified into two groups, namely, the *volcanic* or extrusive rocks, and the *plutonic* or intrusive rocks. The latter resulted from slow cooling of magma and are thus coarser grained, whereas the former resulted from rapid cooling upon eruption and are thus finer grained. *Metamorphic rocks* are formed through the action of heat and pressure operating on sedimentary and igneous rocks. Common metamorphic rocks are gneiss, mica schists, slates, marble, and quartzite. Other metamorphic rocks of less extensive occurrence are amphibolite, amphibole schist, hornblende schist, chlorite schist and tactite. The interaction of the atmosphere and hydrosphere on the crust of the earth and the subsequent sedimentation process have produced *sedimentary rocks*. Of these sediments, 80 % are shales, 15 % sandstones, and 5 % limestones (Clarke and Washington [9]). Sedimentry rocks tend to accumulate near the interface of the crust with the hydrosphere and atmosphere, and cover approximately three-fourths of the land area of the earth. The significant variations in mineralogical and chemical composition of these three rock types (Wedepohl [46]) are among the factors affecting soil formation (Jackson [21]) and the quality of our environments (Tourtelot [44]).

Geological Classification of Parent Materials

Parent materials of soils can be classified into two major groups, namely, *sedentary* and *transported*. Sedentary or residual parent materials lie in their original position above bedrocks for centuries. These materials are widely distributed on all the continents and have usually been subjected to long and often intense weathering. Parent materials can be transported and deposited by gravity, water, ice or wind leading to the formation of transported parent materials. They include *colluvial, alluvial, marine, lacustrine, glacial,* and *aeolian* deposits.

Colluvial debris is composed of the rock fragments detached from the heights above and carried down the slopes mainly by gravitational force. These materials are generally not of great importance in the formation of productive soils. Alluvial sediments are deposited by streams. Three general classes of alluvial deposits are flood plains, alluvial fans and delta. The soils derived from such sediments usually are very fertile. The products of deposition in the oceans, seas and gulfs are marine sediments. These deposits have been worn and weathered by many natural processes and thus usually carry less of the mineral nutrient elements. Despite this, their soils when adequately managed support a great variety of crops.

The materials deposited directly by the ice are glacial tills and are a mixture of rock debris of great diversity, particularly particle size. The soils derived from such parent materials are thus most heterogeneous. Glacial outwash plains are formed by streams heavily laden with glacial sediments. This type of sediment is generally assorted and consequently of great variety in texture.

Wind erosion results in the formation of aeolian deposits. There are three

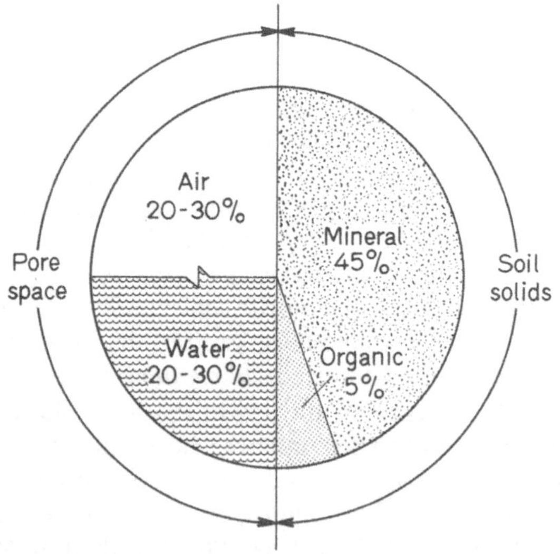

Fig. 1. Volume composition of a silt loam surface soil when in good condition for plant growth (Brady [6])

types of such deposits, namely, loess, volcanic ash and sand dunes. The wind-blown material, termed loess, is deposited in the uplands; the thickest deposits are present in the place where the valleys are widest. Loess has given rise to soils of considerable diversity. Loess soils thus have great differences in fertility and productivity. Soils derived from volcanic ash are light and porous and often have limited agricultural value. Sand dunes support little vegetation under most conditions.

Major Components of Soils

Soils consist of four major components–mineral and organic components, water and air (Fig. 1). The proportion of water and air fluctuates greatly under natural conditions, depending on the climatic and other factors. These four major components of soils exist mainly in an intimately mixed condition. Compared to surface soils, subsoils have a lower content of organic components and air and a higher proportion of mineral components and water.

Mineral Components

The inorganic portion of soils is composed of small rock fragments and a wide variety of crystalline and noncrystalline minerals (Table 2). These materials vary in particle size ranging from gravel to clay fractions (Table 3). The sand and silt fractions of soils are dominated by the *primary minerals* which are inherited from the parent igneous and metamorphic rocks. On the other hand, the clay fraction is enriched in the *secondary minerals* which are formed by low-temperature reactions and inherited in soils from sedimentary rocks or formed *in situ* by weathering.

Primary Minerals

Quartz and Feldspars. The most abundant primary minerals in soils throughout the world are quartz and feldspars, just as they are the dominant rock-forming minerals in the earth's crust. Quartz consists of a continuous framework of silica tetrahedra and is the main form of free silica occurring in soils. The silicate minerals comprise the major weight and volume percentage of most soils. Feldspars are anhydrous three-dimensional aluminosilicates containing varying amounts of Na, K, and Ca and occasionally of other large cations such as Ba. The feldspar minerals are found in virtually all soils and sediments and their nature and quantity vary with the type of parent material and the stage of weathering.

Micas and Accessory Minerals. Muscovite and biotite occur extensively in soils and are important reserves of soil K. They exist in soils primarily through inheritance from the parent rock of the soil. Accessory minerals of soils include a wide variety of minerals which occur in small but significant amounts. These accessory minerals are excellent source minerals for many nutrient elements in soils.

Table 2. Common primary and secondary minerals occurring in soils

Mineral group	Mineral name	Chemical formula
	Primary Minerals	
Silica minerals[a]	Quartz	SiO_2
	Cristobalite	SiO_2
	Tridymite	SiO_2
	Opal (inorganic origin)	$SiO_2 \cdot nH_2O$
Feldspars[b]	Alkali feldspars:	
	Microcline and orthoclase	$KAlSi_3O_8$
	Plagioclase feldspars:	
	Albite	$NaAlSi_3O_8$
	Anorthite	$CaAl_2Si_2O_8$
Micas	Muscovite	$K(Si_3Al)Al_2O_{10}(OH)_2$
	Biotite	$K(Si_3Al)(Mg, Fe^{II})_3O_{10}(OH)_2$
Accessory minerals	Pyroxenes:	
	Enstatite	$MgSiO_3$
	Hypersthene	$(Mg, Fe)SiO_3$
	Diopside	$CaMgSi_2O_6$
	Augite	$Ca(Mg, Fe, Al)(Si, Al)_2O_6$
	Amphiboles:	
	Tremolite	$Ca_2Mg_5Si_8O_{22}(OH)_2$
	Actinolite	$Ca_2Fe_5Si_8O_{22}(OH)_2$
	Hornblende	$(Ca, Na)_2(Mg, Fe, Al)_5(Si, Al)_8O_{22}(OH)_2$
	Olivines:	
	Forsterite	Mg_2SiO_4
	Fayalite	Fe_2SiO_4
	Apatite[c]	$Ca_{10}(F, OH, Cl)_2(PO_4)_6$
	Tourmaline	$Na(Mg Fe)_3Al_6(BO_3)_3Si_6O_{18}(OH)_4$
	Rutile and anatase	TiO_2
	Zircon	$ZrSiO_4$
	Secondary Minerals	
Carbonate minerals	Calcite and aragonite	$CaCO_3$
	Mg-calcite	$Ca_{1-x}Mg_xCO_3, 0 < x < 0.5$
	Dolomite	$CaMg(CO_3)_2$
	Nahcolite	$NaHCO_3$
	Trona	$Na_3CO_3HCO_3 \cdot 2H_2O$
	Soda	$Na_2CO_3 \cdot 10H_2O$
Sulfate, sulfide, and halide minerals	Gypsum	$CaSO_4 \cdot 2H_2O$
	Epsomite	$MgSO_4 \cdot 7H_2O$
	Pyrite	FeS_2
	Jarosite	$KFe_3(OH)_6(SO_4)_2$
	Halite	$NaCl$
Layer silicates[d]	Vermiculite	$X_{0.55}(Al_{1.15}Si_{2.85})(Al_{0.25}Fe^{III}_{0.35}Mg_{2.4})O_{10}(OH)_2 \cdot nH_2O$
	Smectite	$X_{0.4}(Al_{0.15}Si_{3.85})(Al_{1.3}Fe^{III}_{0.45}Mg_{0.25})O_{10}(OH)_2 \cdot nH_2O$
	Chlorite	$(AlSi_3)AlMg_5O_{10}(OH)_8$
	Kaolinite	$Si_2Al_2O_5(OH)_4$
	Halloysite	$Si_2Al_2O_5(OH)_4 \cdot 2H_2O$

Table 2 (continued)

Mineral group	Mineral name	Chemical formula
Crystalline oxides, oxyhydroxides and hydroxides	Gibbsite	α-$Al(OH)_3$
	Boehmite	α-AlOOH
	Goethite	α-FeOOH
	Hematite	α-Fe_2O_3
	Lepidocrocite	γ-FeOOH
	Maghemite	γ-Fe_2O_3
	Pyrolusite	β-MnO_2
Noncrystalline mineral components	Allophane	$Al_2O_3 \cdot 2SiO_2 \cdot nH_2O$
	Imogolite	$Al_2O_3 \cdot SiO_2 \cdot 2.5H_2O$
	Opal (biogenic origin)	$SiO_2 \cdot nH_2O$
	Ferrihydrite	$Fe_5HO_8 \cdot 4H_2O$
	Other noncrystalline hydrous[e] oxides and hydroxides of Al and Fe	

[a] These minerals contain impurities (e.g., Al, Fe, Ti, Na, K, Mg and Ca) which increase in magnitude from quartz to opal due to increased microporosity (Wilding et al. [47])

[b] Alkali feldspar crystals of soils are perthitic in nature and stores of numerous trace elements (Huang [20])

[c] Apatite is the most common primary mineral carrier of phosphorus, but occurs also in calcareous sediments (Jackson [21])

[d] Chemical formulae of layer silicates vary with structural types and isomorphous substitution; imperfectly ordered micas are common in soils derived from sedimentary rocks (Jackson [21]; Gieseking [16])

[e] The noncrystalline state is stabilized by clay minerals (Rich [35] and a series of inorganic and organic ligands (Luciuk and Huang [29]; Schwertmann and Taylor [42]; Kwong and Huang [28]; there is no definite composition)

Table 3. Classification of soil particles according to size (International Society of Soil Science)

Name of separate	Size range (mm)
Clay	<0.002
Silt	0.002–0.02
Fine sand	0.02–0.2
Coarse sand	0.2–2.0
Gravel	>2.0

Secondary Minerals

Carbonates, Sulfur-Bearing Minerals, and Halides. The most abundant carbonate mineral of soils is *calcite* which is inherited from the parent rocks of soils or formed in subsoils of subhumid and more arid regions. Depending on pH and the CO_2 pressure of the soil solution, some calcite is dissolved, providing a source of available Ca. *Gypsum* is the most common soil sulfate mineral and accumulates in semiarid and arid regions, frequently beneath the horizon in which calcium carbonate is accumulated. Gypsum has a relatively high solubility and rate of dissolution compared to silicates. Sulfide minerals, such as *pyrite*, are formed under reducing conditions in the presence of organic matter. The activity of sulfate-reducing bacteria is extremely important in the distribution of sulfide minerals in the soil fabric. The only important halide mineral in soils is *halite* which may form crusts on soil surfaces.

Layer Silicate Minerals. Layer silicates or phyllosilicates play a prominent role in determining the physical and chemical properties of most soils. The two coordinating units are basic to the crystal structure of these minerals. The first is the silicon tetrahedral sheet and the second the octahedral sheet of Al, Fe or Mg. Different combinations of these two types of sheets yield the structures of various layer silicates of importance in soil environments: (1) 1:1 (tetrahedra:octahedra) type minerals, (2) expansible 2:1 type minerals, (3) nonexpansible 2:1 type minerals, and (4) 2:2 type minerals.

The common layer silicates occurring in soils are *mica, vermiculite, smectite, chlorite, kaolinite* and *halloysite.* These specific mineral groups do not always occur independently of the other. Several layer silicates are likely in an intimate mixture. Because of structural similarities, they often occur in a mixed order of stacking in which an individual crystal may consist of two or more layer silicates. These assemblages of layer silicates are commonly referred to as *mixed-layer* or *interstratified* minerals. Furthermore, it is important to recognize that hydroxy forms of Al and Fe play significant roles in modifying the physical and chemical properties of clay minerals, particularly vermiculite and smectite. Hydroxy interlayered minerals are also known as *intergradient layer silicates* and are common in soil environments.

Crystalline Oxides, Hydroxides, and Oxyhydroxides and Noncrystalline Mineral Components

The most common crystalline oxide minerals in soils are gibbsite, hematite, goethite, and pyrolusite. Notable noncrystalline inorganic components are allophane, imogolite and hydrous oxides of Al, Fe, Si, and Mn. They are usually low in amount in soil environments, but their importance in determining soil properties and transformations of nutrients and pollutants in terrestrial systems can never be over-emphasized.

Organic Components

Organic Matter. Organic matter plays both a direct and an indirect role in the determination of pedogenic characteristics and formation of soil horizons. It is also a major factor in soil fertility, the retention and deactivation of anthropogenic organic chemicals introduced into the environment, and the type and amount of organic constituents in the weathering process and in the translocation of clays and sesquioxides (Russell [36]).

The diverse mixture, that includes the *soil biomass*, partially degraded plant, animal and *microbial components* and the soil humic constituents, comprises soil organic matter. The recognizable plant and microbial components constitute 15–25% of the total soil organic carbon in the A horizon of cultivated soils and a large proportion in natural grassland and forest ecosystems. Large plant and animal residues can be removed from the system before analysis. However, most microorganisms and fine root hairs are very small in size. They together with the root exudates are intimately associated with the colloidal components and no known techniques for their separation are presently available. Fractionation of soil humic constituents therefore usually fractionates the microbial components comprising 1–3% of the soil carbon into the different soil fractions regardless of the method of analysis.

Between 52–98% of the organic carbon is associated with the clay and silt sized fractions (Greenland [17]). Much of the remainder is linked to the metal oxides, hydroxides and oxyhydroxides (Schnitzer and Kodama [41]). The classical fractionation technique for soil organic matter includes solubilization and peptization of soil constituents in sodium hydroxide or sodium pyrophosphate. The soluble materials comprise two major fractions, the low molecular weight materials with fairly high functional group acidity *(fulvic acids)* are not precipitated at pH 2, whereas the larger *humic acids* with molecular weights of 20,000 to 100,000 are precipitated. The materials that are so intimately associated with the clays that they are not removed by the basic solution or sequestering agents are known as *humins*. Removal of the clay with agents such as hydrofluoric acid results in the fractionation of the humin into fulvic and humic acids as well as insoluble portions of insect remains, charcoal, etc. (Kononova [25]).

Table 4 shows that humic acids isolated from a wide range of soils have a carbon content ranging from 53.6 to 58.8% with an average of 56.2%. The fulvic acids have a lower carbon content and also show greater variablility in

Table 4. Characteristics of soil organic fractions extracted from a wide range of soil types. Includes the range of values measured. (Adapted from Schnitzer [39])

	Humic acids	Fulvic acids
Element (%)		
C	56.2±2.6	45.7±5.0
H	4.7±1.5	5.4±1.6
N	3.2±2.4	2.1±1.2
S	0.8±0.7	1.9±1.8
O	35.5±2.8	44.8±5.1
Functional groups (meq/g)		
Total acidity	6.7±1.1	10.3±3.9
CO_2H	3.6±2.1	8.2±3.0
Phenolic OH	3.9±1.8	3.0±2.7
Alcoholic OH	2.6±2.4	6.1±3.4
Quinonoid C = O and ketonic C = O	2.9±2.8	2.7±1.5
OCH_3	0.6±0.3	0.8±0.5

composition. They have higher hydrogen, sulphur, and oxygen, but lower nitrogen contents. This is expressed in the functional groups with fulvic acids having a much higher total acidity, carboxylic acid content and alcoholic OH's than the humic acids isolated from the same soil (Schnitzer [39]).

Humic and fulvic acids behave like linear flexible polyelectrolytes that are readily aggregated at low pH with the aid of hydrogen bonding, van der Waal's interactions and interactions between the π electron systems of adjacent molecules (Flaig et al. [15]). Fulvic acids at pH values above 2–3 occur as elongated fibers and bundles of fibers with a relatively open structure. With increases in pH, the fibers tend to mesh into a woven network yielding a sponge-like structure. Humic acids show similar structures but because of the lower solubility in water the same structures are observed over a narrower pH range.

Soil organic constituents can be separated into a number of subgroups depending on the soil climate and inorganic constituents (Kuwatsuka et al. [27]). One of the major criteria of a series of papers which studied 37 different humic acids was the degree of humification which is a measure of the light absorbed at 600 nm per unit of carbon, i. e., the darkness of the humic acid solution. The H:C and the O:C ratio also are of significance as shown in

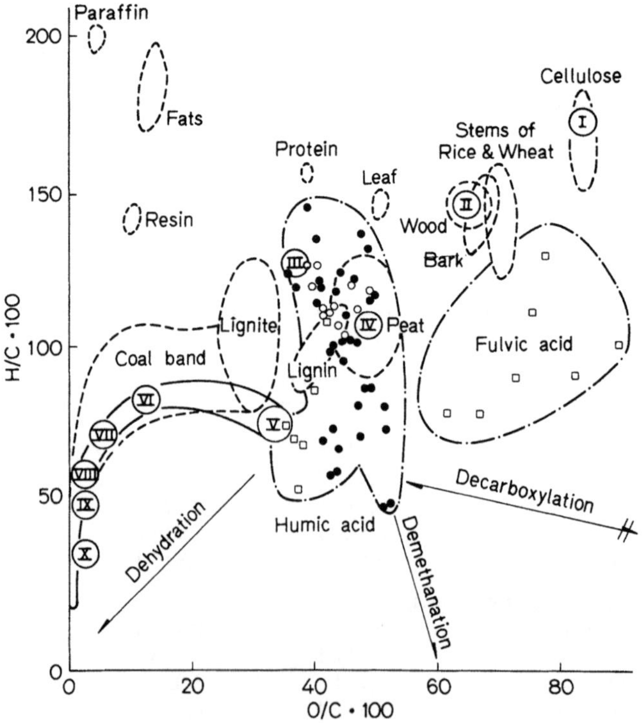

Fig. 2. Atomic H/C versus O/C diagram. I, Cellulose; II, Wood; III, Lignine; IV, Peat; V. Brown coal (Lignite); VI, Low rank bituminous coal; VII, Medium rank bituminous coal; VIII, High rank bituminous coal; IX, Semi-anthracite; X, Anthracite. (From Kuwatsuka et al. [27])

Fig. 2 which compares the oxygen to carbon and hydrogen to carbon ratio of humic and fulvic acids to other organic materials in nature.

The dark amorphous multicomponent humic system never occurs in pure form but is intimately associated with other organic compounds and with the mineral soil colloids. The term *heteropolycondensate* best describes the series of related acidic mycels in which the aromatic rings are bridged by -O-, -NH-, -N-, -S-, linkages. Formation of humic material is not enzymatically controlled but is the result of polycondensation of a wide range of microbially produced and plant degradation products. Physical and chemical characteristics of the environment at the time of formation greatly affect the humic structures and these materials cannot be expected to be uniform in structure but are a series of closely related structures. The type of structure suggested by Stevenson ([43], Fig. 3) presents the known characteristics and components of

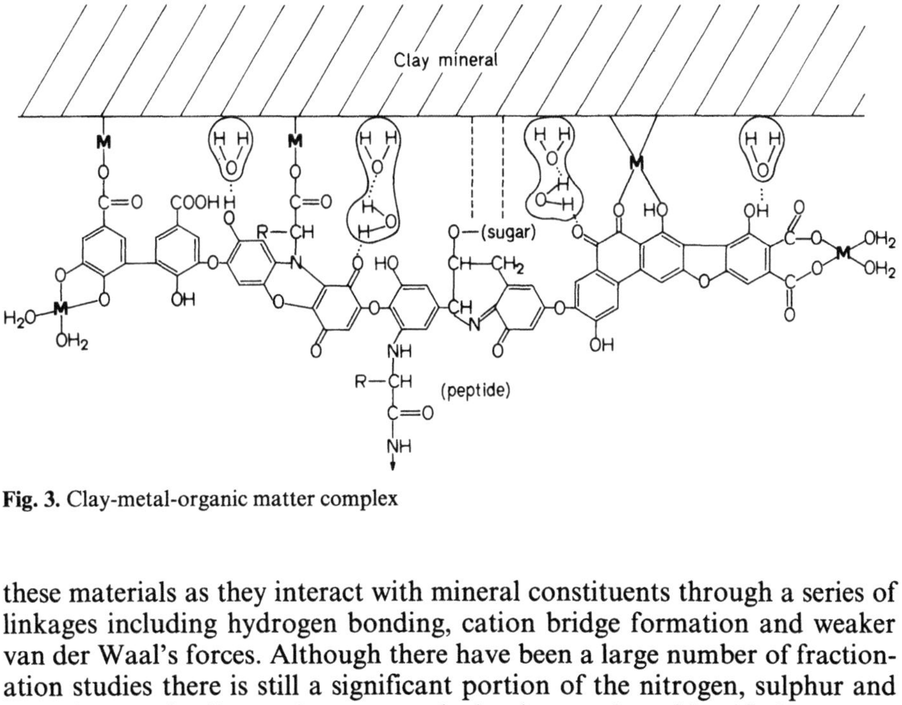

Fig. 3. Clay-metal-organic matter complex

these materials as they interact with mineral constituents through a series of linkages including hydrogen bonding, cation bridge formation and weaker van der Waal's forces. Although there have been a large number of fractionation studies there is still a significant portion of the nitrogen, sulphur and phosphorus of soil organic compounds that has not been identified.

Many of the carbohydrates and alkanes are found within the fulvic acids. However, the proteinacous materials as well as the aromatics are distributed throughout the fulvic and humic acids and humin. Nitrogen is present primarily as amino acids, amino sugars, nucleic acids and cyclic nitrogen within the humic material (Haider et al. [18]). The known groups of phosphorus compounds in soil include the inositol phosphates, nucleic acid derivatives, sugar phosphates, phosphoproteins and phospholipids (Halstead and McKercher [19]). Although some of the organic sulphur is associated with amino acid carbon in the form of C-S linkages, the majority of the sulphur is found as

C–O–S linkages (ester sulphates). These materials are found in a number of compounds such as sulphated polysaccharides which tend to accumulate in the fulvic fractions.

Content and Turnover of Soil Organic Constituents. Studies of Organic matter in various soils have shown that the tropics contain humic materials at great depths (Bohn [5]), and the organic matter contents of tropical soils are usually at least as high as that of temperate woodlands but usually do not attain the levels of temperate grasslands. Plant residues which are the major source of soil organic matter are fairly rapidly decomposed with half-lives varying from days to months (Paul and van Veen [33]). The living organisms in the soil (the biomass) are comprised primarily of bacteria and fungi. The weight of these usually equals in weight the amount of carbon that enters the system in one year. On the average it turns over less than once during a growing season (Jenkinson and Ladd [23]), but the diverse microbial population varies in its growth rate. Bacteria capable of growth on readily available substrates can double their mass within hours, whereas the majority of the soil population is in a resting state and persists for long periods on minimal exogenous carbon supplies.

Microbial by-products and resistant plant residues adsorbed to soil particles have turnover times in terms of years. The humic materials are more resistant. Carbon dating has indicated that the fulvic acids have turnover times in hundreds of years, whereas the humins and humic acids usually

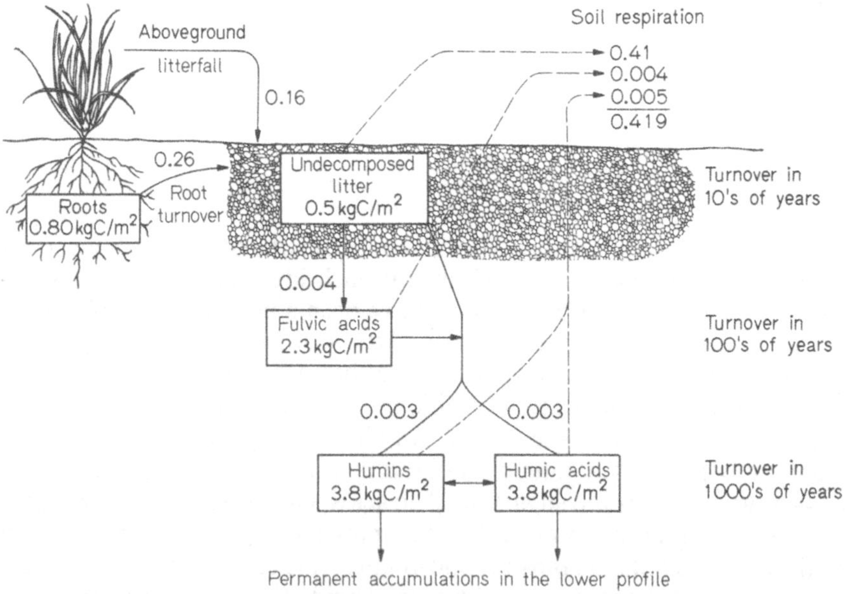

Fig. 4. Detrital carbon dynamics for the 0–20 cm layer of a chernozem grassland soil. Carbon pools (kgC m^{-2}) and annual transfers (kgC m^{-2} yr^{-1}) are indicated. Total profile content to 20 cm is 10.4 kgC m^{-2}. (From Schlesinger [38])

approach a thousand years in their turnover time (Paul and van Veen [33]). Figure 4 shows the distribution of carbon in the various fractions and the annual transfers for a grassland chernozem. Although the humins and humic acids comprise by far the majority of carbon within a system, their very slow turnover rate means that they only contribute a small proportion to the annual cycling of carbon within the soil. In this diagram the undecomposed litter estimate also includes the soil biomass and microbial metabolites which together with the plant residues constitute the active fraction of organic matter which plays the prominent role in replenishing the majority of nutrients for plant growth and in the cycling of nutrient elements within the geocycle on an annual basis.

Water, Air, and Structure

Soil Water. Soil water is held within soil pores with varying degress of tenacity depending upon the nature of soil particles and the amount of water present. Soil water can be divided into three types of physical classes, namely, *gravitational, capillary* and *hygroscopic*. Gravitational water is that in excess of the field capacity (10^4 to 2×10^4 Pa or 0.1–0.2 bar suction) and occupies the larger pores, thereby reducing soil aeration. Capillary water is held in pores of capillary size and exerts suctions between 10^4 and 31×10^5 Pa (0.1 and 31 bars). Water retained tightly by soil particles at suctions greater than 31×10^5 Pa (31 bars) is termed hygroscopic and moves primarily in vapor form.

Based on biological classification, there is a relationship between moisture retention and its availability to plants. Gravitational water is of little use to plants. Moisture retained by the soil at tension greater than 15×10^5 Pa (15 bars) is also largely unavailable to plant growth. Water held by soil particles between the field capacity and the permanent wilting coefficient (15×10^5 Pa) is termed *available water*. Soil water dissolves soluble salts and certain organic components making up the soil solution which is so important as a medium to plant life. Therefore, soil water is affected by osmotic as well as matric tensions.

Soil Air. Soil air is the phase of the soil composed of gases. It differs from the atmosphere in many aspects: (1) it is not continous but is located in the maze of soil pores; (2) it generally has a higher moisture content than the atmosphere; and (3) the content of CO_2 is usually 5 to 50 times higher than the atmosphere. There is a general inverse relationship between O_2 and CO_2 contents and their sum is close to 21% of soil air by volume. The concentration of N_2 in soil air is in the order of 79% by volume and comparable to atmospheric air. Under conditions of poor aeration, N_2O, NO, H_2, CH_4, C_2H_4, and H_2S may be present in soil air (Yoshida [48]; Burford [7]; Bailey and Beauchamp [2]).

The content and composition of soil air is determined by the following factors: (1) the soil-water relationships; (2) the rate of production and consumption of the various gases in the soil; and (3) the rate of exchange between the soil air and atmospheric air (de Jong and Paul [10]). Changes in the content and composition of soil air have significant effects on the physical environment of soils.

Soil Structure. Soil structure results from aggregation of soil particles. Important factors governing the structural formation are the nature of soil parent material and the physical, chemical and biochemical processes of soil formation. Noncrystalline Al and Fe oxides of the weathering products act as cementing agents in binding soil separates (Kroth and Page [26]; Jones and Uehara [24]). Aluminium oxides are much more effective than iron oxides in promoting the stability of soil aggregates (Saini et al. [37]; El Swaify and Emerson [13]).

Organic matter, particularly polysaccharide, is closely associated with soil aggregation. It may promote soil aggregation through the following linkages: clay-(Al, Fe)-organic matter-(Al, Fe)-clay (Edwards and Bremner [12]). Large organic molecules usually form bonds with several mineral particles, increasing aggregate stability.

There are four major types of soil structure, namely, *platy, prism-like, block-like* and *spheroidal.* The important changes in physical properties as a result of cultivation are structural rather than textural. Soil condition and characteristics such as bulk density, porosity, aeration, and water movement are much influenced by soil structure.

Weathering Reactions and Soil Formation and Distribution

Parent rocks of soils are disintegrated by physical weathering into rock fragments and the individual minerals of which they are composed. Rock fragments and the minerals therein are further attacked by chemical and biochemical forces. Weathering of minerals through these processes releases nutrients to nourish plants. Residues from the organisms return to the weathering mass and are transformed to humus. During soil forming processes, primary minerals are altered to varying extents or transformed to crystalline secondary minerals, noncrystalline mineral components (Jackson [22]; Marshall [30]) and organomineral complexes (Schnitzer and Kodama [41]), depending on the nature of parent materials, climate, living organisms, topography and time. Crystalline and noncrystalline mineral colloids, humus and living organisms together with life-sustaining water affect markedly the kind and extent of horizon differentiation that occurs in soil environments.

A *soil horizon* is defined as a layer of soil, approximately parallel to the soil surface, with characteristics produced by soil-forming processes (US Soil Conservation Service [45]). In addition to genetic soil horizons, many soils show stratification due to variations in parent material of lithological discontinuities. Designations of master soil horizons are summarized in Table 5. Each soil is characterized by a given sequence of soil horizons. The sequence is referred to as a *soil profile*. Soils which have been developed by various processes vary in many profile characteristics and differ from region to region.

Table 5. Soil horizon designations (FAO-Unesco [14])

Master horizon	Description
H	An organic horizon formed or forming from accumulations of organic material deposited on the surface that is saturated with water for prolonged periods and contains 30% or more organic matter if the mineral fraction contains more than 60% of clay, 20% or more organic matter if the mineral fraction contains no clay, or intermediate proportions of organic matter for intermediate contents of clay
O	An organic horizon formed or forming from accumulations of organic material deposited on the surface that is not saturated with water for more than a few days and contains 35% or more organic matter
A	A mineral horizon formed or forming at or adjacent to the surface that either: (1) shows an accumulation of humified organic matter intimitately associated with the mineral fraction, or (2) has a morphology acquired by soil formation but lacks the properties of E and B horizons
E	An eluvial mineral horizon showing a concentration of sand and silt fractions high in resistant minerals, resulting from a loss of silicate clay, iron or aluminium or some combination of them
B	A mineral horizon in which rock structure is obliterated or is but faintly evident, characterized by one or more of the following features: (1) an illuvial concentration of silicate clay, iron, aluminium, or humus, alone or in combination; (2) a residual concentration of sesquioxides relative to source materials; (3) an alteration of material from its original condition to the extent that silicate clays are formed, oxides are liberated, or both, or granular, blocky, or prismatic structure is formed
C	A mineral horizon (or layer) of unconsolidated material from which the solum is presumed to have formed and which does not show properties diagnostic of any other master horizons
R	A layer of continuous indurated rock that is sufficiently coherent when moist to make hand digging with a spade impracticable

Table 6. The nomenclature of the soil units (FAO-Unesco [14])

Soil unit[a]	Nomenclature
Fluvisols	From L. *fluvius,* river; connotative of floodplains and alluvial deposits
Gleysols	From Russian local name *gley,* mucky soil mass; connotative of an excess of water
Regosols	From Gr. *rhegos,* blanket; connotative of mantle of loose material overlying the hard core of the earth; soils with weak or no development
Lithosols	From Gr. *lithos,* stone; connotative of soils with hard rock at very shallow depth
Arenosols	From L. *arena,* sand; connotative of weakly developed coarse-textured soils

Table 6 (continued)

Renzina	From Polish *rzedzic*, noise; connotative of noise made by plough over shallow stony soil
Rankers	From Austrian *Rank*, steep slope; connotative of shallow soils from siliceous material
Andosols	From Japanese *An*, dark, and *Do*, soil; connotative of soils formed from materials rich in volcanic glass and commonly having a dark surface horizon
Vertisols	From L. *verto*, turn; connotative of turnover of surface soil
Solonchaks	From Russian *sol*, salt
Solonetz	From Russian *sol*, salt
Yermosols	From Sp. *yermo*, desert
Xerosols	From Gr. *xeros*, dry
Kastanozems	From L. *castaneo*, chestnut, and from Russian *zemlja*, earth, land; connotative of soils rich in organic matter having a brown or chestnut colour
Chernozems	From Russian *chern*, black, and zemlja, earth, land; connotative of soils rich in organic matter having a black colour
Phaeozems	From Gr. *phaios*, dusky, and Russian *zemlja*, earth, land; connotative of soils rich in organic matter having a dark colour
Greyzems	From Anglo-Saxon *grey;* colour formed by blending of white and black; connotative of white silica powder which is present in layers rich in organic matter, and from Russian *zemlja*, earth, land; connotative of soils rich in organic matter having a grey colour
Cambisols	From late L. *cambiare*, change; connotative of changes in colour, structure and consistence resulting from weathering in situ
Luvisols	From L. *luvi*, from *luo*, to wash, lessiver; connotative of illuvial accumulation of clay
Podzols	From Russian *pod*, under, and *zola*, ash; connotative of soils with a strongly bleached horizon
Podzoluvisols	From Podzols and Luvisols
Planosols	From L. *planus*, flat, level; connotative of soils generally developed in level or depressed topography with poor drainage
Acrisols	From L. *acris*, very acid; connotative of low base saturation
Nitosols	From L. *nitidus*, shiny; connotative of shiny ped surfaces
Ferralsols	From L. *ferrum* and aluminium; connotative of a high content of sesquioxides
Histosols	From Gr. *histos*, tissue; connotative of soils rich in fresh or partly decomposed organic matter

[a] These units are generally comparable to the "great group" level

By establishing a common denominator between different soil classification systems, the major soil units, which have been recognized in all parts of the world, both in virgin conditions and under cultivation, are combined into one outline. The *soil units* (Table 6) adopted are selected on the basis of present knowledge of the formation, characteristics and distribution of the soils covering the earth's surface, their importance as resources for production and their significance as factors of the environment.

References

1. Ahrens, L.H.: Distribution of the Elements in our Planet. McGraw-Hill, Toronto 1965; p. 96
2. Bailey, L.D., Beauchamp, E.G.: Can. J. Soil Sci. *53*, 213 (1973)
3. Berry, L.G., Mason, B.: Mineralogy. W.H. Freeman, San Francisco 1959; Chap. 5
4. Birkeland, P.W.: Pedology, Weathering, and Geomorphological Research. Oxford University Press, New York 1974
5. Bohn, H.L.: Soil Sci. Soc. Am. J. *40*, 468 (1976)
6. Brady, N.C.: The Nature and Properties of Soils, 8th ed. MacMillan, New York 1974; Chap. 1
7. Burford, J.R.: J. Environ. Qual. *4*, 55 (1975)
8. Clarke, F.W.: US Geol. Survey Bull. 770, 1924
9. Clarke, F.W., Washington, H.S.: US Geol. Survey Prof. Paper 127, 1924
10. de Jong, E., Paul, E.A.: The Composition of the Soil Atmosphere and its Relationship to Soil Aeration and Respiration; In: Encyclopedia of Earth Science Series, VI. Soil Science and Applied Geology, O.W. Finkle (ed.). Reinhold, New York 1979; p. 15
11. Duchaufour, Ph.: Atlas Ecologique des Sols du Monde. Masson, Paris 1976
12. Edwards, A.P., Bremner, J.M.: J. Soil Sci. *18*, 64 (1967)
13. El Swaify, S.A., Emerson, W.W.: Soil Sci. Soc. Amer. Proc. *39*, 1056 (1975)
14. FAO-Unesco "Soil Map of the World". Unesco, Paris 1974; Vols. I–X
15. Flaig, W., Beutelspacher, H., Rietz, E.: Chemical composition and physical properties of humic substances; In: Soil Components, Vol. 1, Organic Components, J.E. Gieseking (ed.). Springer-Verlag, New York 1975; Chap. 1
16. Gieseking, J.E.: Soil Components, Vol. 2, Inorganic Components. Springer-Verlag, Berlin 1975; Chaps. 3–7
17. Greenland, D.J.: Soil Fert. Commonw. Bur. Soil Sci. *28*, 521 (1965)
18. Haider, K., Martin, J.P., Filip, Z.: Humus biochemistry; In: Soil Biochemistry, Vol. 4, E.A. Paul; A.D. McLaren (ed.). Marcel Dekker, New York 1975; Chap. 6
19. Halstead, R.L., McKercher, R.B.: Biochemistry and cycling of phosphorus; In: Soil Biochemistry, Vol. 4, E.A. Paul; A.D. McLaren (ed.). Marcel Dekker, New York 1975; Chap. 2
20. Huang, P.M.: Feldspars, olivines, pyroxenes and amphiboles; In: Minerals in Soil Environments, J.B. Dixon; S.B. Weed (ed.); Soil Sci. Soc. Amer., Madison, Wisc. 1977; Chap. 15
21. Jackson, M.L.: Chemical composition of soils; In: Chemistry of the Soil, F.E. Bear (ed.). Reinhold, New York 1964; Chap. 2
22. Jackson, M.L.: Int. Congr. Soil Sci. Trans. 9th (Adelaide), *2*, 705 (1968)
23. Jenkinson, D.S., Ladd, J.N.: Microbial biomass in soil – measurement and turnover; In: Soil Biochemistry, Vol. 5, E.A. Paul; J.N. Ladd (ed.). Marcel Dekker, New York 1980; Chap. 10
24. Jones, R.C., Uehara, G.: Soil Sci. Soc. Amer. Proc. *37*, 792 (1973)
25. Kononova, M.M.: Soil Organic Matter. Pergamon Press, New York 1966
26. Kroth, E.M., Page, J.B.: Soil Sci. Soc. Amer. Proc. *11*, 27 (1947)
27. Kuwatsuka, S., Tsutsuki, K., Kumada, K.: Soil Sci. Plant Nutr. *24*, 337 (1978)
28. Kwong, K.F. Ng Kee, Huang, P.M.: Proc. 6th Int. Clay Conf. (Oxford, England) 1978, 527
29. Luciuk, G.M., Huang, P.M.: Soil Sci. Soc. Amer. Proc. *38*, 235 (1974)
30. Marshall, C.E.: The Physical Chemistry and Mineralogy of Soils, Vol. II: Soils in Place. Wiley-Interscience, Toronto 1977; Chap. 2
31. Nat. Acad. Sci. (US): Geochemistry and the Environment, Vol. 1, The Relation of Selected Trace Elements to Health and Disease. Washington, D.C. 1974
32. Nat. Acad. Sci. (US): Geochemistry and the Environment, Vol. II, The Relation of Other Selected Trace Elements to Health and Disease. Washington, D.C. 1977

33. Paul, E.A., van Veen, J.A.: Trans. Int. Soc. Soil Sci., Vol. 3, Symposia Papers, 1978, p. 61
34. Rankama, K., Sahama, Th.G.: Geochemistry. Univers. Toronto Press, Toronto 1950; Chap. 4
35. Rich, C.I.: Clays Clay Minerals *16*, 15 (1968)
36. Russell, E.W.: Soil Conditions and Plant Growth; 10th ed. Longman, London 1973
37. Saini, G.R., MacLean, A.A., Dolyle, J.J.: Can. J. Soil Sci. *46*, 155 (1966)
38. Schlesinger, W.H.: Ann Rev. Ecol. Syst. *8*, 51 (1977)
39. Schnitzer, M.: Recent findings on the characterization of humic substances extracted from soils from widely differing climatic zones; In: Soil Organic Matter Studies, Vol. II; I.A.E.A.: Vienna 1977; IAEA-SM-211/7
40. Schnitzer, M., Khan, S.U.: Soil Organic Matter. Elsevier, Amsterdam 1978
41. Schnitzer, M., Kodama, H.: Reactions of minerals with soil humic substances. In: Minerals in Soil Environments, J.B. Dixon; S.B. Weed (ed.). Soil Sci. Soc. Amer., Madison, Wisc. 1977; Chap. 21
42. Schwertmann, U., Taylor, R.M.: Iron oxides; In: Minerals in Soil Environments, J.B. Dixon; S.B. Weed (ed.). Soil Sci. Soc. Amer., Madison, Wisc. 1977; Chap. 5
43. Stevenson, F.J.: J. Environ. Qual. *1*, 333 (1972)
44. Tourtelot, H.: Chemical compositions of rock types as factors in our environments; In: Environmental Geochemistry in Health and Disease, H.L. Cannon; H.C. Hopps (ed.). Geolog. Soc. Amer., Boulder, Colo. 1971; p. 13
45. US Soil Conservation Service: Soil Taxonomy: A Basic System of Soil Classification for Making and Interpreting Soil Survey. US Soil Survey Staff, Washington, D.C. 1974; Chap. 3
46. Wedepohl, K.H.: Handbook of Geochemistry, Vol. 1. Springer-Verlag, Berlin 1969; Chaps. 7–9
47. Wilding, L.P., Smeck, N.E., Drees, L.R.: Silica in soils: quartz, cristobalite, tridymite and opal; In: Minerals in Soil Environments, J.B. Dixon; S.B. Weed (ed.). Soil Sci. Soc. Amer., Madison, Wisc. 1977; Chap. 14
48. Yoshida, T.: Microbial metabolism of flooded soils; In: Soil Biochemistry, Vol. 3, E.A. Paul; A.D. McLaren (ed.). Marcel Dekker, New York 1975; Chap. 3

The Oxygen Cycle

J. C. G. Walker
College of Engineering, University of Michigan
Ann Arbor, MI 48109, USA

Introduction

The presence of abundant free oxygen in the terrestrial atmosphere is an anomaly in a solar system and universe composed predominantly of hydrogen. It is, of course, a direct consequence of the presence of life on earth, particularly the presence of photosynthetic organisms that use water as electron donor to reduce carbon dioxide to organic matter, producing molecular oxygen as a waste product. The accumulation of this metabolic waste product in the atmosphere is of interest because all aerobic organisms, including all multicellular plants, depend on free oxygen to sustain their vital processes.

There exists a reasonably coherent understanding of the biogeochemical cycles of oxygen, based on the work of a number of authors [1–8]. In broad outline these cycles are as follows (Fig. 1): The only significant source of free oxygen is *photosynthesis*. It is closely linked to a group of processes generally labelled "*respiration and decay*", which recombine the oxygen and organic carbon produced by photosynthesis to restore carbon dioxide and water to the biosphere. The cycle of photosynthesis followed by respiration and decay connects a relatively large reservoir of atmospheric oxygen to a relatively small reservoir of organic carbon at the surface of the earth. Stoichiometric considerations suggest that this cycle can not, by itself, determine the amount of oxygen in the atmosphere. Instead, it controls the size of the reservoir of surface organic carbon. If photosynthesis were to stop, surface organic carbon would be consumed by respiration and decay; atmospheric oxygen would be reduced in the process by only a percent or so.

The abundance of atmospheric oxygen is controlled by a much slower cycle that connects the atmospheric reservoir to a relatively very large reservoir of *reduced minerals* in the crust of the earth. Oxygen is extracted from the atmosphere when these minerals are exposed at the surface and oxidized during the course of chemical weathering. Corresponding to this relatively

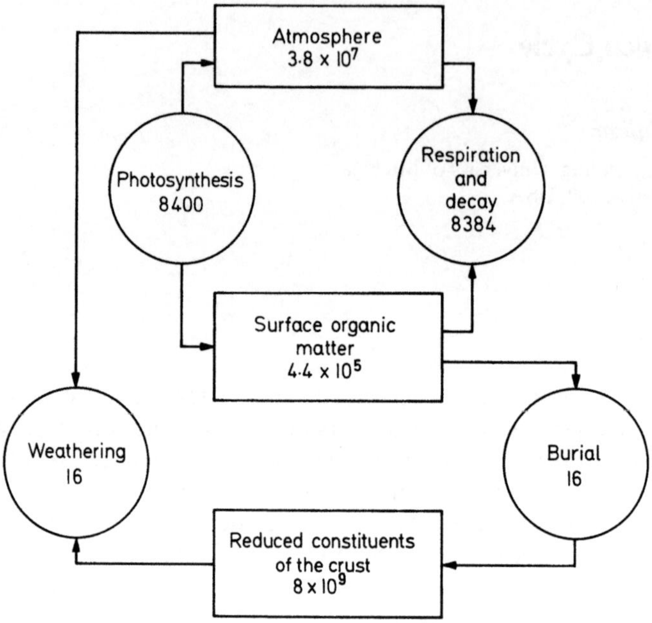

Fig. 1. A simplified representation of the biogeochemical cycles of oxygen. Circles denote processes with rates expressed in 10^{12} moles of oxygen per year or equivalent capacity to combine with oxygen. Rectangles denote stores either of oxygen or of reduced matter with which oxygen can combine. The units are 10^{12} moles of oxygen or capacity to combine with oxygen

slow consumption of oxygen is a net source which arises because not all photosynthetically produced organic matter is consumed, in the short term, by respiration and decay. A small fraction of this organic matter escapes from the biosphere by settling to the bottom of the ocean and becoming incorporated in sedimentary rocks. The oxygen that was released to the atmosphere when this organic matter was synthesized is left behind, to be consumed, eventually, by weathering.

The rate at which oxygen is consumed by weathering may depend on the abundance of atmospheric oxygen. It may, in principle, increase with increasing oxygen partial pressure. On the other hand, the fraction of photosynthetically produced organic matter that escapes oxidation by respiration and decay to become incorporated in sediments should, in principle, decrease with increasing oxygen partial pressure. In principle, therefore, an increase in the amount of oxygen in the atmosphere leads to an increase in the rate of consumption of oxygen and a decrease in the net rate of production. These interactions are thought to control oxygen abundance, causing an equilibrium to be established at the oxygen level that yields equality between the rates of consumption of oxygen by weathering and net production of oxygen corresponding to the burial of organic matter in sediments.

This picture of the biogeochemical cycles of oxygen is plausible, but not entirely satisfying. For one thing, there is no assurance that it is correct in any

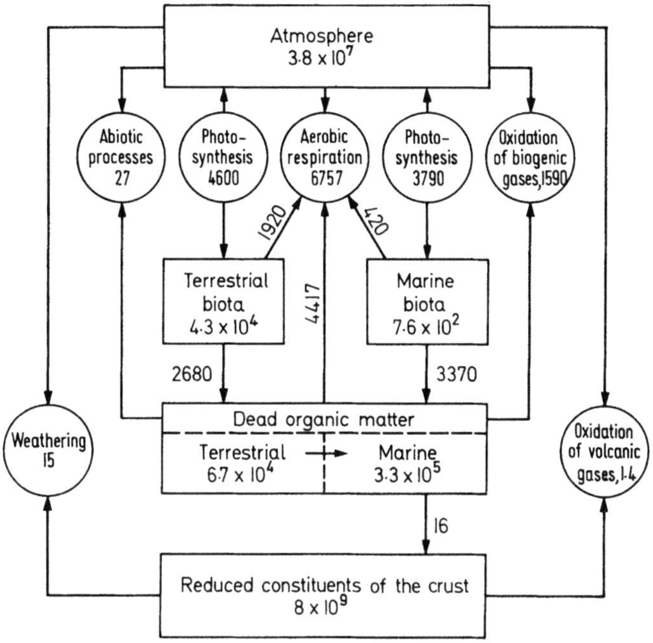

Fig. 2. The biogeochemical cycles of oxygen shown in greater detail, summarizing the numbers presented in Tables 1 and 2

but the broadest outline. The rates of production and consumption of oxygen are not known with sufficient accuracy to determine even whether the system is close to equilibrium. Furthermore, the predictive capability of the model is very low. Because we lack information on how the rates of production and consumption vary with oxygen partial pressure or as a result of external factors we are not able to estimate what the oxygen abundance might have been in the past or might be in the future.

Data that might permit a qualitative improvement in our understanding of what controls the abundance of atmospheric oxygen are lacking, but it is at least possible to re-examine the basic picture in order to expose possible defects or to reveal areas in which improved understanding can be achieved. This is the goal of this paper, in which the fluxes of oxygen and of reduced matter are analyzed in more detail than has been the custom in previous examinations of the biogeochemical cycles of oxygen.

Gains and Losses of Atmospheric Oxygen

Figure 2 provides a summary of the fluxes of oxygen and reduced matter that are described in this section. These fluxes transfer matter into and out of reservoirs, depicted in the figure by rectangles; the circles denote processes.

The relative sizes of the reservoirs are important in enabling a decision to be made about what is controlling what. Reservoir sizes are expressed here in units of 10^{12} moles of oxygen for the atmosphere and capacity to consume oxygen for the other reservoirs. Fluxes are expressed in units of 10^{12} moles O_2 per year or 10^{12} moles O_2 consumption capacity per year in the case of fluxes of reduced matter. Most of the numbers are quite uncertain; fluxes are expressed to several significant figures simply in order to balance the budget.

The size of the atmospheric reservoir is known with considerable precision [5, 9]. The masses of the biota and of dead organic matter are calculated from the corresponding masses of organic carbon quoted by Holland [3], compare [10]. The calculation makes allowance for the reducing power of organic nitrogen according to the following prescription: Photosynthesis produces 1 molecule of oxygen for every atom of organic carbon. A schematic representation of the reaction is

$$CO_2 + H_2O \longrightarrow CH_2O + O_2 \tag{1}$$

where CH_2O describes the average redox state of the carbon in organic matter. Plants and other photosynthetic organisms also incorporate nitrogen during the course of growth, generally reducing it from the level of nitrate to the level of *ammonia*

$$NO_3^- + 2CH_2O \longrightarrow NH_3 + 2CO_2 + OH^- \tag{2}$$

Thus, 2 atoms of carbon are oxidized for every atom of nitrogen incorporated. Plants and photosynthetic organisms constitute most of the mass of the biota. Land plants contain on average 16 carbon atoms for each nitrogen atom, while marine phytoplankton contain on average 7 carbon atoms for every nitrogen atom [11]. Therefore, growth of land plants releases 1.125 moles of oxygen for each mole of carbon incorporated, while growth of marine plankton releases 1.28 moles of oxygen for each mole of carbon incorporated. For convenience I have rounded these factors to 1.15 and 1.3, and then applied them to Holland's figures for the sizes of the surface organic carbon reservoirs and for the rates of production of organic carbon by photosynthesis.

The oxygen consumption capacity of the crust is derived from the estimates of Ronov and Yaroshevsky [12, 13] for the total crustal content of reduced carbon, ferrous iron, and sulfide sulfur. The individual contributions to the total are 4.3×10^{21} moles of oxygen for ferrous iron, 3.2×10^{21} moles of oxygen for reduced carbon, and 5.6×10^{20} moles of oxygen for sulfide sulfur.

Table 1 presents estimates of *annual gains of atmospheric oxygen*. The values for photosynthesis were derived from the estimates of Holland in the manner already described (compare [14]). Net photosynthesis refers to the excess of photosynthesis over respiration by photosynthetic organisms. It describes the growth in biotic mass and atmospheric oxygen that would result if there were no consumption of plant matter by animals and other heterotrophs. The table shows that photosynthesis is by far the most important source of oxygen. A very minor contribution comes from photolysis of nitrous oxide

$$N_2O + h\nu \longrightarrow N_2 + \tfrac{1}{2} O_2. \tag{3}$$

Table 1. Annual gains of atmospheric oxygen (Units of 10^{12} moles O_2 yr^{-1})

1. Net photosynthesis on land	4,600	[3]
2. Net photosynthesis in the ocean	3,790	[3]
3. Photolysis of N_2O	0.36	[19]
4. Escape of H to space	0.007	[5]

The nitrous oxide is a product of microbial denitrification (Burns and Hardy, 1975), so nitrous oxide photolysis, like photosynthesis, must be judged a biological source of oxygen. The only abiotic source is photolysis of water vapor followed by the escape of hydrogen to space [15, 18, 5]. Its contribution to present day oxygen cycles is negligibly small.

Table 2. Annual losses of atmospheric oxygen (Units of 10^{12} moles O_2 yr^{-1})

1. Aerobic respiration	6730	[3]
2. Photochemical oxidation of NH_3	4.3	[19]
3. Microbial oxidation of NH_3 (nitrification)	1200	[11]
4. Photochemical oxidation of CH_4	113	[33]
5. Microbial oxidation of CH_4	226	
6. Photochemical oxidation of terpenes	45.5	[71, 21]
7. Photochemical oxidation of H_2S	1	[72]
8. Microbial oxidation of H_2S	2	
9. Photochemical oxidation of CH_3SCH_3	2.5	[72]
10. Surface reaction of O_3	24	[73]
11. Fixation of nitrogen by lightning	3.3	[44]
12. Oxidation of volcanic H_2	.55	[3]
13. Oxidation of volcanic CO	.15	[3]
14. Oxidation of volcanic SO_2	.7	[3]
15. Chemical weathering of kerogen carbon	7.5	[3]
16. Chemical weathering of sulfide sulfur	3.8	[3]
17. Chemical weathering of ferrous iron	1.3	[3]
18. Combustion of fossil fuel	350.0	[3]
19. Industrial fixation of nitrogen	2.7	[19]

Table 2 presents estimates of the rates of processes that *consume atmospheric oxygen*. Many of the numbers are uncertain, the list is not complete, and it is possible that some consumption reactions have been counted more than once. The purpose of this tabulation is to provide more indication of what happens to oxygen than is provided simply by "respiration and decay" and "weathering". The relative lengths of Tables 1 and 2 show, once again, the anomalous situation of oxygen. In a world that is reducing in overall composition there are many ways to bind oxygen and very few ways to produce it.

For purposes of this presentation "aerobic respiration" is taken to refer to the schematic reaction

$$CH_2O + O_2 \longrightarrow CO_2 + H_2O \tag{4}$$

in which organisms derive energy by oxidizing organic carbon. The value quoted for this process is derived by requiring that the overall budgets of atmospheric oxygen and reduced organic matter be in balance. Specifically excluded from aerobic respiration is oxidation of the nitrogen incorporated in reduced form in organic matter. Most of this nitrogen is returned to the biosphere as *ammonia*, and most of the ammonia is oxidized to nitrate by specific aerobic microbes in the process known as nitrification [19].

$$NH_4^+ + {}^3/_2O_2 \xrightarrow{\text{Nitrosomonas}} NO_2^- + 2H^+ + H_2O \qquad (5)$$

$$NO_2^- + {}^1/_2\,O_2 \xrightarrow{\text{Nitrobacter}} NO_3^-.$$

An estimate of the consequent rate of consumption of oxygen appears as item 3 in Table 2. Most of the nitrate, in turn, is reincorporated into photosynthetic organisms, undergoing reduction in the process and resulting in the increase already described in the relative oxygen yield of photosynthesis.

Because of its high solubility in water, very little of the ammonia released by organisms escapes from soil or water into the atmosphere. According to Dawson [20] and Graedel [21] the *ammonia flux* into the atmosphere is only about 50×10^{12} g yr^{-1}, corresponding to a potential oxygen consumption rate of only 6×10^{12} moles yr^{-1}. Most atmospheric ammonia is returned in reduced form to the surface as a component of rain [22]. The estimate given in Table 2 for the photochemical oxidation of ammonia (item 2) is probably high.

Methane is an abundant product of microbial degradation of organic matter in anaerobic environments [23–25]. It appears to be largely produced by autotrophic methanogenic bacteria which derive energy from the reaction [26]

$$CO_2 + 4H_2 \longrightarrow CH_4 + 2H_2O. \qquad (6)$$

The hydrogen and carbon dioxide are products of microbial fermentation [27]

$$H_2O + CH_2O \longrightarrow 2H_2 + CO_2. \qquad (7)$$

In the atmosphere methane is oxidized by a chain of photochemical reactions that has been studied in detail [28–32]. Intermediate products include hydrogen, carbon monoxide, ozone, and formaldehyde [5]. The ultimate products of photochemical activity are carbon dioxide and water

$$CH_4 + 2O_2 \longrightarrow CO_2 + 2H_2O. \qquad (8)$$

The figure for the corresponding rate of oxygen consumption (item 4) is one of the more certain numbers in Table 2 [33].

Much less certain is the flux of methane that is oxidized by bacteria that derive energy by catalysing the equivalent of reaction (8) within the soil or water. Unlike ammonia, methane is not highly soluble so we might anticipate that a larger fraction of the total methane production is released to the atmosphere. Experiments in anaerobic lakes suggest that about one third of the methane escapes, the rest being oxidized within the water column [34]. This fraction is used in deriving a value for item 5 in Table 2, but the rate of microbial oxidation of methane may be considerably larger.

There is, apparently, a significant flux into the atmosphere of nonmethane hydrocarbons of biological origin [35]. Item 6 gives an estimate of the oxygen consumption resulting from photochemical oxidation of a component of this flux. Graedel [21] has quoted a much larger estimate for the total flux of non-methane hydrocarbons, but much of this total may be of anthropogenic origin, and might therefore belong in a different part of the table.

There is also a flux of *hydrogen sulfide* into the atmosphere, where it suffers photochemical oxidation largely to the level of sulfate (item 7) and then is returned to the surface as a component of precipitation [26, 37].

$$H_2S + 2O_2 \longrightarrow H_2SO_4. \tag{9}$$

The hydrogen sulfide is produced in anaerobic environments by the activities of sulfate-reducing bacteria [38], which derive energy from a process of anaerobic respiration

$$2CH_2O + SO_4^{2-} \xrightarrow{\text{Desulfovibrio}} 2HCO_3^- + H_2S. \tag{10}$$

As in the case of methane and ammonia, probably only a small fraction of the hydrogen sulfide produced by microorganisms is released to the atmosphere. In many environments it reacts, instead, with iron to form insoluble iron sulfide [39–41], an abundant constituent of anaerobic sediments. No oxygen is consumed when this happens. Much of the sulfide that is not immobilized in this fashion is oxidized by bacteria that derive energy from the reaction

$$H_2S + 2O_2 \longrightarrow SO_4^{2-} + 2H^+ \tag{11}$$

as soon as it reaches the aerobic level of the soil or water profile [42]. The estimate of the rate of consumption of oxygen by this process (item 8) is no more than a guess. As in the case of methane, I suspect that the true value is higher. There is also an appreciable flux of dimethyl sulfide (item 9), which I have assumed to be completely oxidized by photochemical processes within the atmosphere.

$$CH_3SCH_3 + 5O_2 \longrightarrow 2CO_2 + 2H_2O + H_2SO_4. \tag{12}$$

No separate entry for the oxidation of *carbon monoxide* is made. Much of this atmospheric constituent is produced during the photochemical oxidation of methane and other hydrocarbons or during the combustion of fossil fuels, so the contribution of CO oxidation to oxygen consumption is included under the corresponding entries of the table.

In preparing the budget summary of Fig. 2 items 2–9 of Table 2 are combined into the process labelled "oxidation of biogenic gases". The combined rates of these processes constitute almost one-quarter of the rate of aerobic respiration. Much of this total, of course, consists of nitrification and so is closely related to the rate of photosynthesis (Fig. 3), but the rate of methane oxidation is not negligible. The possible significance of methane production to the regulation of atmospheric oxygen has been discussed by Watson et al. [43]. We may think of methane production as extracting the

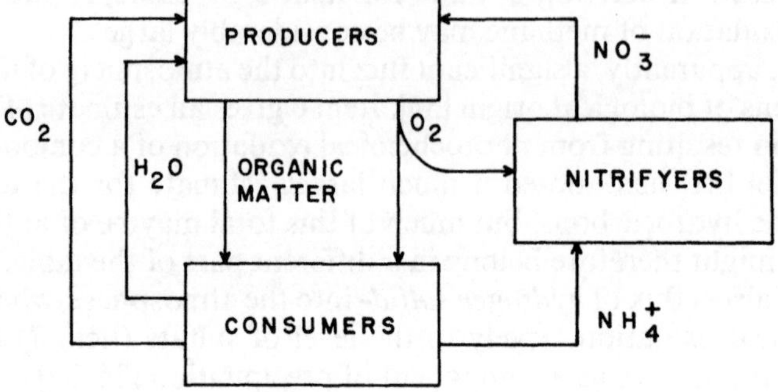

Fig. 3. Destruction of oxygen by nitrifying microorganisms is closely related to the production of organic matter by photosynthetic organisms

reducing power of organic carbon from anaerobic sediments and carrying it back into the aerobic world where it can destroy oxygen.

Items 11 and 12 in Table 2 are included in Fig. 2 under the term abiotic processes. The first is the consumption of *ozone*, produced from oxygen by photochemical reactions in the atmosphere, by reaction with reduced material at the ground. Presumably most of the reduced material oxidized by ozone is dead organic matter. Item 12 is the fixation of nitrogen by lightning. Chemical reactions in the high temperature air produced by a lightning bolt lead to the production of nitric oxide from nitrogen and oxygen [44, 45]

$$N_2 + O_2 \longrightarrow 2NO. \tag{13}$$

Subsequent photochemical processes oxidize the nitric oxide to the level of nitrate, whereupon it is removed from the atmosphere as a component of precipitation and enters the cycle of nitrate assimilation followed by decay and nitrification that was described above [5]. Because the nitrate is ultimately consumed in the anaerobic respiration process called denitrification

$$5CH_2O + 4NO_3^- \longrightarrow 2N_2 + 4HCO_3^- + CO_2 + 3H_2O \tag{14}$$

this process is shown as drawing down the reservoir of dead organic carbon as well as the reservoir of atmospheric oxygen.

Since all of the processes discussed so far act on the reservoir of surface organic carbon they correspond to "respiration and decay" in the simple model of oxygen cycles described in the introduction and illustrated in Fig. 1. While it has been interesting to break respiration and decay down into its component parts, the exercise has revealed no startling new insights. We now turn to the components of the process termed "weathering" in the simple model. These are the processes that act on the large reservoir of reduced constituents of the crust.

The photochemical oxidation of *reduced volcanic gases* (items 12–14) requires little comment. The oxidation state of volcanic gases is determined by redox equilibria in the source regions of the gases, depending largely on the ratio of ferrous to ferric iron in the source magma [46–54]. The values quoted here are derived from the work of Holland [3], who notes that they should probably be taken as upper limits. The computation of the rate of chemical weathering of reduced constituents of the crust (items 15–17) is also due to Holland [3]. The assumption is made that these reduced minerals are completely oxidized to carbon dioxide, sulfate, and ferric iron, in the course of chemical weathering. This assumption is probably valid for ferrous iron and sulfide sulfur, but whether the organic carbon (*kerogen*) in sedimentary rocks is also fully oxidized is a matter of some doubt. This material is typically a highly refractory hydrocarbon polymer, which is not at all easy even to extract from the rocks. No convincing explanation has been given of how kerogen might be oxidized under earth surface conditions either by organisms or by abiotic processes. The numbers in Table 2 might suggest that it is not an important question, but such is not the case. As described in the introduction, it is the slow cycle of burial and weathering of reduced crustal constituents that is believed to control the oxygen content of the atmosphere, not the fast cycle of photosynthesis followed by respiration and decay. And the reduced crustal constituent with the greatest capacity for the consumption of oxygen is kerogen (item 15).

The last two items of Table 2 refer to the *activities of man*. They are included as a matter of interest, but are left out of the budget of Fig. 2. Combustion of fossil fuel (item 18) is an accelerated form of the chemical weathering of kerogen. Industrial fixation of nitrogen is assumed to produce nitrate from the reaction of nitrogen and oxygen, as in the case of lightning fixation (item 11).

Oxidation of Reduced Organic Carbon

As already noted, the rapid cycle of photosynthesis followed by the various processes that together comprise respiration and decay is not what controls the abundance of atmospheric oxygen. This cycle controls the size of the relatively small reservoir of surface organic carbon. In the long run what matters is the balance between consumption of oxygen in the weathering of reduced constituents of the crust and the net production of oxygen that corresponds to the very small fraction (approximately one part in a thousand) of photosynthetically produced organic matter that escapes immediate oxidation to become incorporated in sedimentary rocks. Improved understanding of the biogeochemical cycles of oxygen depends to a large extent on better knowledge of the processes that control the rate of fossilization of reduced organic matter. This section is devoted to a description and discussion of these processes.

There is evidence that almost all organic matter will suffer oxidation if left in an aerobic environment for a geologically reasonable period of time. The organic carbon content of deep sea aerobic sediments is very low, and such carbon as survives appears to be incorporated in amino acids and derivatives that are tightly bound to clay minerals [55]. The contribution of such sediments to the total rate of burial of organic matter is negligibly small [4, 5]. Further evidence of the almost complete oxidation of organic matter exposed to air is the absence of isotopically detectable traces of terrestrial organic matter in oceanic sediments outside the immediate vicinity of river mouths [56]. Evidently very little of the large amount of organic matter synthesized on land escapes oxidation to become incorporated in sediments.

A rough estimate of the time for aerobic respiration to consume organic matter when oxygen is not in short supply can be derived by dividing the rate of aerobic respiration shown in Fig. 2 into the reservoir of dead organic matter. This time is only 100 yr. If organic matter is to escape oxidation it must be moved into a restricted environment from which oxygen is excluded. Such an environment can be established when the entering flux of reduced organic matter exceeds the entering flux of oxygen, for then aerobic respiration will consume all of the entering oxygen and leave an excess of reduced matter. As an example of such a restricted environment the processes which consume reduced matter and oxidants in newly deposited sediments are described here. The same considerations apply to processes in anaerobic waters such as those at the bottom of many lakes, the Black Sea, or the Cariaco trench [57, 58]. This discussion is based on the writings of Berner [40, 59], Heath et al. [55], and Irwin et al. [60].

Organisms package organic matter in solid particles, denser than water, that settle to the bottom of the sea, suffering oxidation on the way. The surviving organic matter enters sea-floor sediments at a concentration that depends on the settling flux and on the rate of accumulation of inorganic constituents of the sediments.

Dissolved oxygen diffuses into the uppermost levels of an anaerobic sediment where it is consumed by aerobic respiration, destroying a stoichiometrically equivalent amount of organic matter in the process. If reduced material is sufficiently abundant, however, the oxygen dissolved in interstitial waters of the sediment is exhausted not far below the surface. Oxidation of organic matter continues, however, by the process of anaerobic respiration known as denitrification. In this process dissolved nitrate serves as electron acceptor, undergoing reduction to nitrogen or nitrous oxide. Denitrification generally persists to somewhat greater depths than aerobic respiration, presumably because the rate coefficient for the reaction of nitrate with organic matter is smaller than the rate coefficient for the reaction of oxygen. Oxidation continues even after the nitrate is exhausted, by another process of anaerobic respiration, sulfate reduction. In this process sulfate serves as electron acceptor, suffering reduction to hydrogen sulfide. The hydrogen sulfide may react with iron to form insoluble iron sulfide or it may diffuse upwards out of the sediment. Destruction of organic matter continues even below the level of sulfate reduction as a result of fermentation. Fermentation converts organic matter into methane and carbon dioxide which diffuse upwards out of the sediment.

This marked depth stratification of the processes of oxidation of organic carbon can be readily understood in terms of the relative rates of the different destruction processes. Aerobic respiration yields more energy per mole of organic matter consumed than any of the other processes and can be expected to proceed most rapidly. Nitrate reduction is a little slower, so nitrate is not exhausted quite as close to the top of the sediments. There is, however, considerable overlap between the levels of aerobic respiration and nitrate reduction, and some microorganisms are able to derive energy from either form of respiration. Sulfate reduction yields considerably less energy than either of the preceeding processes and appears to be much slower. Sulfate can therefore diffuse further into the sediment before it is exhausted. In addition, the sulfate reducing bacteria are obligate anaerobes and, therefore, can not become active until oxygen has disappeared [61]. Under sedimentary conditions fermentation is a slower process even than sulfate reduction. Because it does not require the diffusion into the sediment of any oxidant, it dominates destruction of organic matter at the greatest depths [62]. This situation is illustrated schematically in Fig. 4.

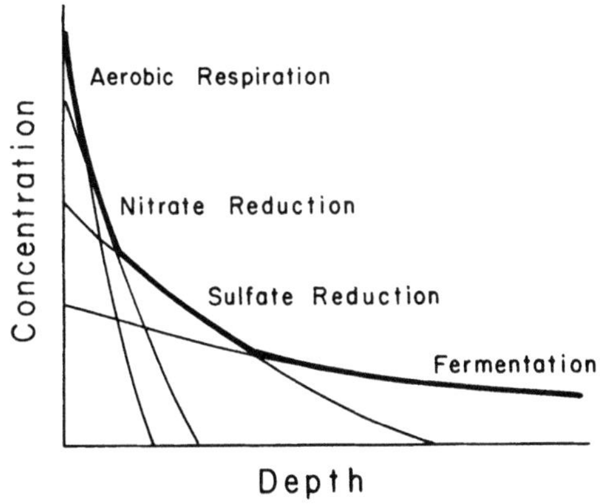

Fig. 4. Schematic representation of the stratification with depth of the processes that consume organic matter in sediments. The heavy line represents the concentration of organic matter. The light lines are proportional to the rates of the different destruction processes and, in principle, to the concentrations of the different oxidants

A key question is what prevents fermentation from continuing until essentially all of the organic matter is exhausted. Measurements on the organic carbon content of selected oceanic sea floor cores show an exponential decrease with depth with a characteristic decay time of some 10,000 yr [55], but that this decay does not continue indefinitely is evidenced by the relatively high organic carbon (kerogen) contents of many ancient sedimentary rocks. It is possible that some fraction of the material that reaches the fermentation level in the sediments is not susceptible to anaerobic decay [59]. Alternatively, perhaps compaction and cementation of the sediments isolates particles of organic matter, denying access to the fermenting bacteria or trapping the products of fermentation so that subsequent diagenesis can convert these products into kerogen.

The factors that govern the oxidation of sedimentary organic matter, which will be called "carbon" for convenience, need to be discussed now. Let "C" be the concentration of carbon as a function of depth in the sediment, with "C_0" the concentration at zero depth. Oxygen and nitrate together will be called "O", and expressed in units of capacity to oxidize carbon. Thus oxidation of one unit of C involves consumption of one unit of O. Suppose that O is exhausted at depth L in the sediment, at which point C has decreased by ΔC (Fig. 5). The layer of sediment above depth L is the reaction region for destruction of organic matter by aerobic respiration and by nitrate reduction.

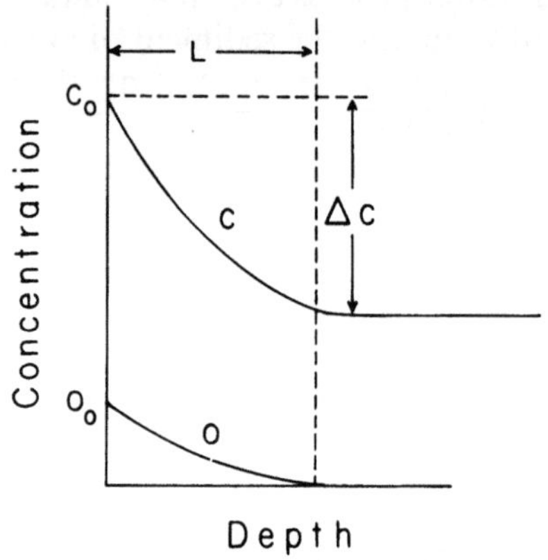

Fig. 5. Schematic representation of the reaction between an oxidant diffusing into sediments and organic carbon being transported to greater depths as a result of sediment accumulation

Carbon is carried into the reaction region at the top and out of the reaction region at the bottom as a result of the accumulation of sediments at velocity W. The rate of loss of carbon within the reaction region, assuming a steady state, is the difference between the flux in and the flux out, or $W\Delta C$. Oxidant is carried into the reaction region by diffusion through the interstitial waters of the sediment. The inward flux is $D\,O_0/L$, where O_0 is the oxidant concentration in the overlying water, D is the diffusion coefficient, and the outward flux is zero because O is zero at the bottom of the reaction region. The inward oxidant flux is equal to the rate of loss of carbon. If τ is the characteristic time for destruction of carbon under more or less aerobic conditions, the integrated loss per unit area within the reaction region is CL/τ.

In equilibrium the integrated loss of C within the reaction region is equal to the inward flux of oxidant and the net flux of carbon.

$$W\Delta C = D\,O_0/L = CL/\tau. \tag{14}$$

Therefore

$$L = (DO_0\tau/C)^{1/2} \tag{15}$$

and

$$\Delta C/C = L/\tau W = (DO_0/C\tau)^{1/2}/W. \tag{16}$$

The depth of penetration of oxidant, L, depends on the diffusion coefficient, D, the reaction time, τ, and the relative concentrations of oxidant and organic matter, O_0/C. The fractional loss of organic matter, $\Delta C/C$, varies inversely with the sediment accumulation velocity, W, and depends on the other factors also.

Obviously, if W is sufficiently small, $\Delta C/C$ can approach or even exceed unity. In this case the sediment remains aerobic, and nearly all organic matter is oxidized. For the sediment to become anaerobic it is necessary that

$$W > (D\,O_0/C\tau)^{1/2} \tag{17}$$

Typical sea water contains 6 ml l^{-1} of dissolved oxygen and 40×10^{-6} g atom l^{-1} of nitrate nitrogen. The corresponding carbon consumption capacity is 0.3 g atom C m^{-3}, mostly due to oxygen. A sediment containing 1 per cent organic carbon by dry weight has about 260 g atom C m^{-3} of pore water [59]. So $O_0/C \sim 10^{-3}/f$, where f is the weight percent organic carbon.

A typical value of the diffusion coefficient for dissolved constituents in sediment pore waters is $D = 6 \times 10^{-6}$ cm^2 s^{-1} [40], and the estimate already given of the lifetime of organic matter in aerobic conditions is $\tau = 100$ yr. Combining these rough numbers

$$W > 0.04 f^{-1/2} \text{ cm yr}^{-1}$$

can be found as the sediment accumulation rate required for anaerobic conditions to exist at depth. Since sediment accumulation rates are typically measured in centimeters per 1,000 yr, this result looks reasonable.

For a sediment that becomes anaerobic, a typical thickness for the uppermost aerobic layer (the reaction region) is

$$L = 4 f^{-1/2} \text{ cm.}$$

This estimate neglects the effective enhancement of diffusion caused by the activities of burrowing animals [63].

Finally, the fractional loss of organic carbon in the aerobic layer can be calculated

$$\Delta C/C = 0.04/W f^{1/2},$$

where W is expressed in cm yr^{-1}. In the Santa Barbara Basin off Southern California W is about 0.2 cm yr^{-1} and f may be as high as 4, so $\Delta C/C$ would be 0.1, if the overlying water were not anoxic.

If a sulfate reaction region is assumed to be underlying the oxygen region, the same arguments apply. Now, however, O_0 refers to the carbon oxidation capacity of the sulfate at the top of the reaction region, and τ is the characteristic time for oxidation of carbon by sulfate. For typical sea water, without allowance for diffusion through the oxygen reaction region, O_0 for sulfate is 60 g atom C m^{-3}, and the results of Berner [40, 59] suggest that $\tau \sim 300$ yr. If carbon is to exhaust sulfate at depth, therefore, it is necessary that

$$W > 0.04 f^{-1/2} \text{ cm yr}^{-1}.$$

The figures quoted above for the Santa Barbara Basin suggest that in this case sulfate comes close to exhausting carbon. The data cited by Berner [59] indicate that both sulfate and carbon persist in appreciable concentrations well below the region of reaction, where sulfate and carbon concentrations are changing with depth. Berner's interpretation is that sulfate exhausts reactive carbon, so that the carbon that persists is not susceptible to consumption by sulfate reducing bacteria.

In a situation where sulfate is exhausted, the thickness of the reaction region is

$$L = 120\, f^{-1/2}\ \mathrm{cm}$$

and the fractional loss of carbon is

$$\Delta C/C = 4/Wf^{1/2}.$$

If reactive iron is present in the sediments in sufficient abundance, the carbon that is lost is replaced by a stoichiometrically equivalent amount of iron sulfide. In this case oxidation of sedimentary organic matter by sulfate reducing bacteria has no effect on the biogeochemical cycles of oxygen. If iron is deficient, however, hydrogen sulfide diffuses upwards into the aerobic zone where it is oxidized, consuming oxygen in the process.

If sediment accumulation rates and initial carbon contents are so high that some reactive organic matter survives destruction by oxygen, nitrate, and sulfate it will decay by fermentation, producing methane in the process. The characteristic time for this decay appears to be some 10,000 yr, as already noted. The methane, joined by any hydrogen sulfide that does not react with iron, diffuses upward until it meets oxygen diffusing into the sediments. There both methane and oxygen are consumed by methane oxidizing bacteria. The interaction can be treated by the methods described in this section, and it can be shown that if the methane flux is sufficiently large, oxygen is completely excluded from the sediments and possibly even from the lower levels of the overlying water column. A full treatment of the conditions for establishment of anaerobic environments should therefore allow for the effects of this return flux of reduced biogenic gases.

These considerations indicate how important the rate of accumulation of sediments and the initial carbon content of the sediments is to the survival of reduced organic carbon. The data of Heath et al. [55] show a very close correlation between carbon burial rate and sediment accumulation rate, the carbon flux varying as accumulation rate raised to the power of 1.4 for a diverse sample of both anaerobic and aerobic sediments. It has, indeed, been suggested that accumulation rate is more important to the preservation of organic matter in sediment than the establishment of anaerobic conditions (compare [58, 64, 67]). It is not entirely clear, however, that accumulation rate alone is the controlling factor because, at least in the Pacific, there is also a correlation between accumulation rate and the productivity of the overlying waters [65], which might determine the rate of delivery of organic matter to the top of the sediments. More work on this problem, both observational and theoretical, is plainly needed.

Controls on Atmospheric Oxygen

The budget presented in this paper indicates that weathering can consume all atmospheric oxygen in a time of only a few million years. On the other hand, Metazoa, which require oxygen to breathe, have been continuously present on earth for at least 600 million years. Evidently, there is a negative feedback mechanism that prevents oxygen from fluctuating wildly in response to changes in, for example, rates of erosion and weathering. Our ideas about this feedback mechanism have been described in the introduction. The detailed examination of the budget of oxygen presented in this paper has revealed no obvious flaws in these ideas. The simple model of preservation of organic matter in sediments has shown, indeed, how preservation increases as oxygen decreases.

A question can be asked, however, about the relative importance of sedimentary carbon and sedimentary sulfide in the overall budget. The question exists because of doubt as to whether sedimentary organic carbon, kerogen, is indeed oxidized when it is exposed to the air by erosion. The argument in favor of oxidation is as follows: Data on the ratios of stable carbon isotopes in carbonate rocks of various ages suggest that the total mass of kerogen in the crust does not change with time [66]. On the other hand, radiocarbon dating of the organic matter in sea floor sediments indicates that this carbon is mostly young, of recent photosynthetic origin. If new organic carbon is entering sediments and if organic carbon is not accumulating in the crust, old carbon must suffer oxidation during the course of chemical weathering. On the other hand, data of Sackett and Poag [68] show that, in Antarctic waters where the input of newly synthesized carbon is small, the organic matter in new sediments is mostly old, recycled kerogen. And data presented by Heath et al. [55] show a significant decrease of carbon content with increasing depth in sediments all over the Pacific Ocean. It seems at least possible that most of the newly synthesized organic matter in the upper layers of the sedimentary column is destroyed by the various processes discussed above, and that the carbon which survives destruction to become a component of sedimentary rocks is old, recycled kerogen which is highly resistant to oxidation. This would open the possibility that most old kerogen is not oxidized during the course of chemical weathering and would offer a ready explanation of why fermentation in anaerobic sediments does not proceed until nearly all organic matter has been consumed.

If most kerogen is not oxidized, the burden of controlling the oxygen content of the atmosphere shifts to the balance between sulfide weathering and deposition [69, 70]. Our basic picture of the controlling processes is not qualitatively changed, but the lifetime of oxygen against consumption in weathering is increased to about 7 million years. This possibility should be kept in mind.

Acknowledgment

I have benefitted from discussion of some of these ideas with H. D. Holland, K. K. Turekian, L. Margulis, and J. W. Schopf. This work was performed at the National Astronomy and Ionosphere Center which is operated by Cornell University under contract with the National Science Foundation.

References

1. Van Valen, L.: The history and stability of atmospheric oxygen. Science *171*, 439 (1971)
2. Holland, H.D.: Ocean water, nutrients, and atmospheric oxygen. In: Proceedings of Symposium on Hydrogeochemistry and Biogeochemistry, Vol. 1. Washington, D.C.: The Clarke Co. 1973, p. 68
3. Holland, H.D.: The Chemistry of the Atmosphere and Oceans. New York: Wiley-Interscience 1978
4. Walker, J.C.G.: Stability of atmospheric oxygen. Am. J. Sci. *274*, 193 (1974)
5. Walker, J.C.G.: Evolution of the Atmosphere. New York: Macmillan Publishing Co. 1977
6. Garrels, R.M., Perry, E.A.: Cycling of carbon, sulfur, and oxygen through geologic time. In: The Sea, Vol. 5. E. Goldberg (Ed.). New York: Wiley-Interscience 1974, p. 303
7. Garrels, R.M., Lerman, A., Mackenzie, F.T.: Controls of atmospheric O_2 and CO_2: Past, present and future. Am. Sci. *64*, 306 (1976)
8. Schidlowski, M., Eichmann, R., Junge, C.E.: Precambrian sedimentary carbonates: Carbon and oxygen isotope geochemistry and implications for the terrestrial oxygen budget. Precambrian Res. *2*, 1 (1975)
9. Verniani, F.: The total mass of the Earth's atmosphere. J. Geophys. Res. *71*, 385 (1966)
10. Degens, E.T.: Carbon in the sea. Nature *279*, 191 (1979)
11. Bowen, H.J.M.: Trace Elements in Biochemistry. New York: Academic Press 1966
12. Ronov, A.B., Yaroshevsky, A.A.: Chemical structure of the Earth's crust. Geochemistry, 1041 (1967). (Trans. from Geokhimiya, No. 11, 1285, 1967)
13. Ronov, A.B., Yaroshevsky, A.A.: Chemical composition of the Earth's crust. In: The Earth's Crust and Upper Mantle. P.J. Hart (Ed.). Washington: Am. Geophys. Union Monograph *13*, 1969, p. 37
14. Holligan, P.M.: The productive oceans. Nature *279*, 191 (1979)
15. Hunten, D.M.: The escape of light gases from planetary atmospheres. J. Atmos. Sci. *30*, 1481 (1973)
16. Liu, S.C., Donahue, T.M.: The aeronomy of hydrogen in the atmosphere of the earth. J. Atmos. Sci. *31*, 1118 (1974)
17. Liu, S.C., Donahue, T.M.: Mesospheric hydrogen related to exospheric escape mechanisms. J. Atmos. Sci. *31*, 1466 (1974)
18. Lui, S.C., Donahue, T.M.: Realistic model of hydrogen constituents in the lower atmosphere and escape flux from the upper atmosphere. J. Atmos. Sci. *31*, 2238 (1974)
19. Burns, R.C., Hardy, R.W.F.: Nitrogen fixation in bacteria and higher plants. New York: Springer-Verlag 1975
20. Dawson, G.A.: Atmospheric ammonia from undisturbed land. J. Geophys. Res. *82*, 3125 (1977)
21. Graedel, T.E.: The kinetic photochemistry of the marine atmosphere. J. Geophys. Res. *84*, 273 (1979)
22. Graedel, T.E.: The oxidation of ammonia, hydrogen sulfide, and methane in nonurban tropospheres. J. Geophys. Res. *82*, 5917 (1977)
23. Deuser, W.G. et al.: Methane in Lake Kivu: New data bearing on its origin. Science *181*, 51 (1973)
24. Zeikus, J.G., Winfrey, M.R.: Temperature limitation of methanogenesis in aquatic sediments. Appl. Environ. Microbiol. *31*, 99 (1976)

25. Dacey, J.W.H., Klug, M.J.: Methane flux from lake sediments through water lilies. Science *203*, 1253 (1979)
26. Wolfe, R.S.: Microbial formation of methane. Advances in Microbial Physiology *6*, 107 (1971)
27. Gray, C.T., Gest, H.: Biological formation of molecular hydrogen. Science *148*, 186 (1965)
28. Levy, H.: Normal atmosphere: Large radical and formaldehyde concentrations predicted. Science *173*, 141 (1971)
29. Levy, H.: Photochemistry of the lower troposphere. Planet. Space Sci. *20*, 919 (1972)
30. Levy, H.: Photochemistry of minor constituents in the troposphere. Planet. Space Sci. *21*, 575 (1973)
31. Levy, H.: Tropospheric budgets for methane, carbon monoxide, and related species, J. Geophys. Res. *78*, 5325 (1973)
32. Logan, J.A. et al.: Atmospheric chemistry: Response to human influence. Phil Trans. Roy. Soc. *290*, 187 (1978)
33. Ehhalt, D.H., Schmidt, U.: Sources and sinks of atmospheric methane. Pure Appl. Geophys. *116*, 452 (1978)
34. Rudd, J.W.M., Hamilton, R.D.: Methane cycling in a eutrophic shield lake and its effects on whole lake metabolism. Limnol. Oceanogr. *23*, 337 (1978)
35. Whitby, R.A., Coffey, P.E.: Measurement of terpenes and other organics in an Adirondack Mountain pine forest. J. Geophys. Res. *82*, 5928 (1977)
36. Friend, J.P.: The global sulfur cycle. In: Chemistry of the Lower Atmosphere. S. I. Rasool (Ed.). New York: Plenum Press 1973, p. 177
37. Garrels, R.M., Mackenzie, F.T., Hunt, C.: Chemical Cycles and the Global Environment. Los Altos, California: William Kaufmann, Inc. 1973
38. Postgate, J.R.: The sulphur cycle. In: Inorganic Sulfur Chemistry. G. Nickless (Ed.). New York: Elsevier 1968, p. 259
39. Berner, R.A.: Sedimentary pyrite formation. Amer. J. Sci. *268*, 1 (1970)
40. Berner, R.A.: Principles of Chemical Sedimentology. New York: McGraw-Hill 1971
41. Berner, R.A.: Sulfate reduction, pyrite formation, and the oceanic sulfur budget. In: The Changing Chemistry of the Oceans. D. Dyrssen and D. Jagner (Ed.). New York: Wiley 1972, p. 347
42. Stanier, R.Y., Douderoff, M., Adelberg, E.A.: The Microbial World, 3rd edition, Englewood Cliffs, New Jersey: Prentice-Hall Inc. 1979
43. Watson, A., Lovelock, J.E., Margulis, L.: Methanogenesis, fires and the regulation of atmospheric oxygen. Bio Systems *10*, 293 (1978)
44. Chameides, W.L. et al.: NO_x production in lightning. J. Atmos. Sci. *34*, 143 (1977)
45. Chameides, W.L.: Effect of variable energy input on nitrogen fixation in instantaneous linear discharges. Nature *277*, 123 (1979)
46. Kennedy, G.C.: Equilibrium between volatiles and iron oxides in igneous rocks. American J. Sci. *246*, 529 (1948)
47. Holland, H.D.: Model for the evolution of the Earth's atmosphere. In: Petrologic Studies: A Volume in Honor of A.F. Buddington. A.E.J. Engel, H.L. James, and B.F. Leonard (Ed.). New York: Geological Society of America 1962, p. 447
48. Holland, H.D.: On the chemical evolution of the terrestrial and cytherean atmospheres. In: The Origin and Evolution of Atmospheres and Oceans. P.J. Brancazio and A.G.W. Cameron (Ed.). New York: John Wiley and Sons 1964, p. 86
50. Heald, E.F., Naughton, J., Barnes, I.L.: The chemistry of volcanic gases, use of equilibrium calculations in the interpretation of volcanic gas samples. J. Geophys. Res. *68*, 545 (1963)
51. Fanale, F.P.: History of Martian volatiles: Implications for organic synthesis. Icarus *15*, 279 (1971)
52. Nordlie, B.E.: Gases-Volcanic. In: The Encyclopedia of Geochemistry and Environmental Science. R.W. Fairbridge (Ed.). New York: Van Nostrand 1972, p. 387
53. Cruikshank, D.P., Morrison, D., Lennon, K.: Volcanic gases: Hydrogen burning at Kilauea Volcano, Hawaii. Science *182*, 277 (1973)
54. Allard, P., Tazieff, H., Dajlevic, D.: Observations of seafloor spreading in Afar during the November 1978 fissure eruption. Nature *279*, 30 (1979)
55. Heath, G.R., Moore, T.C., Dauphin, J.P.: Organic carbon in deep-sea sediments. In: The

Fate of Fossil Fuel CO_2 in the Oceans. N.R. Anderson and A. Malahoff (Ed.). New York: Plenum Press 1978, p. 605
56. Degens, E.T.: Biogeochemistry of stable carbon isotopes. In: Organic Geochemistry; Methods and Results. G. Eglinton and M.T.J. Murphy (Ed.). Berlin: Springer-Verlag 1969, p. 304
57. Richards, F.A.: Anoxic basins and fjords. In: Chemical Oceanography, Vol. 1. J.P. Riley and G. Skirrow (Ed.). New York: Academic Press 1965, p. 611
58. Deuser, W.G.: Organic-carbon budget of the Black Sea. Deep-Sea Res. *18*, 995 (1971)
59. Berner, R.A.: An idealized model of dissolved sulfate distribution in recent sediments. Geochim. Cosmochim. Acta *28*, 1497 (1964)
60. Irwin, H., Curtis, C., Coleman, M.: Isotopic evidence for source of diagenetic carbonates formed during burial of organic-rich sediments. Nature *269*, 209 (1977)
61. Morris, J.G.: The physiology of obligate anaerobiosis. In: Advances in Microbial Physiology, Vol. 12, A.H. Rose and D.W. Tempest (Ed.). New York: Academic Press 1975, p. 169
62. Martens, C.S., Berner, R.A.: Methane production in the interstitial water of sulfate-depleted marine sediments. Science *185*, 1167 (1974)
63. Rhoads, D.C.: The influence of deposit-feeding benthos on water turbidity and nutrient recycling. Amer. Jour. Sci. *273*, 1 (1973)
64. Reimer, T.O., Barghoorn, E.S., Margulis, L.: Primary productivity in an early Archean microbial ecosystem. Precambrian Research *9*, 93 (1979)
65. Koblentz-Mishke, O.J., Volkovinsky, V.V., Kabanova, J.G.: Plankton primary production of the world ocean. In: Scientific Exploration of the South Pacific. W.S. Wooster (Ed.). Washington: National Academy of Sciences 1970, p. 183
66. Broecker, W.S.: A boundary condition on the evolution of atmospheric oxygen. J. Geophys. Res. *75*, 3553 (1970)
67. Richards, F.A.: The enhanced preservation of organic matter in anoxic marine environments. In: Symposium on Organic Matter in Natural Waters. D.W. Hood (Ed.). Occas. Pub. Inst. Mar. Sci. Univ. Alaska *1*, 399 (1970)
68. Sackett, W.M., Poag, C.W., Eadie, B.J.: Kerogen recycling in Ross Sea, Antarctica. Science *185*, 1045 (1974)
69. Holland, H.D.: Systematics of the isotopic composition of sulfur in the oceans during the Phanerozoic and its implications for atmospheric oxygen. Geochim. Cosmochim. Acta *37*, 2605 (1973)
70. Schidlowski, M., Junge, C.E., Pietrek, H.: Sulfur isotope variations in marine sulfate evaporites and the Phanerozoic oxygen budget. J. Geophys. Res. *82*, 2557 (1977)
71. Rasmussen, R.A., Went, F.W.: Volatile organic material of plant origin in the atmosphere. Proc. Nat. Acad. Sci. *53*, 215 (1965)
72. Graedel, T.E.: Reduced sulfur emission from the open oceans. Geophys. Res. Lett. *6*, 329 (1979)
73. Chameides, W.L., Stedman, D.H.: Tropospheric ozone: Coupling transport and photochemistry. J. Geophys. Res. *82*, 1787

The Sulfur Cycle

A. J. B. Zehnder
Federal Institute for Water Resources and Water Pollution Control (EAWAG)
CH-8600 Dübendorf, Switzerland

S. H. Zinder
Division of Environmental and Nutritional Sciences
School of Public Health, University of California
Los Angeles, CA 90024, USA

Introduction

The geochemistry of sulfur is to a great extent affected by the biosphere. Living organisms transform sulfur from one oxidation state to another, thus changing the chemical properties of sulfur compounds and their distribution in the atmosphere, soils, waters and sediments. Sulfur compounds change considerably the physiochemical properties of a system (pε, pH). The highly specific catalytic and kinetic effects of enzyme systems of organisms, may create environments with a distinct chemical character.

In biogeochemistry, sulfur shows one of the most complex cycles because sulfur exists in most oxidation states between -2 and $+6$ and forms a large variety of organic and inorganic species. The activities of man in modern civilization steadily modify the natural flux of sulfur and add to the complexity of the cycle of this element.

The aim of this chapter is to summarize the important physical, chemical and biological facts of the sulfur cycle. A substantial amount of information is available about the forms of sulfur and its organic geochemistry with special emphasis on microbial transformations [61, 147, 168, 209, 219–221]. The microbiology is covered in numerous monographs and reviews [18, 34, 46, 65, 92, 106, 111, 114, 155–158, 162, 163, 179, 184, 188, 198, 213, 216]. Challenger [40] gives a historical account of sulfur research in biochemistry and natural products. Metabolic pathways of sulfur are summarized by Young and Maw [215], Maw [130], Peck [149], and Roy and Trudinger [175]. A comprehensive review of the sulfur system in marine environments is given by Goldhaber and Kaplan [74]. Bremner and Steele [29] describe the role of microorganisms in the atmospheric sulfur cycle. The properties of sulfur compounds and recovery of sulfur in relation to energy production and environmental protection is extensively treated by Meyer [132], and the fluxes and reservoirs in the sulfur cycle are discussed by numerous authors [26, 47, 51–53, 67, 71, 79, 83, 98, 100–102, 107, 129, 136, 161, 170, 171]. The analytical methods for determination and identification of sulfur compounds in the environment are summarized for soil by Banwart and Bremner [10] and for the atmosphere by Tanner et al. [196].

Global Sulfur Cycle

The masses used to calculate the global sulfur cycle are given in Table 1. Sulfur concentrations in the air are so small that new analytical methods had to be developed before accurate concentration measurements and hence rate transfer measurements became possible. In the ocean, the relatively high amount of sulfate makes it difficult to track the changes of intermediates which occur at very low concentrations. In the sediments, problems are caused by inhomogeneity of strata and deposits which exhibit erratic changes in total sulfur

Table 1. Mass of the spheres

Sphere	Estimated mass (10^{20} g)
Atmosphere [a]	
Mass of the troposphere (to 11 km)	40
Total mass	52
Pedosphere [b]	16
Hydrosphere [c]	
Rivers and lakes	2
Groundwaters	81
Polar ice caps, icebergs, glaciers	278
Sea	13,480
Lithosphere	
Sediments [d]	3,000
Sedimentary rocks [e]	29,000
Metamorphic rocks [e]	76,200
Igneous rocks [e]	189,300

[a] Values from Butcher and Charlson [33]

[b] The pedosphere is the uppermost part of the lithosphere which is colonized by living organisms. The average thickness of the pedosphere is 5 m. The surface was estimated to be 130 million km^2 (surface of the continents minus the surface covered by high mountains and glaciers). Average density 2.5 g cm^{-3}

[c] Partition according to Baumgartner and Reichel [15]

[d] Total amount from Holser and Kaplan [87]

[e] These values were calculated on the basis of the partition of the lithosphere by Ronov and Yaroschevskiy [172] except that the estimated total sedimentary mass of $32,000 \times 10^{20}$ g from Garrels and MacKenzie [70] was used instead of the usual $16,500 \times 10^{20}$ g [87]. Poldervaart [154] estimated the total sedimentary mass (sediments+sedimentary rocks) as $17,000 \times 10^{20}$ g, Horn [88] obtained $20,400 \times 10^{20}$ g. Gregor [80] leaning heavily on Ronov's work estimated the phanerozoic sedimentary mass at about $18,000 \times 10^{20}$ g whereas Ronov himself [173] puts the total mass at about $24,000 \times 10^{20}$ g

content and in the ratio of the various sulfur forms. In addition, some steps in the sulfur cycle proceed only in geological time periods and are therefore far beyond a meaningful extrapolation from laboratory experiments. Thus, the residence time for sulfate from pyrite in the ocean is about 38 million years; that of sulfate from gypsum 55 million years [72]. In sedimentary rocks, a sulfur atom remains for 261 and for 263 million years when it is bound in calcium sulfate or pyrite respectively [72].

Contents of the Reservoirs

Concentration of Various Sulfur Compounds in the Atmosphere

Hydrogen Sulfide and Volatile Organic Sulfur Compounds

The reactivity of reduced sulfur with oxygen creates problems which are difficult to overcome. Thus, earlier measurements of hydrogen sulfide in the atmosphere probably are not very accurate [107]. Our knowledge about the natural concentration levels in the atmosphere of the so-called reduced sulfur compounds, hydrogen sulfide, carbon disulfide, dimethyl sulfide (DMS), dimethyl disulfide (DMDS), methyl mercaptane (methane thiol) and others is very fragmentary. Smith et al. [183] found in the atmosphere over Wales a hydrogen sulfide concentration of 0.14–0.43 μg S m^{-3} and Junge [100] suggested 1.8–18 μg S m^{-3} on the basis of measurements mostly made in Europe. Over Panama and the Amazonas, Lodge and Pate [119] measured only traces of H_2S, and Breeding et al. [27] reported values of 0.03–0.06 μg S m^{-3} in Colorado. Natusch et al. [138] found values of approximately 0.07 μg S m^{-3} for the same US state.

Dimethyl sulfide was suggested to be an important sulfur carrier since it is produced by some species of marine algae [39], freshwater algae [16] and bacteria [5, 217]. Lovelock et al. [122] found that water in the North Atlantic is saturated with DMS ($1.2 \cdot 10^{-11}$ g·ml^{-1}), but they failed to find DMS in samples of the open air (the distribution coefficient of DMS between air and seawater is 0.30, this would at equilibrium give an atmospheric concentration of 1.6 μg S m^{-3}). Beside the oceans, DMS was found to be produced by most different soils [11, 12, 58, 122]. Maroulis and Bandy [131] reported for the air over Wallop Island (Virginia, Eastern Shore, near the Maryland border, USA) a mean DMS concentration of 0.063–0.088 μg S m^{-3}. They showed that higher values for DMS are obtained when the wind blows from the open ocean. DMS is very likely being produced in aerobic and microaerobic parts of soils and oceans because under strict anoxic conditions, it would be rapidly metabolized to methane, carbon dioxide and inorganic sulfur components, most likely sulfide [218].

Recently Sandalls and Penkett [176] detected in the atmosphere over Harwell, England, *carbon disulfide* (CS_2) at a mean concentration of 0.54 μg S m^{-3} and its photolysis product [214] carbonyl sulfide (COS) of 0.73 μg S m^{-3}.

0.34 ng S m^{-3} *sulfur hexafluoride* (SF_6) has been observed at Yosemite, California [180].

Sulfur Dioxide. Large amounts of sulfur dioxide emissions come from fuel burning. Most of man's sulfur pollution of the atmosphere (~95%) occurs in form of sulfur dioxide. Therefore, the concentration of sulfur dioxide in the atmosphere varies enormously, depending on whether it is measured over industrialized areas or over land exclusively covered with vegetation. As far back as 1954, sulfur dioxide was measured in "pure" marine air in Hawaii [97] at an average concentration of about 0.35 µg SO_2–S m^{-3}. Georgii [73] obtained over the Atlantic 1–4 µg S m^{-3}. The sulfur dioxide concentration had its peak at latitude 40 °N and decreased gradually to 0.2–0.5 µg S m^{-3} between latitude 20° N and 10° N, reaching virtually zero at the equator. All these measurements were made along the meridian 39° W. Georgii attributes the higher values to pollution from industrial areas which are most concentrated between latitude 60° N and 40° N in Europe and 50° N and 35° N in North America. Over Colorado he found values from 0.5 to 2 µg S m^{-3}. From flights over the oceans in areas remote from land, Nguyen et al. [139] reported a rather low mean sulfur dioxide concentration (Table 2). In the Antarctica, 21

Table 2. Atmospheric concentration of sulfur dioxide and sulfate aerosols (µg^{-S}m^{-3}) [139]

Location	SO_2	SO_4^{2-}	SO_2/SO_4^{2-}
Antarctic (60° S - 70° S)	0.13	1.57	0.08
Sub-Antarctic (40° S - 60° S)	0.18	1.68	0.11
South Pacific Ocean (20° S - 40° S)	0.12	1.15	0.10
North Atlantic Ocean (50° N - 8° N)	-	2.33	-
Mediterranean Sea	2.27	8.43	0.27

out of 37 measurements were below the detection limit (0.3 ppb per volume or 0.4 µg SO_2–S m^{-3}) of the method used by Fisher et al. [54]. Prahm et al. [159] obtained an average of 0.09 µg SO_2–S m^{-3} from measurements at the Faroe Islands in air which had been over the Atlantic for several days, according to trajectory analysis. The same group found 0.27 µg SO_2–S m^{-3} for air masses passing over the British Isles.

No data for land areas remote from industrial influence seem to be available except those reported by Lodge et al. [120]: The atmosphere over the Amazonas, Brazil, contains about 0.7 µg SO_2–S m^{-3} and over Panama 1.4 µg S m^{-3}. Many data are available for Europe, North America and Japan, but man-made emissions contribute here probably the major part to the concentration of SO_2. For example in Rome, the average over a day is 60–75 µg SO_2–S m^{-3} in the air with peaks of 500 µg SO_2–S m^{-3} [32] and in Cincinnati 30 µg SO_2–S m^{-3} with a peak concentration of over 80 µg SO_2–S m^{-3} [189].

Sulfate (Sulfur in Aerosols). The major sources of sulfur sulfate in the atmosphere are believed to include:

(i) the oxidation of sulfur gases such as SO_2, H_2S and, perhaps organic sulfides to particulate SO_4^{2-};

(ii) the ejection of particulate sulfates as a result of bubble rupture at the air-sea interface, leading to a sulfur enrichment in marine aerosols and
(iii) influx of sulfate-containing particles from continental weathering sources.

The contribution and spatial distribution of each of these sources are poorly known but are currently the subject of active research.

The ratio of SO_2/SO_4^{2-} over the ocean compared with values for remote, relatively unpolluted land areas provides a rough indication to which extent SO_2 is being converted into sulfate. Land values range from 1.3 to 10 [73, 164] whereas values in oceanic areas range around 0.1 as shown in Table 2 [139]. Prahm et al. [159] found in a series of measurements an average of 0.14 μg SO_4–S m^{-3} in aerosols of the atlantic air and a ratio of SO_2/SO_4^{2-} of 0.6. Over land, the sulfate concentration rose to 1.1 μg SO_4–S m^{-3}, and the ratio was 0.3. The effect of pollution on the sulfate concentration was measured by Appel et al. [6] over the Los Angeles area. The 24-h average fluctuated between 6 and 22 μg SO_4–S m^{-3} reaching its highest concentration of 150 μg S m^{-3} around noon over Downtown Los Angeles. Sulfate existed primarily as ammonium salt with some evidence of unneutralized sulfates in the vicinity of sulfur oxide emissions. The measured mass median diameter of the sulfate aerosols amounted to 0.3–0.4 μm [6].

The stratosphere contains a layer of particles with maximum concentrations at altitudes of about 16–18 km. Junge and Mason found in 1961 that aerosol particles probably consist of sulfate from this layer which is now called the Junge layer. The total sulfur concentration in this layer seems to vary considerably. Castleman et al. [38] and Nielson [140] measured the sulfur isotope distribution to identify volcanic sulfur as a major stratospheric sulfur source. Junge found a concentration from 1960 through 1961 of 0.01–0.05 μg S per standard m^3 [101]. In the period from 1968 through 1969, the ambient sulfate concentration was about 10 times higher than those measured in 1961. However, in 1971, following a prolonged period of low volcanic activity, the values showed a return to the concentrations measured by Junge before the 1963 eruption of the Agung volcano in Bali [107]. The presence of sulfate aerosols can contribute to opacity [26], and possibly to the "green house effect" of the atmosphere [132]. The total contribution of sulfur to this effect was estimated to be about 0.03% [206]. This is negligible, no matter what present distribution model is chosen, compared to the effect of CO_2 (0.8%), nitric oxides (0.7%) or the fluorocarbons (0.5%) [132].

Sulfur Compounds in the Pedosphere

Field soils contain between 0.01 and 0.5% sulfur, but most soils average between 0.01 and 0.06% [194]. Most sulfur in soils throughout the world occurs in organic rather than in inorganic state [60]. Very little sulfur is found as elemental sulfur (S^0), sulfide (S^{2-}) or inorganic sulfate (SO_4^{2-}). Of 208 different soils investigated, an average of only 5.2% of the total sulfur was present as SO_4^{2-}, and little sulfur, if any, was found in the elemental or sulfide form. Their concentration scarcely reached 3% [56]. Tabatabai and Bremner

[194] were unable to detect sulfur in the latter two forms in an investigation on the sulfur contents of 39 representative samples of soils from Iowa. The distribution of sulfur between the organic and inorganic fraction in peat is quite different from that of soil. In the freshwater swamp Okefenokee, about 20% of the 0.2% (weight %) total sulfur was present in its inorganic form. In the organic fraction of the Everglades peat (marine), 40% of the total 1.7% consists of inorganic sulfur [37]. High sulfur coal was prohbably formed from similar peat as is found nowadays in the Everglades.

The organic forms of sulfur in soils constitute of different groups of compounds. One group, probably the most obvious one, includes the sulfur-containing amino acids and sulfonates in which sulfur is directly bound to carbon (e.g. cysteine, methionine, taurine). This fraction is tentatively defined as soil sulfur which is reduced to S^{2-} by Raney Nickel [62]. However, there is also evidence suggesting that some soils may contain carbon-bound sulfur which is not reduced to S^{2-} by Raney Nickel [56] Carbon-bound sulfur amounts from 10% [194] to 70% [64] of the total sulfur content of a soil. Humic acid may also contain some carbon-bound sulfur [124]. In another important group of sulfur-containing compounds, the organic esters of sulfuric acid, sulfur is bound to oxygen in the form of $C{-}O{-}SO_3^-$. These types of sulfoconjugates include the sulfate esters of phenols (the arylsulfates), of aliphatic alcohols (the alklysulfates and cholin sulfate), of carbohydrates (i.e. chondroitin sulfate) and of amino acids (i.e. tyrosine sulfate). The percentage of total sulfur in form of sulfate esters oscillates around 50% in surface soils (e.g. [45, 64, 194]). With increasing depth, the percentage of sulfur present as ester sulfate increases compared to the carbonbound sulfur, but in any depth sulfur shows a significant correlation between organic carbon and total nitrogen [194]. The sulfated thioglycosides (oxine-o-sulfate esters) and the sulfonates which occur in form of an $N{-}O{-}SO_3^-$ and an $N{-}SO_3^-$ linkage respectively might be potentially important, yet they are less well studied.

Sulfur in the Hydrosphere

Marine Systems. Sulfate it the most abundant sulfur species in the ocean. Its concentration is in the average 28.7 m*M* or 2.71 g/kg seawater at a salinity of 35‰. This totals 1.4×10^9 teragrams in the oceans. In the last million year the sulfate content remained more or less constant. However, over the history of the oceans there has probably been a change of the sulfur content. Since the origin of the ocean, sulfate steadily increased to about 40 mmoles per liter in the Ordovician (palaeozoic era) 400 million years ago. Then it continously decreased to the amount found today (Table 3).

Besides sulfate, hydrogen sulfide, DMS and carbon disulfide were found in marine systems. The waters of the North Atlantic are saturated with DMS (1.2×10^{-8} g l^{-1} [122] and contain 5.2×10^{-10} g l^{-1} carbon disulfide [23]. In stagnant waters and near the coast, the concentration of carbon disulfide is higher (Table 4). Lovelock [123] concludes that carbon disulfide originates from anaerobic parts on the seafloor and as it is stable for at least 10 days in seawater containing oxygen it may participate to the transfer of sulfur from

Table 3. Change in sulfate concentrations in the oceans over the last 1.5 billion years [a]

Eras and periods	Time before now ($\times 10^6$ yr)	SO_4^{2-} concentrations (mmol per liter ocean water)
Tertiary		
Pliocene [b]	3.5	29.0
Eocene	45	30.0
Mesozoic		
Cretaceous	90	30.8
Triassic	210	34.3
Palaeozoic		
Permian	251	35.7
Cambrian	530	40.3
Precambrian		
Upper	1,000	31.3
Middle [c]	1,500	18.0

[a] The sulfate concentrations in this table were calculated by using the sedimentation rates and the quantities of sulfate in relation to the relative sedimentary thickness from Conway [44]
[b] The times given in this table refer to the approximate middle of each period
[c] Means toward the end of the "Middle-Precambrian" era

Table 4. Carbon disulfide in seawater [123]

Origin	Concentration (ppm)
Atlantic off Ireland	7.8×10^{-7}
Stagnant bay water	5.4×10^{-6}
Mud at sea bottom	2.95×10^{-5}
Open Atlantic (50° N, 65° S)	5.2×10^{-7}

the seafloor to the surface waters. In trapped or deep sea water bodies oxygen may be lost to organic matter faster than diffusion or convection can replenish it. Thus, an environment is created, where sulfate can be reduced to sulfide by microorganisms. At present time, such reduction is only observed in the Black Sea, the Cariaco Trench, in fjords, lagoons and estuaries which suffer from stagnation. The various observed sulfide concentrations in seawater are given in Table 5. The total amount of hydrogen sulfide sulfur in the hydrosphere is very small and hence does not contribute significantly to the sulfur reservoir in the hydrosphere. The formation of the volatile hydrogen sulfide, however, is of importance for the transfer of sulfur from the hydrosphere to the atmosphere.

Freshwater Systems. Sulfur in natural waters comes from a wide variety of natural and anthropogenic sources and sulfur is a major pollutant in freshwaters. Beeton [17] and Nriagu [142] could show that between 1850 and 1967 the sulfate levels in Lake Erie increased at a rate of about 2 mg/l each decade.

Table 5. Comparison of the sulfide concentrations in some isolated marine systems

System	Depth (m)	S^{2-} tot[a] (10^{-3} ppm)	References
Intermittently anoxic			
Gulf of Cariaco[b]	60–80	800	Richards [165], Gade [68]
Dramsfjörd	55	580	Richards and Benson [166]
(Oslofjord, Norway)	105[c]	1,220	
Saanich Inlet	160	300	Richards [167]
(Vancouver Island, B.C)	240[c]	640	
Baltic Sea[d]	240[c]	2,000	Fonselius [57]
(Gotland Deep)			
Permanently anoxic			
Black Sea	150	60	Skopintsev et al. [181]
	11,500	20,900	
Dead Sea	100	350	Nissenbaum and Kaplan [141]
(Ein Gedi)	300	560	
Kaoe Bay	441	430	Van Riel [204]
(East Indies)			
Cariaco Trench	400	130	Richards and Benson [166]
	1,200[c]	500	
Lake Nitinat	50	3,200	Richards [167]
(Vancouver Island B.C)	210[c]	10,800	

[a] Sum of $H_2S + HS^- + S^2$
[b] There may either be annual or intermittent flushing of the basin in February [165]
[c] Just above the sediments
[d] The intrusion of new bottom water does not follow a yearly cycle

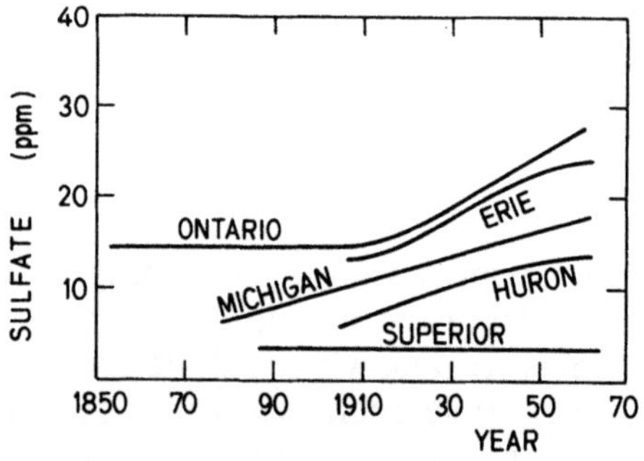

Fig. 1. Historical changes in the concentrations of sulfate in each of the Great Lakes. (After Beeton [17])

Probably, also due to man's activities are the recent increases in the sulfate concentrations of the other great Lakes (Fig. 1). Livingstone [118] gives for an average river water a sulfate concentration of 11.2 mg l^{-1}. About 28% of that concentration derives from pollution [23], as can be seen in Table 6. Groundwaters are important in the mobilization and the transfer of sulfate from one sphere to another. Table 7 summarizes some data of the sulfate content of groundwaters draining the various rock types.

Table 6. Source of sulfate in world average river water [23]

Source	SO_4^{2-} (ppm)	% of total
Rock weathering		
Pyrite	1.3	12
$CaSO_4$ minerals	2.6	23
Volcanic activity	0.8	7
Cyclic sulfur		
Sea spray	0.7	6
Marine H_2S	2.7	24
Pollution	3.1	28
Total[a]	11.2	100

[a] Livingstone [118]. This value might be 7.5% to high [108]

Table 7. Sulfate contents of groundwaters [143, 207]

Source	Rock	Concentration range (ppm)	Mean concentration (ppm)
Igneous	granitic	0.1- 468	6.0
	diorite, andesite, syenite	0.1- 115	(?)
	basic and ultrabasic	0 - 61	5.0
Metamorphic	quartzite, marble	2.0- 130	13
	slate, schist, gneiss	0 - 132	31
Sedimentary	sandstone	1.4- 362	22
	siltstone, clay and shale	1.5-2,420	54
	limestone and dolomite	0.2- 707	20
	unconsolidated sand and gravel	1.0-1,910	25

Sulfur Content of the Lithosphere

The sulfur content of the lithosphere has been studied by Ricke [169] and Holser and Kaplan [87]. The data are much less certain than for the other spheres. Table 8 summerizes some mean sulfur concentrations in the sediment and different rocks. Table 9 gives a summary of the mineable inventory of sulfur. Sulfate deposits are very large and are used as source of gypsum. For mining purposes four types of sulfur containing materials are important: Elemental sulfur deposits, sulfur in fossil deposits, pyrite deposits and volcanic deposits.

In Sicily, sulfur deposits are spread over an area of about 2,600 km^2. So far, some 50 million tons have been mined, most of it before 1900. Sulfur is also found around salt domes. Highest sulfur concentrations are found in domes with a thick anhydrite and calcite rock cap and with large petroleum deposits in the flanking sediments [132]. Ivanov [92] gives a very detailed description and analysis for most of the well known sedimentary sulfur deposits, their history of research and a review of their geological nature.

Table 8. Sulfur content of the lithosphere

Rock type	Estimated mass of rock (10^{20} g)	Mean sulfur concentration (ppm)[a]	Estimated reservoir (10^6 Tg)
Igneous rocks[b]			
Basalt, gabbros, amphibolites, eclogites	127,000	750	9,530
Granites	29,500	300	885
Granodiorites, diorites	31,100	280	871
Alkali massifs	1,100	1,130	124
Dunites, peridotite	600	100	6
Metamorphic rocks[b]			
Gneisses	59,600	540	3,220
Schists	14,100	1,050	1,480
Marbles	2,500	1,200	300
Sedimentary rocks[c]			
Sandstone	7,250	900	650
Shale	15,080	2,700	4,070
Limestone	6,380	1,300	830
Evaporites	290	170,000	4,070
Sediments	3,000	250	75

[a] For igneous and metamorphic rocks according to Schneider [177]; for sedimentary rocks from Holser and Kaplan [87]
[b] From Ronov and Yaroshevskiy [172]
[c] Total mass from Garrels and Mackenzie [70] relative partition according to Holser and Kaplan [87]

Table 9. World's mineable reserves of sulfur (Tg) [89]

Petroleum	Natural gas	Native ore	Dome	Sulfide ores	Pyrite	Coal
334	680	555	150	265	375	12,000

Hydrogen sulfides and sulfur from petroleum constitute the most important source of elemental sulfur. From the viewpoint of sulfur, the sour gas deposits in France and Canada are economically the most interesting sources. The world reserves are not yet fully known, because the estimates of »proved reserves« have doubled every ten years, together with production. Crude oil contains from 0.05% to 14% sulfur, but the great majority of oil contains between 0.1%–3%. Coal contains pyrite, sulfate and organic sulfur at a concentration of 1%–14%. This corresponds to a sulfur reserve of over 20,000 million tons which is ten times more than all other proven sulfur reserves together.

Large pyrite reserves exist in over 30 countries. US reserves are about 80 million tons, and Spain and Portugal have over 500 million tons. The total world reserves might reach about 2,000 million tons.

The largest, presently known volcanic sulfur deposits are located in the Andes Mountains in South America. This area contains over 100 deposits representing a total of over 100 million tons of elemental sulfur. Only about 5% of the available sulfur have been mined from these locations. Other volcanic deposits in Asia, Europe and around the Pacific Ocean are too small to have commercial potential.

Transfer Mechanisms, Fluxes, and Rates

The global sulfur cycle (Fig. 2) is based on a preindustrial steady state in the pedosphere [79], i.e. in the absence of anthropogenic sources, the scheme of fluxes of sulfur from one reservoir to another constitutes the main part of a global sulfur cycle. The estimated values of the different fluxes vary considerably from author to author (Table 10). Geochemical data about "global" processes are based on difficult-to-test assumptions. They are obtained by computing the sum of a variety of different regional and local sulfur cycles, comparing processes in different areas, weighing the relative importance of different models and adding extrapolated data. Thus, only few steps in the sulfur cycle can be directly determined; the remaining links must be deduced by balancing the cycle. In this presentation, the atmospheric position (the most active portion) and the fluxes in the hydrosphere are balanced (Table 10). For the atmosphere it was assumed that the burden in its reservoir is neither increasing nor decreasing. The results of recent studies upon hydrothermal processes in the sea floor [49a] allow to balance the sulfur fluxes in the oceans. The other two spheres, pedosphere and lithosphere, however, are not balanced (Table 10) because some transfer mechanisms and rates are still unknown. In the following the input into each sphere are discussed separately.

Atmosphere

Estimations of the yearly atmospheric flux (input = output) varies between 149 and 360 Tg yr^{-1} (Table 10).

Biological Processes. The input of volatile sulfur compound by biological processes are used to balance the flow of sulfur in the atmosphere. Models of the global sulfur cycle indicate that a substantial proportion of atmospheric sulfur is derived from H_2S which was emitted to the atmosphere through microbial reduction of sulfate in anoxic aquatic environments. However, no measurements of H_2S emissions from aquatic systems have yet been reported. Conway [44] suggests that large amounts of H_2S in the atmosphere have their origin in the microbial reduction of sulfate. This hypothesis has been accepted in numerous discussion of the global sulfur cycle [47, 51, 52, 67, 83, 101, 107, 135] and has been critized by others [122, 160]. In recent articles Hitchcock [84, 85] has shown that biological H_2S production gives rise to measurable levels of particulate sulfate in the urban and nonurban atmospheres, and that these sources are most active in summer and fall, when they appear to exceed local anthropogenic sources. Maroulis and Bandy [131] measured dimethylsulfide in the air above the Atlantic. These measurements are not conclusive enough

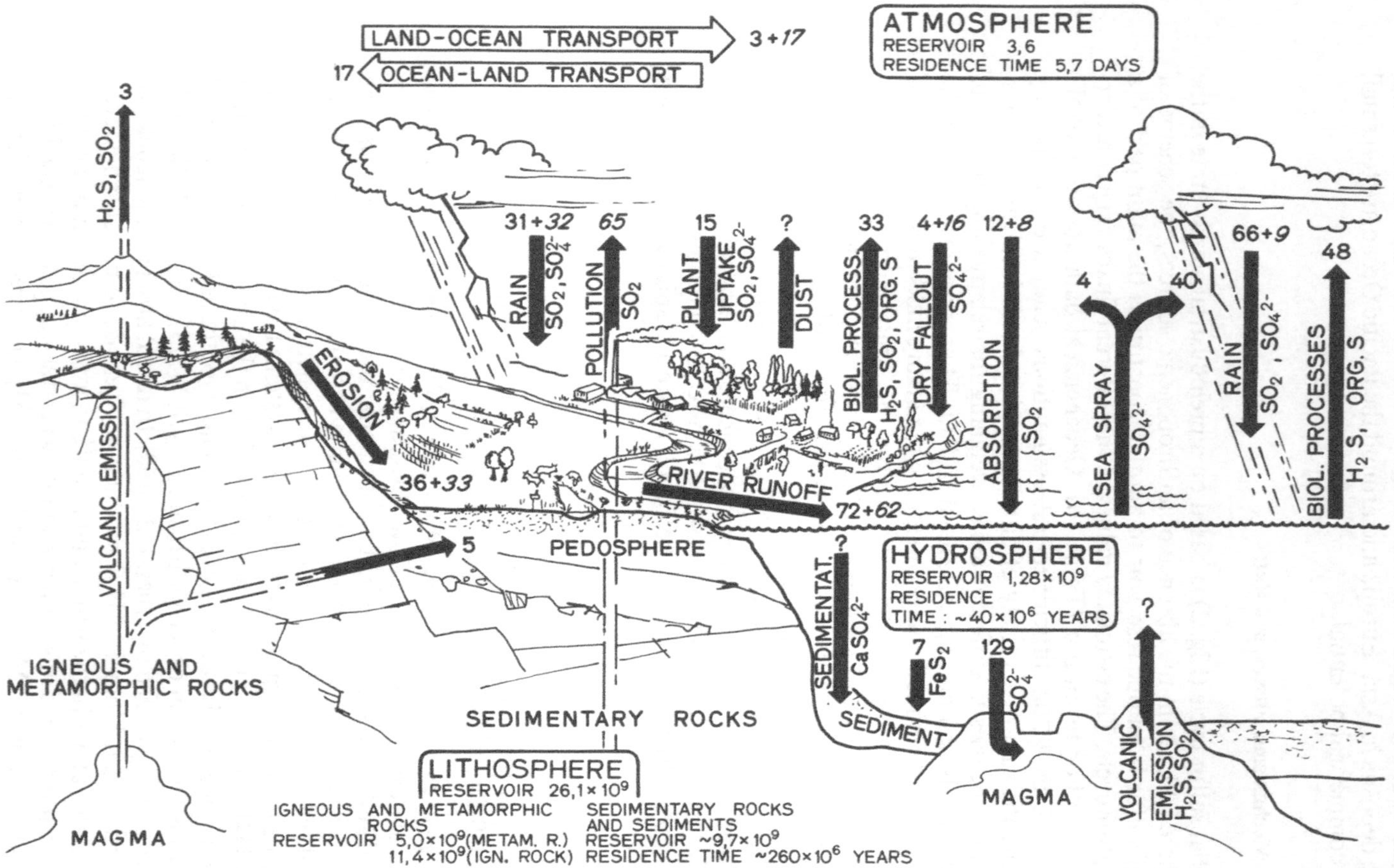

Fig. 2. The global sulfur cycle. The fluxes shown are given in millions of tonnes sulfur per year of Tg S yr^{-1}. Roman typed numbers denote the transfers as estimated to have prevailed before civilization had a significant influence on the sulfur cycle. The italic numbers give the amounts of what man had added by his various activities

Table 10. Annual fluxes of the sulfur in the environment (in Tg yr^{-1})

Sources of sulfur	Eriksson [51]	Junge [100, 101]	Robinson and Robbins [170]	Kellog et al. [107]	Friend [67]	Garrels et al. [71]	Granat et al. [79]	This study
Biological processes, land (H_2S, SO_2, org. S)	110	70	68	} 89	58	56	3	33
Biological processes, ocean (H_2S, org. S)	170	160	30	}	48	35	34	48
Pollution (SO_2, SO_4^{2-})	40	40	70	50	65	50	65	65
Sea spray (SO_4^{2-})	40	–	44	43	44	44	44	44
Volcanic emission, atmosphere (H_2S, SO_2)	–	–	–	~ 1	2	–	3	3
Volcanic emission, pedosphere	–	–	–	–	–	–	5	5
Rain over ocean[a] (SO_2, SO_4^{2-})	}	60	71	72	71	73	63	75
Rain over land[b] (SO_2, SO_4^{2-})	} 165	56	70	86	86	} 87	43	63
Dray fallout over land (SO_4^{2-})	}	15	20	10	20	}	28	20
Absorption by ocean (SO_2)	100	70	25	-	25	-	10	20
Plant uptake (H_2S, SO_2, SO_4^{2-})	75	70	26	15	15	25	–	15
River runoff (SO_4^{2-})	80	95	73	–	136	117	122	132
Sedimentation (pyrite)	–	–	–	–	36[c]	–	–	7
Sedimentation ($CaSO_4$)	–	–	–	–	64	–	–	–
Ridge crest hydrothermal activity (SO_4^{2-})	–	–	–	–	–	–	–	129
Erosion pyrite } Rockweathering	15	}	11	–	26	36	}	}
Erosion $CaSO_4$ }		} 25				24	} 66	} 69
Fertilizer application	10	}	14	–	42		}	}
Atmospheric balance:								
Land → sea	- 10		+26	+ 5	+ 8	+ 10	+ 18	+20
Sea → land	5		4	4	4	7	17	17
Balance in the atmosphere	+20	- 1	0	0	0	0	0	0
Balance in the pedosphere	- 25	+ 1	0	+111	- 5	- 1	+ 20	+ 7
Balance in the hydrosphere	+70	+85	+91	+ 29	+40	+111	+117	0
Balance in the lithosphere	- 65	- 65	-95	- 51	- 35	- 110	- 139	- 8

[a] Including 35 Tg sea spray for Eriksson, 0 for Junge, 39 for Kellog et al. and 40 for the remaining authors
[b] These values contain 4 Tg sea spray (except Eriksson 5, and Junge 0)
[c] Sedimentation, partially as dead organic material

to allow an estimate to be made whether this flux to the atmosphere is one of the most important as was suggested by Lovelock et al. [122] and Rasmussen [160]. Extrapolations of available emission rate data [85] show that marine algae and hence the aquatic environment contribute 0.05 Tg S yr^{-1} in form of dimethylsulfide.

Although it has been assumed that substantial amounts of H_2S are released from soils to the atmosphere by soil microorganisms, Banwart and Bremner [12–14] and Bremner [28] have indicated that this assumption may not be valid and that most of the sulfur volatilized from soils through microbial activity is in form of organic compounds such as dimethylsulfide, dimethyldisulfide and methylmercaptane. However, there seems to be a possibility of H_2S production by soils if plant residues and other sulfur containing materials (e.g. manure, sewage sludge) are added to soils [29] and if the soils lack the cations needed to precipitate the H_2S produced under these conditions. In Table 11 a very rough estimate is given for the production of volatile

Table 11. Estimated volatilized sulfur by unamended soils

Soil	Estimated surface area (10^{12} m^2)[a]	Average S emission [10^{-12}g/g(soil) ×day][b]	Calculated global S emission (Tg S g^{-1})[c]
Aerobic	89.5	16.7	0.5
Waterlogged	26.5	283	2.2

[a] The total aerobic soil surface is composed of the regions covered by woods of the temperate zones (27.5 million km^2), the tundra (8 million km^2), the bush desert (18.0 million km^2), div. graslands (24 million km^2) and the cultivated land (14 million km^2). The waterlogged soil was assumed to include the tropical rain forests (17 million km^2) other rain forests (7,5 million km^2) and bugs and marshes (2 million km^2)

[b] Values from Banwart and Bremner [12]

[c] The most active part of the soil was estimated to be the upper 40 cm with a volume density (including gas bubbles and water) of 2 g cm^{-3} [25]

sulfur compounds by soils. The values in this table may be taken as the very minimum even though Banwart and Bremner [13] measured the production at 30 °C and the slow down of microbial activities during the winter time is not considered. The formation of particulate sulfate from H_2S as discussed by Hitchcock [85] requires further research in order to obtain a more precise idea about the influence of microbiological processes on the atmospheric load of sulfur.

Pollution from Fossil Fuel Combustion. Friend [67] estimated the emission of pollution sulfur to be 65 Tg yr^{-1} in 1965, an estimate very close to that of 64 Tg yr^{-1} given by Robinson and Robbins [170].

Volcanic Sulfur. According to Granat et al. [79] the volcanoes contribute 3 Tg S yr^{-1} to the sulfur budget of the atmosphere. This value is in reasonable agreement with that of 3.75 Tg yr^{-1}, by Cadle [35], of 2 Tg yr^{-1} by Friend [67], of 3.5 Tg yr^{-1} by Stoiber and Jepsen [190] and of 0.75 Tg yr^{-1} given by Kellogg et al. [107].

Sea Spray. This is a widely accepted estimate given in the work of Eriksson [52].

Dust. The mineralogical composition of the dust in a given geographic area depends on the minerals occuring in the soil and weathered rocks. The mineral components of global dust are given by Windom [211]. Dust can be transported over very large distances. Sporadically, powerful volcanic eruptions inject material into the stratosphere, so dispersing the ash globally (e.g. eruption of the Krakatoa in 1883). Lamb [112] provides a comprehensive summary of recent and historical cases of dispersal of volcanic dust through the atmosphere. High-altitude winds transport dust between the continents. Clastic material from the Sahara desert are carried by winds across the Atlantic and are the external sources of the soils on Bermuda and Barbados [93]. A result of a long-distance dust transport is the deposition of reddish colored dust on the snow fields in the Alps, where it is brought by winds from the Sahara. This is a fairly common phenomenon and can be seen best in February and March when the wind direction and duration are most favorable and snow is still present in relative low altitudes. Based on the global soil dust emission of 200 Tg yr^{-1} cited by Butcher and Charlson [33], Granat et al. [79] estimate that only 0.2 Tg S yr^{-1} are transported by dust. However, dust may regionally provide a significant contribution to the sulfur cycle (e.g. the black earth in Russia was brought in from the North as dust during the ice-age.

Pedosphere

Erosion and Weathering. Weathering of ancient sediments, containing chiefly gypsum ($CaSO_4 \cdot 2\,H_2O$) and pyrite (FeS_2), yields substantial amounts of sulfur. They are sources of both reduced and oxidized sulfur in a ratio of roughly 2:3. Gypsum weathers by simple solution and enters streams as Ca^{2+} and SO_4^{2-}-ions. Pyrite, on the other hand consumes oxygen to oxydize to Fe_2O_3 and H_2SO_4. Recently the possibility has emerged that oxidation of pyrite in soils may not be complete, and that SO_2 may be released directly to the atmosphere by a reaction of the type [71]:

$$2\,FeS_2 + 5.5\,O_2 = Fe_2O_3 + 4\,SO_2.$$

All the sulfur from weathering of rocks is assumed to enter finally streams as sulfate, but may be partially retained for a period of time by the pedosphere.

Weathering can be estimated as denudation rate. According to Gregor [81] the annual denudation rate is 10,500 Tg. Of this about 9,000 Tg are assumed to be carried by rivers 100 Tg by glaciers and 6 Tg by wind. Weathering of the continental crust is not equal for igneous + metamorphic and sedimentary rocks because of their different resistance to weathering and their different geographical occurance. Conway [44] gives a mean value for the ratio of igneous + metamorphic to sedimentary rock weathering of 0.2. The denudation rates in Switzerland have been estimated to be 0.05 mm yr^{-1} for cristalline rocks and 0.25 mm yr^{-1} for sedimentary rock terrains [150]. As a global average Lerman [115] postulates 0.006 mm yr^{-1} for cristalline formations and 0.02 mm yr^{-1} for sedimentary rocks. By using the ratio of Conway and Peters-Kümmerly and 0.076% as the average sulfur content of igneous and metamorphic rocks and 0.41% for sedimentary rocks (Table 8) one can calculate that the sulfur content of the average weathered rock is:

$$0.8 \times 0.41 + 0.2 \times 0.076 = 0.34\%.$$

This gives with the annual prehuman denudation rate from Gregor [81] a mean annual value for preindustrial weathering of 36 Tg sulfur, a value quite similar to 33 Tg reported by Granat et al. [79].

The estimate of the antropogenic portion which includes the input by fertilizer application was taken from Garrels et a. [71] and from Granat et al. [79].

Wet Deposition of Sulfur. Granat et al. [79] give a detailed discussion of the sulfate sulfur data in the wet deposition. These authors decided to choose a value of 0.34 g m^{-2} yr^{-1} for the total wet deposition of sulfur (including aerosols) over land areas. Based on the data in Table 12 we estimate 0.4 g S m^{-2} yr^{-1} to be the amount of excess sulfur in the wet deposition over the continents (according to Granat et al. [79] the term excess sulfur is used for the fraction that is estimated to emanate from sources other than seaspray). The rain deposits therefore 63 Tg S yr^{-1} of which 4 Tg S yr^{-1} is contributed by sea-spray.

Table 12. Sulfur flux in rain over the continents

Region	Area [a] (10^6 km^2)	Water depth [a] (precipitation in mm)	Sulfate [b] concentration in the rain (mg S l^{-1})	Sulfate deposition (Tg S yr^{-1})
Europe	10.0	657	3.5 [c]	23
Asia	44.1	696	0.6	18.4
Africa	29.8	696	0.3	6.2
Australia [d]	7.6	447	0.4	1.5
N. America	24.1	645	0.3	4.8
S. America	17.9	1,564	0.1	3.6
Antarctica	14.1	169	0.5	1.1
Total land	148.9			58.6

[a] Baumgartner and Reichel [15]
[b] Granat et al. [79] except Europe
[c] Typical mean value for W. Europe [145]
[d] Without islands

Dry Deposition and Plant Uptake. Robinson and Robbins [170] have reasoned that analog to nuclear fallout the dry deposition of sulfate sulfur is about 20%–25% of its total deposition (83 Tg yr^{-1}). Granat and Söderlund [78] found over Scandinavia that for sulfur dioxide the dry deposition is roughly equal to its wet deposition. Plants take up 15 Tg S yr^{-1} mostly as SO_2 (see later). This amount was estimated by Kellogg et al. [107] but without giving details of the calculation.

Volcanic Sulfur. Rough data exists for the volcanic sulfur input into the pedosphere [67, 190]. The elemental sulfur emitted into the gas phase by volcanoes mixes with the atmosphere and partially precipitates around the volcanoes. A generally accepted figure for sulfur emission to the pedosphere is 4–5 Tg S yr^{-1}.

Hydrosphere

River Runoff. The total yearly and global runoff is 4×10^{16} l [15]. For this study the Antarctica and the islands around Australia are excluded for the following reasons:

(i) The runoffs of the former is very small and the sulfur concentration in the ice is rather low.
(ii) The major part of the sulfur in the runoffs of the latter derives directly from the sea-spray. Consequently we use for our calculation 3.56×10^{16} l yr^{-1} (Table 13).

Table 13. Transport of sulfur by rivers

Region	Runoff[a] (10^{18} g)	SO_4^{2-} [b] (ppm)	Total S (Tg)
Europe	2.8	24	22
Asia	12.2	8.4	34
Africa	3.4	13.5	15
Australia[c]	0.2	2.6	0.2
N. America	5.9	20	39
S. America	11.1	4.8	53
Total	35.6		163.2

[a] Baumgartner and Reichel [15]
[b] Livingstone [118]
[c] Without islands

The mean sulfate concentration in the world rivers is believed to be 11.2 ppm [118]. Thus, the resulting flux of sulfur is 132 Tg yr^{-1}. A slightly higher value is obtained when the transport of sulfur by rivers is calculated for each continent separately (Table 13). The difference between these figures is probably not significant because the soil storage may be changing. 60 Tg or 45% of the river-borne sulfate sulfur was taken as anthropogenic in its origin [71].

Rain. 0.08 mg l^{-1} excess sulfur sulfate was found in ice cores of remote areas like the Antarctica [121]. In more polluted parts like Iceland up to 0.25 mg · l^{-1} excess sulfur were measured in rainwater [79]. A value of 0.09 mg l^{-1} seems to be a reasonable approach to represent the excess sulfur concentration. With precipitation of 38.5×10^{16} l over the oceans, rain transports yearly 35 Tg excess sulfur to the oceans. Excess sulfur see above under "Wet Deposition of sulfur".

Absorption of SO_2 *by the Oceans.* Ocean water absorbs SO_2 well around pH 8. With increasing proton activity (below pH 7) the absorbing ability becomes lower. Because of the high pH and the good carbonate buffer capacity the oceans act as good sinks for atmospheric SO_2 and may never be a direct source of it [19, 30, 82, 117]. There is a lack of flux data on the absorption of SO_2 by the oceans. Since approximately the same amount of sulfur is deposited over the land and the oceans we can estimate that roughly 20 Tg SO_2–S yr^{-1} might be absorbed by the ocean waters.

Lithosphere

Sedimentation. Sulfur enters into the sediment in form of sulfate (gypsum), sulfide (pyrite) or as organically bound sulfide in dead biogenic material. The portion of the organic sulfur in the sedimentary sulfur flux is negligible, since most of it is biologically transformed into sulfide and consequently into pyrite. The pyrite is formed in nature by the reaction of amorphous iron monosulfides with elemental sulfur [22]. This process is thought to take years or even decades. Recently Howarth [91] has shown, however, that pyrite can be formed in salt marshes within a few days (see later).

The generation of dissolved sulfide species from sulfate depends largely on the quantity and nature of the organic matter present in sediments (see below), on the quantity and availability of iron, and on the diffusion of sulfate from the overlaying water mass into the sediments (Table 14). It is therefore not surprising that the vertical distribution of dissolved sulfate and sulfide, and the quantity of precipitated iron sulfide are quite variable, even in a rather restricted area such as the basins of the central part of the Gulf of California [20].

One of the major sinks for sulfate is the evaporite. Marine evaporites have been found in sedimentary rocks of all geologic periods from the recent to the late Precambrian [109]. These evaporites can extract dissolved salts from seawater at an alarmingly rapid rate. Consider a shallow basin that annually receives 1 m of seawater, which consequently evaporates completely to dryness. Due to evaporation in such a basin the ocean would annually lose 0.27 g SO_4^{2-} per cm^2 basins area. Today the rate of formation of sulphate evaporite minerals is negligible, but the present situation has not endured more than a few million years [126].

Holser and Kaplan [87] base their estimated flux of 100 Tg S yr^{-1} into the lithosphere on an assumed steady state in the rate of sediment formation. Berner [24], however, calculates at best 7 Tg S yr^{-1}. Water from hydrothermal

Table 14. Rates of sulfur uptake by some marine sediments [24]

Locality	Uptake ($mg\ S\ cm^{-2}yr^{-1}$)	Reference
Black Sea, deep basin	0.05–0.25	Ross et al. [174]; Ostroumov et al. [148]
	0.35	Berner [24]
	0.05–0.85	Sorokin [185]
Santa Barbara Basin	0.8	Kaplan et al. [104]
	1.3	Kaplan et al. [104]; Berner [24]
Gulf of California	0.5–0.8	Van Andel [203]; Berner [20]
Long Island Sound subtidal	0.4	Berner [22]
Polluted depression in Maine fjord	11.3	Berner [24]
Lindsley Pond, Conn. (freshwater mud)	2.0	Stuiver [192]

vents in the mid ocean ridges is depleted in sulfate [128a]. Edmond et al. [49a] propose that the ridge crest hydrothermal activity extracts at least 122 Tg S yr^{-1} from sea water by a thermally driven process (>200 °C) in which sulfate is reduced by ferrous iron and precipitated as pyrite. To balance the hydrosphere we estimate that every year 129 Tg S is lost by the ocean and incorporated into the newly formed crust.

Equilibrium Chemistry of Sulfur

Of a wide variety of inorganic sulfur compounds there are only few involved in biogenic reactions in the environment (Table 15). Although some of the important environmental processes involving sulfur cannot always be expected to attain chemical equilibrium, a model is often useful to determine limiting conditions. Figure 3a shows that in an aqueous system there are three possible dominant species of sulfur, namely the sulfates (HSO_4^- and SO_4^{2-}), the sulfides (H_2S, HS^-, and S^{2-}) and elemental sulfur. In natural environments the form of sulfur species is closely related to the iron. Figure 3b gives the thermodynamic boundaries of the iron-sulfur system. Stable oxidation states for aqueous sulfur species that are reported in water analyses are mostly forms of S^{6+} (sulfates) or of S^{2-} (sulfides). In hydrothermal or in very polluted waters in anoxic sediment or in flooded soils intermediate oxidation states such as polysulfides, thiosulfate, thionates and sulfites may occure. The equilibrium distribution for some metastable sulfur species is shown in Fig. 3c. These unstable sulfur compound play an important role in the biological sulfur cycle (see below).

Table 15. Most important inorganic sulfur species in the environment

Redox state of sulfur	Name of compounds	Formula
S^{+6}	Sulfur trioxide	SO_3
	Sulfate	H_2SO_4, HSO_4^-, SO_4^{2-}
S^{+4}	Sulfur dioxide	SO_2
	Sulfite	H_2SO_3, HSO_3^-, SO_3^{2-}
S^{+5} to $S^{+1.67}$ [a]	Dithionite	$S_2O_4^{2-}$
	Polythionates	$S_nO_6^{2-}$ (n = 2–6)
S^{+2} [a]	Thiosulfate	$H_2S_2O_3$, $HS_2O_3^-$, $S_2O_3^{2-}$
S^o	Elemental sulfur	S_8
$S^{-0.33}$ to S^{-1} [a]	Disulfide	S_2^{2-}
	Polysulfides	S_n^{2-} (n = 3–6)
S^{-2}	Sulfide	H_2S, HS^-, S^{2-}

[a] In these these species more than one oxidation state of sulfur exists, which results in an overall fractional and intermediate valence state

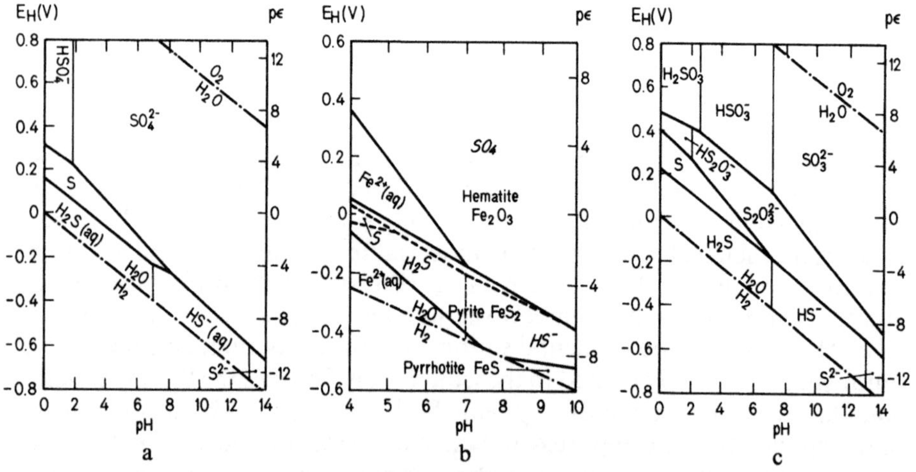

Fig. 3a–c. Equilibrium distribution of sulfur species in water at 25 °C and 1 atmosphere of total pressure. **a** Activity of the dissolved total sulfur species 10^{-1}. **b** Activity of total dissolved sulfur species 10^{-3} and 10^{-4} for the total dissolved iron species. The equilibrium conditions for the sulfur system alone are indicated by the dashed lines and the formulas in italics. **c** Stability fields for metastable sulfur species in absence of sulfate. The Gibbs free energy values for the construction of these diagrams were taken from Garrels and Christ [69]

Physical and Chemical Conversion of Sulfur Compounds in the Environment

The reduction of highly oxidized sulfur compounds is at the current redox state of the environment and at low pressures and temperatures an exclusive privilege of the biosphere (see below). The oxidation of reduced sulfur compounds may as well be a pure chemical process or biologically mediated. In this section only the chemical and physical conversion of reduced sulfur compounds will be discussed.

Atmospheric Reactions

The atmospheric chemistry of reduced sulfur compounds has not been studied as extensively as that of the oxidized forms (e.g. SO_2). The following brief summary is mainly based on the studies of Graedel [77] and Urone and Schroeder [202].

Oxidation of Reduced Sulfur Compounds. The abstraction of a hydrogen atom from H_2S by the HO free radical is the dominant atmospheric reaction and at the same time the rate limiting step in the overall conversion of H_2S to SO_2. Its rate constant is $6.4 \times 10^4\ s^{-1}$ giving a mean residence time for H_2S in the troposphere of 16 hours.

$$H_2S \xrightarrow{HO} HS \xrightarrow{O_2(^3\Sigma)} SO \xrightarrow[\text{or } NO_2]{O_2(^3\Sigma)} SO_2$$

HO readicals also react with methyl mercaptan (CH_3SH) and carbonyl sulfide (COS). Their estimated life times are 0.2 and greater than 60 days respectively. Dimethylsulfide ($CH_3S\ CH_3$) has a proposed atmospheric life time of 3 days, carbon disulfide (CS_2) one of 40 days [77]. Both are attacked by atomic oxygen. The intermediates formed by DMS are:

$$CH_3\,S\,CH_3 + O \longrightarrow \begin{cases} CH_3^{\bullet} + CH_3SO^{\bullet} \\ CH_3O^{\bullet} + CH_3S^{\bullet} \\ CH_3-\underset{\underset{O}{\|}}{S}-CH_3 \end{cases}$$

Oxidation of Sulfur Dioxide. Sulfur dioxide can be oxidized in the atmosphere by photochemical reactions, free radical reactions or by chemical reactions with ozone or nitrogen oxides. Sulfur dioxide has three principal absorption bands at approximately 210, 290, and 380 nm. The width of the 290 nm band allows the SO_2 to absorb sunlight up to about 330 nm and to raise the molecule to an excited singlet ($^1SO_2$) state. The transition to a triplet state ($^3SO_2$) with a subsequent further excitation at 380 nm creates the major reactive photochemical species of sulfur dioxide in the atmosphere [178, 191].

$$SO_2 + h\nu \longrightarrow {}^3SO_2 \begin{cases} + O_2 & \longrightarrow SO_3 + O \\ + O + SO_2 & \longrightarrow SO_3 + SO_2 \\ + RH & \longrightarrow R\,SO_2H \end{cases}$$

SO_2 reacts also directly with ozone to form sulfur trioxide and oxygen:

$$SO_2 + O_3 \longrightarrow SO_3 + O_2$$

Sulfur trioxide readily combines with water to sulfuric acid. In polluted atmospheres the free radical HO_2 may be present and cause with SO_2 the following reaction:

$$SO_2 + HO_2 \longrightarrow SO_3 + OH$$

In presence of nitrogen dioxide and sunlight the rate of SO_2 oxidation increases tenfold [201] over that in clean air. Nitrogen dioxide acts as an oxygen carrier in a cyclic process:

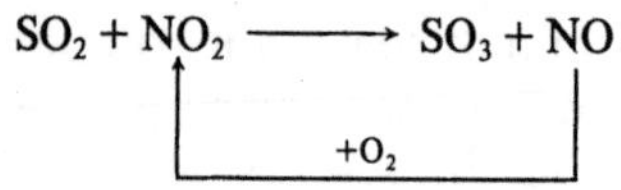

Sulfur dioxide has also a corrosive effect especially on marble and limestone:

$$SO_2 + CaCO_3 + \tfrac{1}{2}\,O_2 \xrightarrow{H_2O} CaSO_4 + CO_2$$

or sulfuric acid formed from SO_2 can dissolve calcium carbonate. Irreparable damages to historical buildings and monuments are caused by these two reactions.

Aqueous Reactions

Sulfide and oxygen may coexist in aqueous solutions for relatively long periods of time. The lifetimes of sulfide species in oxygen saturated seawater (pH 8) are in the order of a few hours (Table 16). Despite numerous attempts in recent years no unified mechanism for the autoxidation of hydrogen sulfide has been found. Often, the results obtained by various investigators were contradictory. Hoffmann and Lim [86] examined the catalytic role of metalphtalocyanine complexes as a model for the naturally occuring porphyrin pigments. Their findings that traces of such complexes affect significantly the autoxidation rate of sulfide may explain some of the discrepant results of others. Table 16 summarizes the current state of kinetic and stoichiometric findings on the autoxidation of sulfide in aquatic environments.

Sulfur Oxidation in the Lithosphere

The weathering of pyrite is probably the most important oxidation process of sulfur in the lithosphere. The pyrite oxidation has been extensively studied for its acidification of coal mine drainage waters. Stumm and Morgan [193] propose the following reaction scheme:

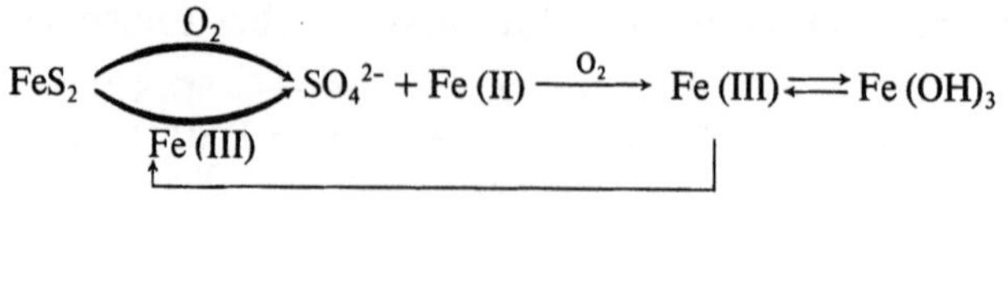

Table 16. Autoxidation of sulfide in natural waters

Observed products	pH	T (°C)	Halflife time[a] (min)	Order of reaction[b]	References
Medium: buffered distilled water					
SO_3^{2-}, $S_2O_3^{2-}$, SO_4^{2-}	7–14	25	1,000	1	Alferova and Titova [2]
S_8,[c] $S_2O_3^{2-}$, SO_4^{2-}	11–14	-	130	1	Avrahami and Golding [7]
S_x^{2-}, SO_3^{2-}, S_8, $S_2O_3^{2-}$, SO_4^{2-}	7.9	25	3,000	1.34	Chen and Morris [41]
S_8, SO_3^{2-}, $S_2O_3^{2-}$, SO_4^{2-}	8.6	25	2,200	1	Demirjian [48]
SO_3^{2-}, $S_2O_3^{2-}$, SO_4^{2-}[d]	8.8	25	e	1	Hoffmann and Lim [86]
S_x^{2-}, SO_3^{2-}, S_8, $S_2O_3^{2-}$, $S_4O_6^{2-}$, SO_4^{2-}	8.3	25	816	1	Lim [116]
SO_3^{2-}, $S_2O_3^{2-}$, SO_4^{2-}	8.0	25	114	1	Moussavi [137]
SO_3^{2-}, $S_2O_3^{2-}$, SO_4^{2-}	7.6	25	880	1	O'Brian and Birkner [144]
Medium: seawater					
S_8, SO_3^-, $S_2O_3^{2-}$, SO_4^{2-}	8.0	23	280	1	Almgren and Hagström [3]
SO_3^{2-}, $S_2O_3^{2-}$, SO_4^{2-}	7.8	9.8	175	1	Cline and Richards [43]
S_8, SO_3^{2-}, $S_2O_3^{2-}$, SO_4^{2-}	8.2	15–22	3,300	1	Skopintsev et al. [182]
S_8, $S_2O_3^{2-}$, SO_4^{2-}	7.7	9.0	600	-	Sorokin [186]

[a] Half life-time of sulfide
[b] With respect to sulfide
[c] Occasionally
[d] Catalyzed reaction. Catalysts: CU(II) –4,4′, 4″, 4‴ - tetrasulfophtalocyanine (10^{-6} *M*)
or Ni(II) TSP (10^{-6} *M*)
or Co(II) TSP (10^{-7} *M*)
[e] The cobalt catalyst at ppb levels increases the rate of autoxidation by a factor of 10^4, the copper and nickel complexes are less effective

At a low pH this oxidation is controlled by the conversion of Fe(II) to Fe(III).

On a global basis much of the atmospheric oxygen would have been consumed over the past several million of years by the weathering of pyrite. However, a part of the sulfate in the oceans is reduced by organisms using organic material derived from photosynthesis. The CO_2 released during this reduction can restore the oxygen content of the atmosphere again, through photosynthesis [72].

Biological Transformations of Sulfur in the Environment

Sulfur Compounds in Biological Systems

Sulfur is an important bioelement which represents approximately 0.5% of the dry weight of plants and microorganisms and 1.3% of animal tissues [147]. A large variety of sulfur containing compounds are found in living cells. Some among the most representative are listed in Table 17 and are described below.

Table 17. Biological sulfur compounds

Name and occurence	Structure
1. Common amino acids	
Cysteine in all organisms	$HS{-}CH_2CH(NH_2)CO_2H$
Methionine in all organisms	$CH_3{-}S{-}CH_2CH_2CH(NH_2)CO_2H$
2. Cofactors	
Thiamine in all organisms	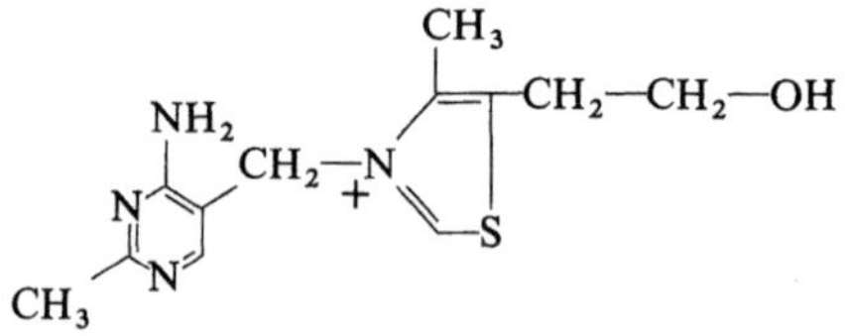
Biotin in all organisms	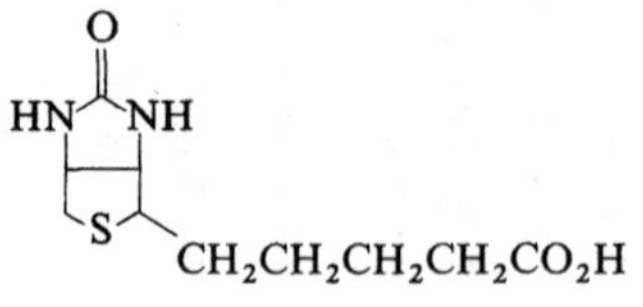

Table 17 (continued)

Name and occurence	Structure
Coenzyme A in all organisms	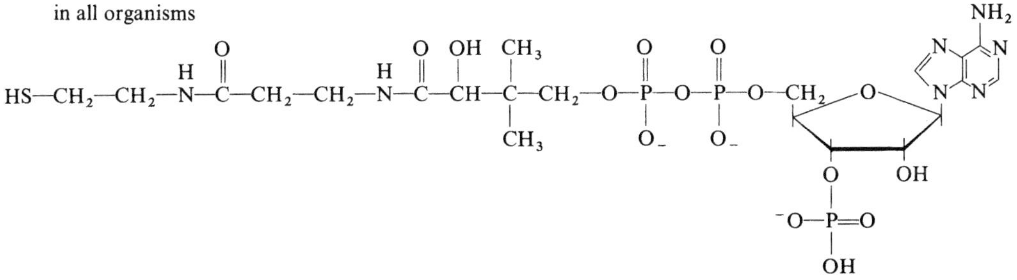
Lipoic acid in all organisms	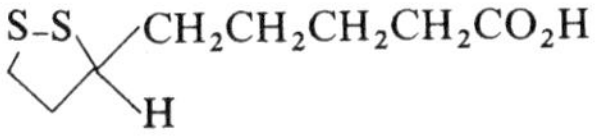
Ferredoxin in bacteria and plants	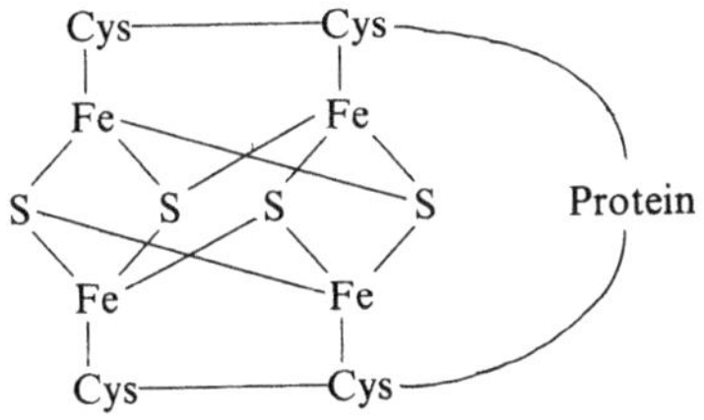
Coenzyme M Methane forming bacteria	HS—CH_2—CH_2—SO_3H
3. Sulfate esters	
Chrondroitin sulfate in animals	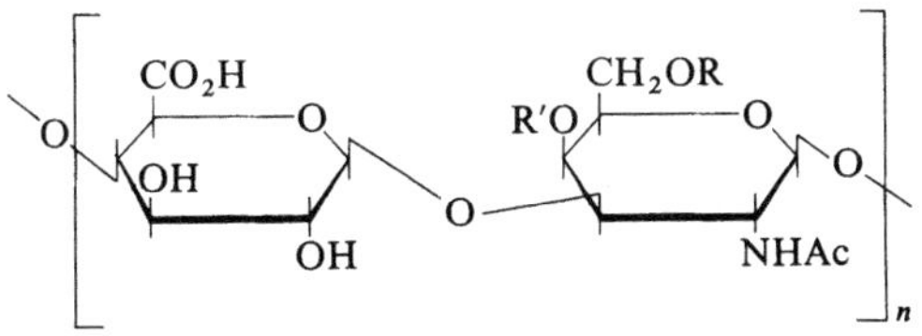 Chondroitin sulfate A: R=H, R′=SO_3H Chondroitin sulfate C: R=SO_3H, R′=H 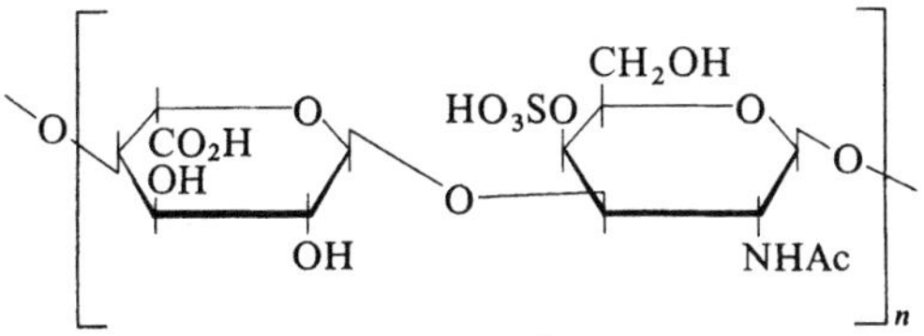 Condroitin sulfate B (Dermatan sulfate)

Table 17 (continued)

Name and occurence	Structure
Tyrosin sulfate in animal excreta and soils	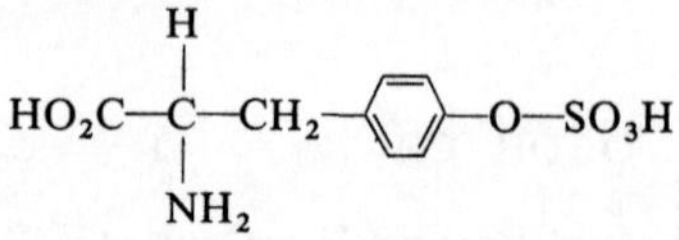
Choline sulfate in fungi and bacteria	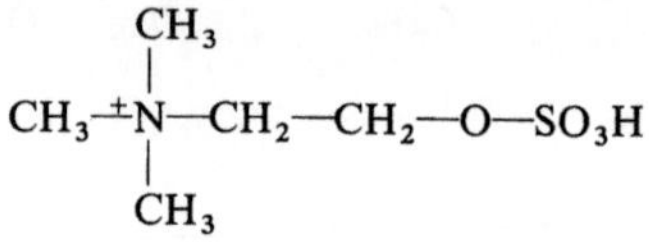
4. Sulfonates	
Cysteic acid in animal excreta and soils	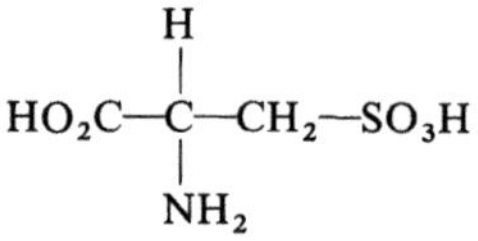
Taurine in animal excreta and soils	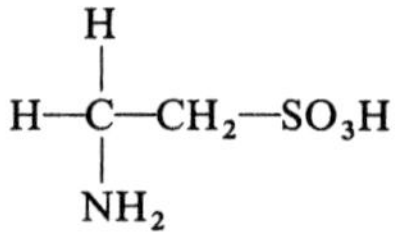
Sulfolipid in plants and photosynthetic bacteria	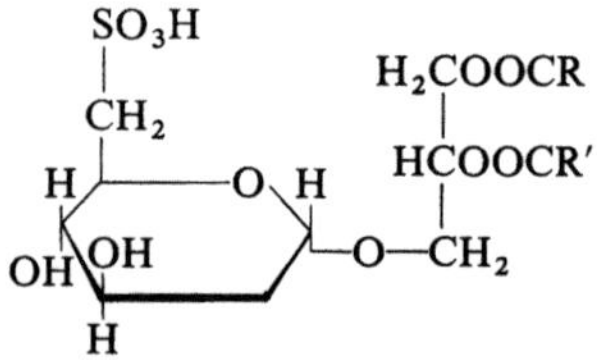
5. Miscellaneous sulfur compounds	
Dimethyl-β-propiothetin in algea and fungi	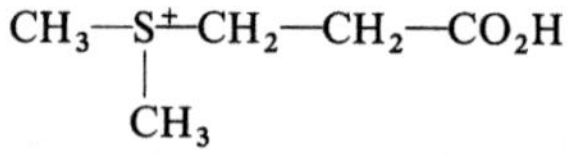
Phosphatidyl sulfocholine in diatoms	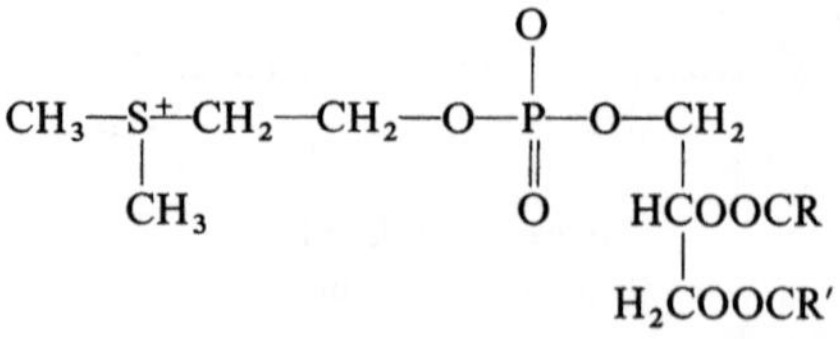

Table 17 (continued)

Name and occurence	Structure
Dimethyl sulfone in human urine	$CH_3-\overset{\overset{\displaystyle O}{\|}}{\underset{\underset{\displaystyle O}{\|}}{S}}-CH_3$
Elemental sulfur in sulfur and photosynthetic bacteria, fungi	S°
Humin sulfur in soils	Complex heterocyclic rings
Diallyl disulfide in plants, especially in garlic	$CH_2{=}CHCH_2-S-S-CH_2CH{=}CH_2$
Propane thial-S-oxide in onions	$CH_3-CH_2-\underset{\underset{\displaystyle H}{\|}}{C}{=}\overset{+}{S}-O^-$

Amino Acids. Most of the sulfur found in living organisms is in the form of the amino acids cysteine and methionine, which are important constituents of protein. Sulfur is in the -2 oxidation state in these amino acids. Cysteine and methionine are readily biologically degraded to form sulfide or volatile organic sulfur compounds under aerobic or anaerobic conditions [1, 103, 216], or can be oxidized to sulfate under aerobic conditions [216].

Cofactors. A number of cofactors containe sulfur. Cofactors are normally only present in trace amounts in cells. For example the vitamins thiamine (vitamin B_1), biotin and coenzyme A. Biotin may be found in the sulfoxide form. Non-heme iron-sulfur proteins, such as the ferredoxins act as acceptors and donors for low potential, high energy electrons. They are important in the metabolism of many anaerobic bacteria and serve as first transfer system for electrons from the excited chlorophyll in the photosynthetic apparatus. Coenzyme M a carrier for one carbon compounds has been found only in methane producing microorganisms [9].

Sulfate Esters. Polysaccharides containing ester-linked sulfate, such as chondroitin sulfate, heparin sulfate, and dermatin sulfate are universally present in connective tissues of animals and in their vitreous and synovial fluids as well. It has been suggested that their function is to bind water in interstitial spaces and to hold cells together in a jellylike matrix. Aryl sulfates such as tyrosine sulfate and steroid sulfates are found in animal excreta [55]. Many fungi and bacteria are capable of forming choline sulfate [55]. Often, over 50% of the sulfur in soils is ester-bonded sulfate [55], and it has been proposed that much of the sulfur in coal is derived from sulfate esters found in peat [36]. Sulfate esters are hydrolyzed by enzymes known as sulfatases.

Sulfonates. Cysteic acid and taurine are both common degradation products of cysteine. Sulfolipids are found in photosynthetic membranes in plants and bacteria [55].

Miscellaneous Sulfur Compounds. Dimethyl-β-propiothetin is an example of a class of sulfonium compounds called thetins found in fungi and algae [39]. The sulfonium analogue of lecithin, phosphatidyl-sulfocholine has been detected in diatoms [4]. Humans excrete approximately 6 mg/day of dimethyl sulfone in their urine [210]. Dimethyl sulfide is a common product of the sulfonium degradation. Elemental sulfur has been detected in fungal spores [151], and is often found as granules in a variety of colorless and photosynthetic sulfur bacteria. Sulfur in soil may be incorporated in complex heterocyclic humic material [55]. Plants related to the onion secrete a variety of sulfur containg oils such as allyl disulfide, and it is thought that these compounds discourage pathogens and parasites [205]. Propanethial-S-oxide (syn.) a volatile substance in onions induces tears. The irritation is caused by sulfuric acid one of the breakdown products of this compound in the eye fluid.

Biological Reduction of Sulfur Compounds

Assimilation

Most organisms are capable of using sulfate as a sole sulfur source for growth; they reduce it to hydrogen sulfide intracellularly and replace the hydroxyl group of serine or homoserine with the sulfhydryl group to form from the former cysteine and methionine from the latter [175]. Some microorganisms cannot assimilate sulfate but need sulfide or forms of intermediate oxidation states such as thiosulfate or sulfite. Other microbes may even have an outright requirement for sulfur containing cofactors or amino acids. Most organisms prefer to assimilate amino acids rather than sulfate.

Assimilatory sulfate reduction is important in the formation of organic sulfur compounds and its main characteristic is that under normal conditions almost no sulfide is released from the cells [127]. The assimilation of radioactive sulfate by phytoplankton and bacteria has been used as methods to measure primary productivity in aquatic systems [133] and bacterial growth [94]. In soils, after 168 days of incubation 18%–25% of added sulfate ends up in insoluble carbon bound sulfur (assumed to be in organisms and perhaps some in humic materials); another 10%–20% went into soluble carbon bound sulfur and most of the remaining was found in sulfate esters [63].

Dissimilation of Inorganic Sulfur Compounds

There is no known purely chemical mechanism for the reduction of sulfate by organic matter at normal earth temperatures and pressures, so that microorganisms are completely responsible for this process [74]. Reduction of sulfate to sulfide occurs virtually in all anoxic habitats containing organic matter and sulfate; e.g. marine and freshwater sediments [74, 158], flooded soils, digesting sewage sludge, rumen and intestines [90] and anaerobic layers of hot springs algal mats [49]. Dissimilatory sulfate reduction is the process by which sulfate is used as the electron acceptor for the oxidation of organic matter, much as oxygen is used in the aerobic respiration. Hydrogen sulfide as the endproduct of the sulfate reduction is excreated by the cells as opposed to the assimilatory reaction. The known strains of microorganisms which carry out sulfate reduction belong to three different genera: *Desulfovibrio*, *Desulfotomaculum* and *Desulfuromonas*. They are all obligate anaerobes and use only a limited range of organic carbon sources (Table 18). Molecular hydrogen can often serve as electron donor, however, autotrophic growth of any of the known strains has never been observed [195, 208]. A variety of new strains have been isolated and are now in the stage of description. They can utilize higher fatty acids up to stearate as well as their corresponding alcohols. These substrates are either oxidized by a β-oxidation to CO_2 or only to acetate (N. Pfennig, personal communication). Besides sulfate, other inorganic forms of sulfur can act as electron acceptors; so are sulfite and thiosulfate widely used substrates and some strains even metabolize elemental sulfur [113, 153, 175].

Sulfate reducing bacteria, as a group can tolerate a wide variety of environmental extremes, such as high temperature, high salinity and high pressure [219], however a low pH limits sulfate reduction. None of the isolated sulfate reducing bacteria can grow below pH 5 [199]. Thiosulfate, polythionates and sulfite are common intermediates of the autoxidation (Table 16) and perhaps biological oxidation of hydrogen sulfide. These thermodynamically metastable sulfur compounds (Fig. 3c) may find their way back into the anaerobic environment and are there reconverted to hydrogen sulfide. Besides sulfate reducers, *Proteus mirabilis* [146], some marine heterotrophs [200] and a spiral-shaped organism from lake sediments [212, 218] are capable to catalyze these reactions and to utilize the liberated energy.

Reduction of Organic Sulfur Compounds

Dimethyl sulfoxide occurs in paper mill wastes and is thought to be a product of the atmospheric oxidation of dimethyl sulfide ([77, 122] see above). Many microorganisms both procaryotic and eucaryotic are capable of reducing dimethyl sulfoxide, to dimethyl sulfide and some of them can even grow on this reaction [218]. Other naturally occuring sulfoxides derive from methionine and biotin, but they are easily re-reduced again by microorganisms [42, 187]. Certain pesticides containing thionophosphorus and thioether groups are oxidized to sulfoxides which can be highly toxic to man and animals. Their microbial reduction may serve as a detoxification.

Table 18. Representative bacteria involved in the reduction or oxidation of inorganic sulfur compounds [a]

Organisms	Reaction catalysed	Morphology	Habitat	Comments
Reduction:				
Desulfovibrio	SO_4^{2-} + lactate $\longrightarrow$ H_2S + acetate + CO_2	Spiral	Anaerobic sediments and soils	Can also use ethanol as electron donor; SO_3^{2-}, $S_2O_3^{2-}$, S^0 as e^-acceptors
Desulfotomaculum	SO_4^{2-} + lactate $\longrightarrow$ H_2S + acetate + CO_2	Sporeforming rod	Anaerobic sediments and soils	Metabolism similar to *Desulfovibrio*
D.acetoxidans	SO_4^{2-} + acetate $\longrightarrow$ $H_2S + CO_2$	Sporeforming rod	Anaerobic sediments and soils	Strain of *Desulfotomaculum* with special characteristics
Desulfuromonas	S^0 + acetate $\longrightarrow$ $H_2S + CO_2$	Rod	Anaerobic sediments and soils	Does not reduce other sulfur compounds
Oxidation:				
Thiobacillus	$H_2S + O_2 \longrightarrow H_2SO_4$	Rod	Soils and water	Can also oxidize $S_2O_3^{2-}$, S^0 and polythionates *T. denitrificans* can use nitrate as an oxidant. *T. ferrooxidans* oxidizes also Fe^{2+} and metal sulfides.
Sulfobacillus	$H_2S + O_2 \longrightarrow H_2SO_4$	Sporeforming rod	Ores	*S. thermosulfidooxidans* plays a role in oxidation of sulfide ores and their warming in deposits
Sulfolobus	$S^0 + O_2 \longrightarrow H_2SO_4$	Lobed sphere	Acid hot-springs	Oxidizes also H_2S and Fe^{2+}
Beggiatoa	$H_2S + O_2 \longrightarrow H_2SO_4$	Filament	Soils and water	S^0 is usually an intermediate and accumulates inside cells
Chromatium	$H_2S + CO_2 \xrightarrow{light} H_2SO_4$ + organics	Rod	Anaerobic waters	Purple sulfur bacterium, S^0 often an intermediate (see also carbon cycle)
Chlorobium	$H_2S + CO_2 \xrightarrow{light} H_2SO_4$ + organics	Rod	Anaerobic waters	Green sulfur bacterium, S^0 often an intermediate (see also carbon cycle)

[a] Many other bacteria are involved. This list is a sampling of the better characterized species

Sulfones are very difficult to reduce chemically and no direct biological reduction of such compounds has been reported up to now.

Geology of Sulfate Reduction

The process of sulfate reduction is of great geological importance in marine sediments because of the large areas and the high sulfate concentration (28 m*M*) involved. Figure 4 is an example of depth profiles for sulfate, and

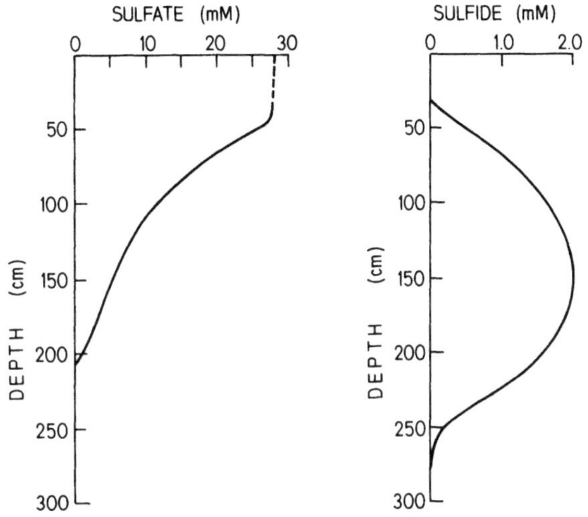

Fig. 4. Plot of dissolved sulfate and sulfide in pore waters of a core from Carmen Basin, Gulf of California. (After Goldhaber and Kaplan [75])

dissolved and acid-volatile sulfide in a marine sediment. Sulfate is depleted with depth until it falls to zero at 200 cm, in this particular sediment. The steady-state maximum concentration of sulfide is less than the amount of sulfate depleted. This is because of losses of sulfide due to diffusion and to formation of pyrite. The kinetics of sulfate reduction are affected by such parameters as sedimentation rate, organic content of sediments, diffusion rates of sulfate and sulfide, temperature, pressure and bioturbation of the sediments by animals. Kinetic models have been presented by Berner [21], Goldhaber and Kaplan [74, 75], and Jørgensen [95].

Althought there may be some formation of black amorphous ferrous sulfide, makinawite (FeS) or greigite (Fe_3S_4), the ultimate fate of most of the sulfide in marine sediments is to form pyrite (FeS_2). Two possible mechanisms have been suggested. One involves the reaction among ferrous iron, sulfide and elemental sulfur as follows:

$$Fe^{2+} + H_2S + S^0 \longrightarrow FeS_2 + 2H^+$$

in which intermediates may include various forms of FeS. An alternate mechanism involves the reaction of sulfide with oxidized iron in geotite:

$$2FeOOH + 2H_2S \longrightarrow FeS_2 + Fe^{2+} + 2H_2O + 2OH^-.$$

Both reactions have been carried out in the laboratory [74] but the latter reaction seems to occur more readily at conditions closest to the natural [91]. The presence of metal ions other than iron can lead to the precipitation of other metal-sulfide minerals [197]. Most sedimentary pyrite is found in raspberry-like clusters of microscopic spheres known as framboids. It has been postulated that bacterial cells serve as a matrix for crystallization of these framboids but such aggregates have been produced in the laboratory without the presence of bacterial cells [74].

Biological Oxidation of Sulfur Compounds

A great variety of microorganisms are capable of oxidizing inorganic sulfur compounds. The sulfur oxidizing bacteria are usually divided into two groups, the colorless and the colored bacteria. The colorless sulfur bacteria which use oxygen as oxidant owe their name to the fact that they lack photosynthetic pigments. The colored or phototrophic sulfur bacteria use the sulfur compounds to reduce their photosynthetic system just as green plants use water (see carbon cycle). Thus, they are capable to mediate the oxidation of sulfur compounds in anaerobic environments where oxidation would otherwise not occur.

Our knowledge on the oxidation of organic sulfur compounds is rather limited. When cystine, the dimer of cystein, is added to soils, sulfate is rapidly formed sometimes with cysteic acid as intermediate [66]. Frederick et al. [59] observed significant production of sulfate in soils amended with taurine, sodium taurochlorate and thiamin. The degradation of sulfonates is especially interesting in the context of environmental pollution caused by synthetic surfactants mainly alkylbenzenesulfonates. *Pseudomonas* strains are able to split of sulfite from sulfonates (e.g. [50]). Sulfite is then readily oxidized to sulfate. The resistance of sulfonates to microbial oxidation is due to the difficulties of its permeation into the cells rather than to the slow kinetics of the desulfonating enzymes systems (K. Wuhrmann, personal communication).

Oxygen as Oxidant

Sulfide and oxygen react chemically in aqueous systems (see above) but the presence of microorganisms can accelerate this process. Usually this kind of microbial sulfur oxidation is found in habitats in which sulfide and oxygen meet such as in waters overlaying anaerobic sediments, the thermoclines or chemoclines of lakes and anaerobic-aerobic interfaces in soil microenvironments. In these zones sulfide is oxidized by the colorless sulfur bacteria (Table 18) e.g. the large and morphologically distinct *Beggiatoa* or *Thiothrix* and *Thiobacillus*.

Beggiatoa forms tufts or mats at sulfide-oxygen interfaces. An interesting habitat of this organism is the rhizosphere of plants growing in flooded soils. There it might reduce hydrogen sulfide levels near the roots, thereby alleviating toxicity to the plants [96].

Thiothrix has never been grown in culture and its physiological properties are obscure. However, its adverse economic activity in sewage treatment systems are well known. If activated sludge receives sewage containing hydrogen sulfide, filamentous masses of *Thiothrix* can form causing poor settling of the sludge, called bulking.

Organisms of the genus *Thiobacillus* are probably the best studied colorless sulfur bacteria, specifically *Thiobacillus ferroxidans*. If metal-sulfide minerals, such as pyrite are unearthed and exposed to air, as is the case during strip mining of coal this bacterium can oxidize the sulfur in these otherwise stable minerals to sulfuric acid causing the nuisance condition known as acid mine drainage [125]. This same reaction can be taken advantage of when one uses these bacteria to leach metal ions out of low-grade metal ores.

Sulfolobus acidocaldarius is the inhabitant of acid hot springs where it is responsible for the oxidation of elemental sulfur to sulfuric acid [31]. Recently the thermophilic *Sulfobacillus thermosulfidooxidans* with properties similar to *T. ferroxidans* has been isolated from a copper-zinc-pyrite deposit [76].

Many common heterotrophic bacteria can also oxidize sulfur compounds [110], however, their role in varius ecosystems remains obscure. It might be that they are even more important than the autotrophic thiobacilli in the sulfur cycle of certain habitats.

Biological Photooxidation

Under anaerobic conditions, photosynthetic bacteria and some cyanobacteria (blue green algae) oxidize sulfide in presence of light. Electrons from sulfide oxidation are used to assimilate carbon dioxide (see also carbon cycle). Most photosynthetic bacteria can oxidize sulfide to sulfate, but elemental sulfur often accumulates as intermediate [152]. Large deposits of elemental sulfur found in the Gulf of Mexico, are believed to be the result of elemental sulfur production by photosynthetic bacteria [74]. The colored sulfur bacteria can be divided into the purple sulfur bacteria (e.g. *Chromatium*) and the green sulfur bacteria (e.g. *Chlorobium*). Their names derive from the distinct color of these bacteria. It is thought that before the advent of oxygen-producing photosyntheses on earth, the photosynthetic bacteria were widespread and responsible for primary production of organic carbon. The sulfate and organic matter produced by the photosynthetic bacteria were then recycled to sulfide and carbon dioxide by the sulfate-reducing bacteria. This type of cycle has been called sulfuretum [8]. Its conception has been extended by Postgate [158] who included the colorless sulfur bacteria into the system. Sulfureta represent biological ecosystems that are now relatively restricted on this planet. They range in size from grains of sands in a polluted estuary to whole seas (e.g. Dead and Black Seas).

Microbial Production of Volatile Sulfur Compounds

Hydrogen sulfide can be biologically produced by sulfate-reduction or by anaerobic degradation of organic sulfur compounds such as cysteine and methionine [1]. It has been proposed that a significant fraction of atmospheric sulfur is derived from hydrogen sulfide emanating from marine and freshwater sediments [107], especially those with shallow water overlying them. However, biological and chemical reactions between sulfide and oxygen in aqueous systems are rapid and make it unlikely that much hydrogen sulfide escapes.

More recently, it has been proposed that volatile organic sulfur compounds such as methane thiol (methyl mercaptan), dimethyl sulfide, and carbon disulfide contribute to the atmospheric sulfur cycle [29, 122, 123]. Methane thiol is generally produced during microbial degradation of methionine, while dimethyl sulfide is a minor product of methionine degradation and a major product of sulfonium compound degradation [29, 103]. Carbon disulfide has been found to be a product of cysteine degradation in soils [11, 12]. Low levels of dimethyl sulfide and carbon disulfide have been detected in ocean and lake waters [122, 123, 217] and dimethyl sulfide has been found to emanate from some soils [58, 122]. Biological consumption of volatile organic sulfur compounds may regulate the rate at which they emanate into the atmosphere from soils and waters. Zinder and Brock [217] were unable to detect any volatile organic sulfur compounds in lake sediments and found that added ^{14}C-labelled methane thiol and dimethyl sulfide were rapidly metabolized by organisms in the sediments to $^{14}CH_4$ and $^{14}CO_2$ (and the sulfur in the compounds was presumably metabolized to H_2S). Thus, any methane thiol or dimethyl sulfide produced in these sediments would be consumed before it could escape. It is unclear as to whether any biological metabolism of volatile organic sulfur compounds takes place in soils. Adsorption to minerals may play a more important role in regulating release of volatile organic sulfur compounds in soil systems than previously thought [29].

Sulfur Isotope Fractionation by Biological Processes

The two most common stable isotopes of sulfur are ^{32}S and ^{34}S. The ratio of ^{32}S to ^{34}S in earth materials is 22.6:1 [70]. In both the chemical and biological reduction of sulfate to sulfide, $^{32}SO_4^{2-}$ is reduced slightly more rapidly than $^{34}SO_4^{2-}$, apparently due to slightly less energy required to break the ^{32}S–O bond [74]. This leads to production of sulfide from sulfate which is enriched in ^{32}S compared to the original sulfate as long as the sulfate is not extensively depleted (Fig. 5). If the reaction goes to completion then the sulfide would necessarily have the same isotopic composition as the starting sulfate. This isotope fractionation is usually quantified in parts per mil as

$$\delta^{34}S = \left(\frac{(^{34}S/^{32}S)\ \text{sample}}{(^{34}S/^{32}S)\ \text{standard}} - 1 \right) \times 1000\ ‰,$$

in which the standard is arbitrarily defined as meteoritic troilite (Fig. 6).

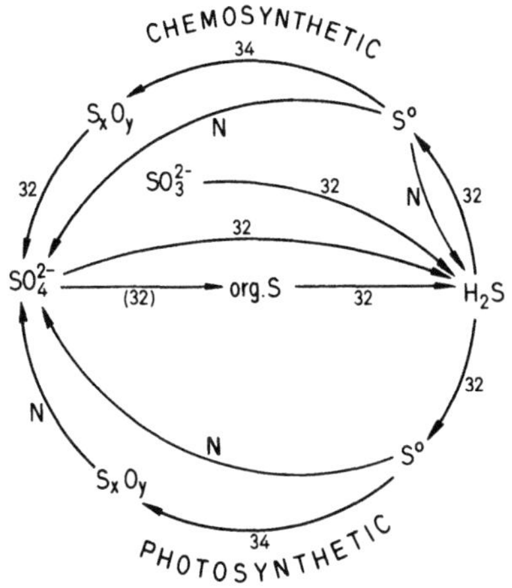

Fig. 5. Enrichment patterns of ^{32}S and ^{34}S by biological processes. N is designated for no fractionation. (According to Goldhaber and Kaplan [74])

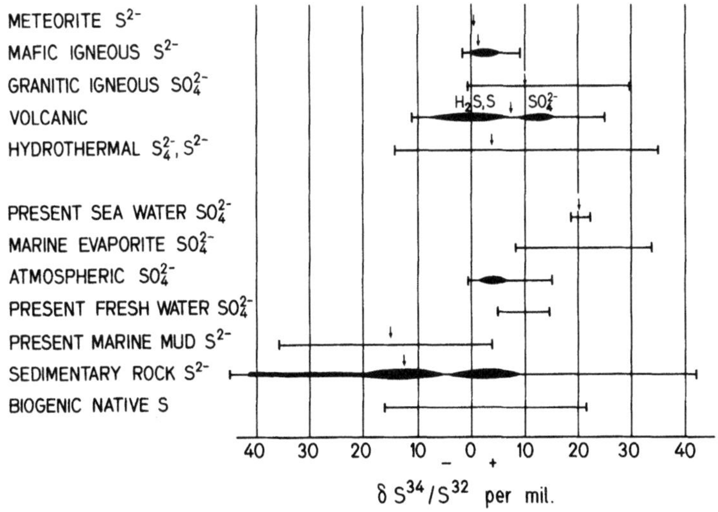

Fig. 6. Ranges in $\delta\,^{34}S$. (After Holser and Kaplan [87])

Cultures of sulfate-reducing bacteria typically produce sulfide with a $\delta\,^{34}S$ of − 10 to − 30‰ relative to that of the sulfate [95, 105, 128]. Some variations may be due to culture conditions such as growth rate and carbon source. On the contrary, assimilatory sulfate reduction does not usually result in a very large fractionation by cultures (Table 19 [74, 127]). Volcanic and mafic igneous rock have $\delta\,^{34}S$ values close to 0‰ while those for oceanic sulfate are about +20‰ and those for sedimentary sulfate at about −20‰ with wide variation (Fig. 6, [74]). Thus, the enrichment of ^{32}S in sedimentary sulfur has

Table 19. Maximum fractionations of the sulfur isotope ratio during the metabolism in microorganisms [a]

Reaction			Organism involved	$\Delta\delta^{34}S$ [b] (‰)
SO_4^{2-}	⟶	H_2S	*Desulfovibrio desulfuricans*	−46.0
SO_4^{2-}	⟶	organic S	*Escherichia coli* *Saccharomyces cerevisiae*	− 2.8
SO_3^{2-}	⟶	H_2S	*D. desulfuricans* *S. cerevisiae*	−14.3 −41.0
Cysteine	⟶	H_2S	*Proteus vulgaris*	− 5.1
H_2S	⟶	S^o	*Thiobacillus concretivorus* *Chromatium*	− 2.5 −10.0
H_2S	⟶	S_xO_y	*T. concretivorus* *Chromatium*	+19.0 +11,2
H_2S	⟶	SO_4^{2-}	*T. concretivorus* *Chromatium*	−18.0 0

[a] After Goldhaber and Kaplan [75]
[b] Given as the enrichment relative to the $^{34}S/^{32}S$ of the starting compound

caused depletion of the lighter isotope in oceanic sulfate. The explanation for this fractionation is that sulfate-reducing bacteria reduce some of the sulfate in the upper layers of the sediment, thereby enriching it with ^{34}S. Some of this ^{34}S-enriched sulfate then exchanges with sulfate in the waters overlying the sediment via diffusion and bioturbation, thereby increasing the ^{34}S content of the seawater sulfate, while at the same time, some of the isotopically light sulfide produced is trapped in the sediments as insoluble metal sulfides. Models for isotope fractionation in sediments have been presented by Goldhaber and Kaplan [74] and Jørgenson [95], among others. Studies of the isotopic composition of sulfide in Precambrian sedimentary rocks show that the isotope effects expected from bacterial sulfate reduction are not present in rocks as old as 3.7×10^9 yr and indicate an emergence of bacterial sulfate reduction about 3×10^9 yr ago [134].

Acknowledgement

The preparation of this chapter was aided by a grant from the Federal Institute for Water Resources and Water Pollution Control (EAWAG), Switzerland, to A.J.B.Z. and by a grant number RD-33 from the Southern California Edison Company to S.H.Z. The manuscript was typed by Barbara Stier and the figures were drafted by Heidi Bolliger.

References

1. Alexander, M.: Adv. Appl. Microbiol. *18*, 1 (1974)
2. Alferova, L.A., Titova, G.A.: Zh. Prikl. Khim. *42*, 192 (1969)
3. Almgren, T., Hagström, I.: Water Res. *8*, 395 (1974)
4. Anderson, R., Kates, M.: Volcani, B.E.: Nature *263*, 51 (1976)
5. Ando, H. et. al.: Japan J. Microbiol. *1*, 335 (1957)
6. Appel, B.R. et al.: Environ. Sci. Technol. *12*, 418 (1978)
7. Avrahami, M., Golding, R.M.: J. Chem. Soc. A *1968*, 647
8. Baas-Becking, L.G.M.: Ann. Bot. *39*, 613 (1925)
9. Balch, W.E., Wolfe, R.S.: J. Bacteriol. *137*, 256 (1979)
10. Banwart, W.L., Bremner, J.M.: Soil Biol. Biochem. *6*, 113 (1974)
11. Banwart, W.L., Bremner, J.M.: J. Environ. Qual. *4*, 363 (1975)
12. Banwart, W.L., Bremner, J.M.: Soil Biol. Biochem. *7*, 359 (1975)
13. Banwart, W.L., Bremner, J.M.: ibid. *8*, 19 (1976)
14. Banwart, W.L., Bremner, J.M.: ibid. *8*, 439 (1976)
15. Baumgartner, A., Reichel, E.: The World Water Balance. R. Oldenburg Verlag, München 1975
16. Bechard, M.J., Rayburn, W.R.: J. Phycol. (Suppl.) *10*, 10 (1974)
17. Beeton, A.M.: Limnol. Oceanogr. *10*, 240 (1965)
18. Beerstecher, E.: Petroleum Microbiology. Elsevier, New York 1954
19. Beilke, S., Lamb, D.: Tellus *26*, 268 (1974)
20. Berner, R.A.: Mar. Geol. *1*, 117 (1964)
21. Berner. R.A.: Geochim. Cosmochim. Acta *28*, 1497 (1964)
22. Berner, R.A.: Am. J. Sci. *268*, 1 (1970)
23. Berner, R.A.: J. Geophys. Res. *76*, 6597 (1971)
24. Berner, R.A.: The changing Chemistry of the Oceans, Nobel Symposium 20 (Dryssen, D., Jagner, D., Eds.) Almquist and Wiksell, Stockholm 1972, p. 347
25. Bölling, W.H.: Bodenkennziffern und Klassifizierung von Böden. Springer Verlag, Vienna, New York 1971
26. Bolin, B., Charlson, R.J.: Ambio *5*, 47 (1976)
27. Breeding, R.J. et al.: J. Geophys. Res. *79*, 7057 (1973)
28. Bremner, J.M.: Proc. Symp. Soil Organic Matter Studies; Braunschweig/Germany, Sept. 6–10, 1976; Internat. Atomic Energy Agency, Vienna 1977; Vol. II, p. 229
29. Bremner, J.M., Steele C.G.: Adv. Microbial Ecology, (Alexander, M., Ed.) Plenum, New York, 1978 Vol. 2, Chap. 4
30. Broecker, W.S., Peng, T.H.: Tellus *26*, 21 (1974)
31. Brock, T.D. et al.: Arch. Mikrobiol. *84*, 54 (1972)
32. Bruner, F. et al.: Analyt. Chem. *44*, 2070 (1972)
33. Butcher, S.S., Charlson, R.J.: An Introduction to Air Chemistry. Academic Press, New York 1972
34. Butlin, K.R.: Research *6*, 184 (1953)
35. Cadle, R.D.: J. Geophys. Res. *80*, 1650 (1975)
36. Casagrande, D.J., Siefert, K.: Science *195*, 675 (1977)
37. Casagrande, D.J., Gronli, F., Patel, V.: Environmental Biogeochemistry and Geomicrobiology (Krumbein, W.E., Ed.). Ann. Arbor Science, Ann Arbor, 1978; Vol. 2, Chap. 40
38. Castleman, A.W. jr., Munkelwitz, H.R., Manowitz, B.: Tellus *26*, 222 (1974)
39. Challenger, F.: Adv. Enzymol. *12*, 429 (1951)
40. Challenger, F.: Aspects of the Organic Chemistry of Sulfur. Butterworth Scientific, London 1959
41. Chen, K.Y., Morris, J.C.: Environ Sci. Tech. *6*, 529 (1972)
42. Cleary, P.P., Dykhuizen, D.: Biochem. Biophys. Res. Commun. *56*, 629 (1974)
43. Cline, J.D., Richards, F.A.: Environ Sci. Technol. *3*, 838 (1969)
44. Conway, E.J.: Proc. R. Ir. Acad. *48*, 161 (1943)
45. Cooper, P.J.M.: Soil Biol. Biochem. *4*, 333 (1972)

46. Davis, J.B.: Petroleum Microbiology. Elsevier, New York 1967
47. Deevey, E.S., jr.: Carbon in the Biosphere (Woodwell, G.M.; Pecan, E.V., Eds.) US Atomic Energy Commission, Washington 1973, p. 182
48. Demiryian, Y.A.: Ph. D. Dissertation, Univ. Michigan, Ann Arbor 1971
49. Doemel, W.N., Brock, T.D.: Limnol. Oceanogr. *21*, 433 (1976)
49a. Edmond, J. M., et al.: Earth Planet. Sci. Lett. *46*, 1 (1979)
50. Endo, K., Kondo, H., Ishimoto, M.: J. Biochem. *82*, 1397 (1977)
51. Eriksson, E.: Tellus *12*, 63 (1960)
52. Eriksson, E.: J. Geophys. Res. *68*, 4001 (1963)
53. Eriksson, E., Rosswall, T.: Nitrogen, Phosphorus, and Sulfur – Global Cycles (Svensson, B.H., Söderlund, R. Eds.) SCOPE Rep. 7, Ecological Bulletins *22*, 11 (1976)
54. Fisher, W.H. et al.: Environ. Sci. Technol. *2*, 464 (1968)
55. Fitzgerald, J.W.: Bacteriol. Rev. *40*, 698 (1976)
56. Fitzgerald, J.W. in: Sulfur in the Environment (Nriagu, J.O., Ed.) Wiley-Interscience, New York 1978; Part II, Chap. 10
57. Fonselius, S.H.: Ambio Spec. Rep. No. 1, (1972)
58. Francis, A.J., Duxbury, J.M., Alexander, M.: Soil Biol. Biochem. *7*, 51 (1975)
59. Frederick, L.R., Starkey, R.L., Segal, W.: Proc. Soil Sci. Soc. Am. *21*, 287 (1957)
60. Freney, J.R., Stevenson, F.J.: Soil Sci. *110*, 307 (1966)
61. Freney, J.R. in: Soil Biochemistry (McLaren A.D.; Peterson, G.H., Eds.) Marcel Dekker, New York 1967, p. 229
62. Freney, J.R., Melville, G.E., Williams, C.H.: Soil Sci. *109*, 310 (1970)
63. Freney, J.R., Melville, G.E., Williams, C.H.: Soil Biol. Biochem. *3*, 133 (1971)
64. Freney, J.R., Melville, G.E., Williams, C.H.: ibid. *7*, 217 (1975)
65. Freney, J.R., Swaby, R.J. in: Sulphur in Australasian Agriculture (McLachlan, K.D., Ed.) Sydney University Press, Sydney 1975, p. 31
66. Freney, J.R.: Aust. J. Biol. Sci. *13*, 387 (1976)
67. Friend, J.P. in: Chemistry of the Lower Atmosphere (Rasool, S.I., Ed.) Plenum, New York 1973, Chapt. 4
68. Gade, H.G.: Bol. Inst. Oceanogr. Univ. Oriente *1*, 287 (1961)
69. Garrels, R.M., Christ, C.L.: Solution, Minerals and Equilibria. Harper and Row, New York and John Weatherhill, Inc., Tokyo 1965
70. Garrels, R.M., Mackenzie, F.T.: Evolution of Sedimentary Rocks. Norton, New York 1971
71. Garrels, R.M., Mackenzie, F.T., Hunt, C.: Chemical Cycles and the Global Environment. William Kaufmann, Los Altos, Cal. 1973
72. Garrels, R.M., Perry, E.A. jr. in: The Sea (Goldberg E.D., Ed.). Wiley-Interscience, New York 1974; Vol. 5, Chap. 9
73. Georgii, H.W.: J. Geophys. Res. *75*, 2365 (1970)
74. Goldhaber, M.B., Kaplan, I.R. in: The Sea (Goldberg, E.D., Ed.). John Wiley, New York 1974, Vol. 5, Chap. 17
75. Goldhaber, M.B., Kaplan, I.R.: Soil Sci. *119*, 42 (1975)
76. Golovacheva, R.S., Karavaiko, G.I.: Mikrobiologiya *47*, 815 (1978)
77. Graedel, T.E.: Rev. Geophys. Space Phys. *15*, 421 (1977)
78. Granat, L., Söderlund, R.: Rep. AC. 32; Internat. Meteorological Inst., Stockholm 1975
79. Granat, L., Rodhe, H., Hallberg, R.O.: Nitrogen, Phosphorus, and Sulphur-Global Cycles (Svensson, B.H.; Söderlund, R., Eds.) SCOPE Report 7, Ecological Bulletins *22*, 89 (1976)
80. Gregor, C.B.: Kon. Ned. Akad, Wetensch. Proc. *71*, 22 (1968)
81. Gregor, B.: Nature *228*, 273 (1970)
82. Hicks, B.B., Liss, P.S.: Tellus *28*, 348 (1976)
83. Hill, F.B.: Carbon in the Biosphere (Woodwell, G.M.; Pecan, E.V., Eds.). US Atomic Energy Commission, Washington 1973, p. 159
84. Hitchcock, D.R.: 2nd Ann. Conf. Water Reuse, Chicago, Illinois, May 1975
85. Hitchcock, D.R.: Environmental Biogeochemistry (Nriagu, J.O., Ed.). Ann. Arbor Science, Ann Arbor 1976; Vol. 1, Chap. 24
86. Hoffmann, M.R., Lim, B.C.H.: Environ. Sci. Technol. *13*, 1406 (1979)
87. Holser, W.T., Kaplan, I.R.: Chem. Geol. *1*, 93 (1966)
88. Horn, M.K., 1966, cited by Garrels and Mackenzie (1971)

89. Horseman, F.: World Sulphur Supply and Demand (1960–1980) for UN Ind. Devel. Org. UNESCO, New York (1970)
90. Howard, B.H., Hungate, R.E.: Appl. Environ. Microbiol. *32*, 598 (1976)
91. Howarth, R.W.: Science *203*, 49 (1979)
92. Ivanov, M.V.: Microbiological Processes in the Formation of Sulfur Deposits, Israel Program for Sci. Translations, Jerusalem 1968
93. Jackson, M.L. et al.: Soil Sci. *116*, 135 (1973)
94. Jassby, A.D.: Ecology *56*, 627 (1975)
95. Jørgensen, B.B.: Geochim. Cosmochim. Acta *43*, 363 (1979)
96. Joshi, M.M., Hollis, J.P.: Science *195*, 179 (1977)
97. Junge, C.E.: Tellus *9*, 528 (1957)
98. Junge, C.E.: J. Geophys. Res. *65*, 227 (1960)
99. Junge, C.E., Mason, J.E.: J. Geophys. Res. *66*, 2163 (1961)
100. Junge, C.E.: Air Chemistry and Radioactivity. Academic Press, New York 1963
101. Junge, C.E.: J. Geophys. Res. *68*, 3975 (1963)
102. Junge, C.E.: Q. J l R. Met. Soc. *98*, 723 (1972)
103. Kadota, H., Ishida, Y.: Ann. Rev. Microbiol. *26*, 127 (1972)
104. Kaplan, I.R., Emery, K.O., Rittenberg, S.C.: Geochim. Cosmochim. Acta *27*, 297 (1963)
105. Kaplan, I.R., Rittenberg, S.C.: J. Gen. Microbiol. *34*, 195 (1964)
106. Kelley, D.P.: Aust. J. Sci. *31*, 165 (1968)
107. Kellogg, W.W. et al.: Science *175*, 587 (1972)
108. Kempe, S.: The Global Carbon Cycle (Bolin, B., Degens, E.T., Kempe, S., Ketner, P., Eds.) SCOPE Report 13. John Wiley, Chichester 1979, Chap. 12
109. Kozary, M.T., Dunlop, J.C., Humphrey, W.: Saline Deposits, A Symposium Based on Papers from the Internal Conference on Saline Deposits. Houston, Texas 1962, Geol. Soc. Am. Spec. Paper 88, p. 43 (1968)
110. Kuenen, J.G.: Plant Soil *43*, 49 (1975)
111. Kuznetsov, S.I., Ivanov, M.V., Lyalikova, N.N.: Introduction to Geological Microbiology. Mc Graw-Hill, New York 1963
112. Lamb, H.H.: Philos. Trans. Royal Soc. London A *266*, 425 (1970)
113. LeGall, J., Postgate, J.R.: Adv. Microbiol. Physiol. *10*, 81 (1973)
114. LeGall, J.: The Aquatic Environment: Microbial Transformations and Water Management Implications (Guarraia, L.J., Ballentine, R.K., Eds.), US Environmental Protection Agency, EPA 430/G–73–008, Washington 1974, p. 75
115. Lerman, A.: Geochemical Processes, Water and Sediment Environments. Wiley-Interscience, New York 1979
116. Lim, B.C.H.: M.Sc. Dissertation, Univ. Minnesota, Minneapolis 1978
117. Liss, P.S., Slater, P.G.: Nature *247*, 181 (1974)
118. Livingstone, D.A.: US Geol. Surv. Prof. Pap. 440–G, 1963
119. Lodge, J.P. jr., Pate, J.B.: Science *153*, 408 (1966)
120. Lodge, J.P. jr. et al.: Tellus *26*, 250 (1974)
121. Lorius, C. et al.: Tellus *21*, 136 (1969)
122. Lovelock, J.E., Maggs, R.J., Rasmussen, R.A.: Nature *237*, 452 (1972)
123. Lovelock, J.E.: Nature *248*, 625 (1974)
124. Lowe, L.E.: Canad. J. Soil Sci. *49*, 129 (1969)
125. Lundgren, D.G., Vestal, J.R., Tabita, F.R.: Water Pollution Microbiology (Mitchell, R., Ed.) Wiley, New York 1972; Vol. 1, Chap. 4
126. Mackenzie, F.T.: Chemical Oceanography, 2nd ed. (Riley, J.P., Skirrow, G., Eds.). Academic Press, London 1975; Vol. 1, Chapt. 5
127. McCready, R.G.L., Kaplan, I.R., Din, G.A.: Geochim. Cosmochim. Acta *38*, 1239 (1974)
128. McCready, R.G.L.: Geochim. Cosmochim. Acta *39*, 1395 (1975)
128a. McDuff, R. E., Edmond, J. M.: EOS *60*, No. 46, O158 (1979)
129. Manahan, S.E.: Environmental Chemistry, 2nd. ed., Willard Grant Press, Boston 1977; Chap. 12
130. Maw, G.A.: Sulfur in Organic and Inorganic Chemistry (Senning, A., Ed.). Marcel Dekker, New York 1972; Vol. 2
131. Maroulis, P.J., Bandy, A.R.: Science *196*, 647 (1977)

132. Meyer, B.: Sulfur, Energy and Environment. Elsevier, Amsterdam 1977
133. Monheimer, R.H.: Can. J. Microbiol. *20*, 825 (1974)
134. Monster, J. et al.: Geochim. Cosmochim. Acta *43*, 405 (1979)
135. Moss, M.R.: J. Environ. Stud. *9*, 209 (1976)
136. Moss, M.R.: Sulfur in the Environment (Nriagu, J.O., Ed.). Wiley-Interscience, New York 1978; Part I, Chap. 2
137. Moussavi, M.: Ph. D. Dissertation, Univ. Southern California, Los Angeles 1974
138. Natusch, D.F.S., et al.: Analyt. Chem. *44*, 2067 (1972)
139. Nguyen, Ba Cuong, Bonsang, B., Lambert, G.: Tellus *26*, 241 (1974)
140. Nielsen, H.: Tellus *26*, 213 (1974)
141. Nissenbaum, A., Kaplan, I.R.: Environmental Biogeochemistry (Nriagu, J.O., Ed.), Ann Arbor Science, Ann Arbor 1976; Vol. 1, Chap. 22
142. Nriagu, J.O.: Isotope Ratios as Pollutant Source and Behavior Indicators, Internat. Atomic Energy Agency, Vienna (1975), p. 77
143. Nriagu, J.O., Hem, J.D.: Sulfur in the Environment (Nriagu, J.O., Ed.). Wiley-Interscience, New York 1978; Part II, Chap. 6
144. O'Brien, D.J., Birkner, F.B.: Environ. Sci. Technol. *11*, 1114 (1977)
145. OECD Program on Long Range Transport of Air Pollutants, OECD, Paris 1977
146. Oltmann, L.F., Van der Beck, E.G., Stouthamer, A.H.: Plant Soil *43*, 153 (1975)
147. Orr, W.: Handbook of Geochemistry (Wedepohl, K.H., Ed.). Springer Verlag, Heidelberg 1974, Vol. II–2; Chap. 16–L
148. Ostroumov, E.A., Volkov, I.I., Fomina, L.C.: Akad. Nauk. S.S.S.R., Inst. Okeanol. Trudy *50*, 93 (1961)
149. Peck, H.D.: Bacteriol. Rev. *26*, 67 (1962)
150. Peters – Kümmerly, B.: Geogr. Helv. *28*, 137 (1973)
151. Pezet, R., Pont, V.: Science, *196*, 428 (1977)
152. Pfennig, N.: Plant Soil *43*, 1 (1975)
153. Pfennig, N., Biebl, H.: Arch. Microbiol. *110*, 3 (1976)
154. Poldervaart, A.: Geol. Soc. Am. Spec. Pap. *62*, 119 (1955)
155. Postgate, J.R.: Ann. Rev. Microbiol. *13*, 505 (1959)
156. Postgate, J.R.: Progr. in Industrial Microbiology (Hockenhüll, D.J.D., Ed.). Heywood and Co., London 1960; Vol. 2
157. Postgate, J.R.: Bacteriol. Rev. *29*, 425 (1965)
158. Postgate, J.R.: Inorganic Sulphur Chemistry (Nickless G., Ed.). Elsevier, Amsterdam 1968, Chap. 8
159. Prahm, L.P., Torp, U., Stern, R.M.: Tellus *28*, 355 (1976)
160. Rasmussen, R.A.: Tellus *26*, 254 (1974)
161. Rasmussen, K.H., Taheri, M., Kabel, R.L.: Water Air Soil Pollut. *4*, 33 (1975)
162. Rheinheimer, G.: Marine Ecology (Kinne, O., Ed.). Wiley-Interscience, London 1972; Vol. 1, Part 3, p. 1459
163. Rheinheimer, G.: Aquatic Microbiology. Wiley, London 1974
164. Rhode, H.: J. Geophys. Res. *77*, 4494 (1972)
165. Richards, F.A.: Deep-Sea Res. *7*, 163 (1960)
166. Richards, F.A., Benson, B.B.: Deep-Sea Res. *7*, 254 (1961)
167. Richards, F.A.: Chemical Oceanography (Riley, J.P., Skirrow, G., Eds.). Academic Press, New York 1965; Vol. 1, Chap. 13
168. Richmond, D.V.: Phytochemistry (Miller, L.P., Ed.). Van Nostrand Reinhold, New York 1973; Vol. III; p. 41
169. Ricke, W.: Geochim. Cosmochim. Acta *21*, 35 (1960)
170. Robinson, E., Robbins, R.: Emissions, Concentrations and Fate of Gaseous Atmospheric Pollutants, Stanford Res. Inst., Menlo Park, Cal. 1968
171. Robinson, E., Robbins, R.: J. Air Pollut. Control Ass. *20*, 303 (1970)
172. Ronov, A.B., Yaroshevskiy, A.A.: Geokhim. *11*, 1285 (1967)
173. Ronov, A.B.: Sedimentology *10*, 25 (1968)
174. Ross, D.A., Degens, E.T., MacIlvain, J.: Science *170*, 163 (1970)
175. Roy, A.B., Trudinger, P.A.: The Biochemistry of Inorganic Compounds of Sulfur. Cambridge University Press, Cambridge 1970

176. Sandalls, F.J., Penkett, S.A.: Atmos. Environ. *11*, 197 (1977)
177. Schneider, A.: Handbook of Geochemistry (Wedepohl, K.H., Ed.) Springer Verlag, Berlin 1978; Vol. II–2, Chap. 16 E/M
178. Sidebottom, H.W. et al.: Environ. Sci. Technol. *6*, 72 (1972)
179. Siegel, L.M.: Metabolism of Sulfur Compounds, 3rd ed. (Greenberg, D., Ed.). Academic Press, New York 1975; Vol. VII, p. 217
180. Singh, H.B.: Geophys. Res. Lett. *4*, 101 (1977)
181. Skopintsev, B.A. et al.: C.R. Acad. Sci. U.S.S.R. *119*, 121 (1958)
182. Skopintsev, B.A., Karpov, A.V., Vershinena, D.A.: Sov. Oceanogr. *4*, 55 (1964)
183. Smith, A.F., Jenkins, D.G., Cunningworth, D.E.: J. Appl. Chem. (London) *11*, 137 (1961)
184. Sokolova, G.A., Karavaiko, G.I.: Physiology and Geochemical Activity of Thiobacilli, Israel Program Sci. Translations, S. Manson, Jerusalem 1968
185. Sorokin, Y.I.: Mikrobiologiya *31*, 402 (1962)
186. Sorokin, Y.I.: Okeanologiya *11*, 423 (1971)
187. Sourkes, T.L., Trano, Y.: Arch. Biochem. Biophys. *42*, 321 (1953)
188. Starkey, R.L.: Ind. Eng. Chem. *48*, 1429 (1956)
189. Stevens, R.K. et al.: Analyt. Chem. *43*, 827 (1971)
190. Stoiber, R.E., Jepsen, A.: Science *182*, 577 (1973)
191. Strickler, S.J., Howell, D.B.: J. Chem. Phys. *49*, 1947 (1968)
192. Stuiver, M.: Geochim. Cosmochim. Acta *31*, 2151 (1967)
193. Stumm, W., Morgan, J.J.: Aquatic Chemistry. Wiley, New York 1970
194. Tabatabai, M.A., Bremner, J.M.: Soil Sci. *114*, 380 (1972)
195. Thauer, R.K., Jungermann, K., Decker, K.: Bacteriol. Rev. *41*, 100 (1977)
196. Tanner, R.L., Forrest, J., Newman, L.: Sulfur in the Environment (Nriagu, J.O., Ed.). Wiley-Interscience, New York 1978; Part I, Chap. 10
197. Trudinger, P.A., Lambert, I.B., Skyring, G.W.: Econ. Geol. *67*, 1114 (1972)
198. Trudinger, P.A.: Sulphur in Australasian Agriculture (McLachlan, K.D., Ed.). Sydney University Press, Sydney 1975; p. 11
199. Tuttle, J.H. et al.: J. Bacteriol. *97*, 594 (1969)
200. Tuttle, J.H., Jannasch, H.W.: J. Bacteriol. *115*, 732 (1973)
201. Urone, P. et al.: Environ. Sci. Technol. *2*, 611 (1968)
202. Urone, P., Schroeder, W.H.: Sulfur in the Environment (Nriagu, J.O., Ed.). Wiley-Interscience, New York 1978; Part 1, Chap. 8
203. van Andel, T.H.: Am. Assn. Petroleum Geol. Memoir 3, 1964, 216
204. van Riel, P.M.: Snellius.-Exped. East. Part Neth.-E.-Indies *2*, 1 (1945)
205. Walker, J.C., Morell, S., Foster, H.H.: Am. J. Botany *24*, 536 (1937)
206. Wang, W.C. et al.: Science *194*, 685 (1976)
207. White, D.E., Hem, J.D., Waring, G.A.: Chemical Composition of Subsurface Waters, US Geological Survey, Prof. Paper 440-F, 1963
208. Widdel, F., Pfennig, N.: Arch. Microbiol. *112*, 119 (1977)
209. Williams, C.H.: Sulphur in Australasian Agriculture (McLachlan, K.D., Ed.). Sydney University Press, Sydney 1975; p. 21
210. Williams, K.I.H., Burstein, S.H., Layne, D.S.: Arch. Biochem. Biophys. *113*, 251 (1966)
211. Windom, H.L.: Geol. Soc. Am. Bull. *80*, 761 (1969)
212. Wolfe, R.S., Pfennig, N.: Appl. Environ. Microbiol. *33*, 427 (1977)
213. Wood, E.J.F.: Marine Microbial Ecology. Chapman and Hall, London 1965
214. Wood, W.P., Heicklen, J.: J. Phys. Chem. *75*, 854 (1971)
215. Young, L., Maw, G.A.: The Metabolism of Sulfur Compounds. Methuen, London 1958
216. Zinder, S.H., Brock, T.D.: Sulfur in the Environment (Nriagu, J.O., Ed.). Wiley-Interscience, New York 1978; Part II, Chap. 11
217. Zinder, S.H., Brock, T.D.: Nature *273*, 226 (1978)
218. Zinder, S.H., Brock, T.D.: Arch. Microbiol. *116*, 35 (1978)
219. ZoBell, C.E.: Prod. Mon. *22*, 12 (1958)
220. ZoBell, C.E.: Organic Geochemistry (Berger, I.A., Ed.). Macmillan, New York 1963; p. 543
221. ZoBell, C.E.: Estuarine Microbial Ecology (Stevenson, H.L.; Colwell, R.R., Eds.). Univ. South Carolina Press, Columbia 1973; p. 9

The Phosphorus Cycle

J. Emsley
Department of Chemistry, King's College, Strand, London, U.K.

Introduction

The investigator of phosphorus and its role in the environment is well served by a large body of data covering all aspects of the subject and spanning the natural, earth and life sciences as well as the technologies of mining, manufacture and agriculture. There are two general texts devoted solely to phosphorus and the environment [1, 2] and a detailed chemistry of phosphorus, including its environmental and biochemical roles is also available [3].

While it may be true that to understand how phosphorus behaves in the environment one must understand its chemistry, it is also true that the chemistry of phosphorus in the natural state is limited to the salts and esters of only one of its many acids – phosphoric acid.[1] Strictly speaking there are several phosphoric acids depending upon the degree of polymerization:

phosphoric acid itself, H_3PO_4,
di- or pyro-phosphoric acid, $H_4P_2O_7$, and
triphosphoric acid, $H_5P_3O_{10}$,

are the main ones with the general formula $HO(P(O)(OH))_nOH$, where $n=1,2,3$. In nature the metal salts are all of the simple acid, but esters of the di- and tri-acids are important in living cells. The best known of these is ATP, adenosine triphosphate, used by cells to transfer the energy necessary for many biochemical reactions.

In *inorganic phosphates* we find the anion PO_4^{3-} (I)

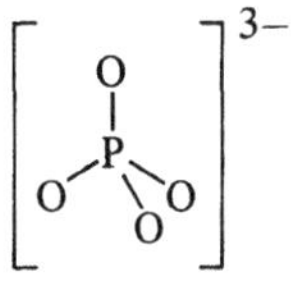

(I)

1 Not strictly true – there are a few naturally occurring compounds with phosphorus-carbon and phosphorus-nitrogen bonds

The oxygen atoms may carry hydrogen as in the ions HPO_4^{2-} and $H_2PO_4^-$ and the acid itself H_3PO_4. These forms are present in aqueous solution, their relative concentrations depending on the pH. They all are covered by the general term "inorganic phosphate", sometimes abbreviated as P_i.

The organic phosphates can be much more complicated according to the one, two or three organic groups attached to the oxygen atoms. Thus there are mono-, di-, and tri-esters: $(RO)PO_3H_2$, $(RO)_2PO_2H$, and $(RO)_3PO$. The mono-

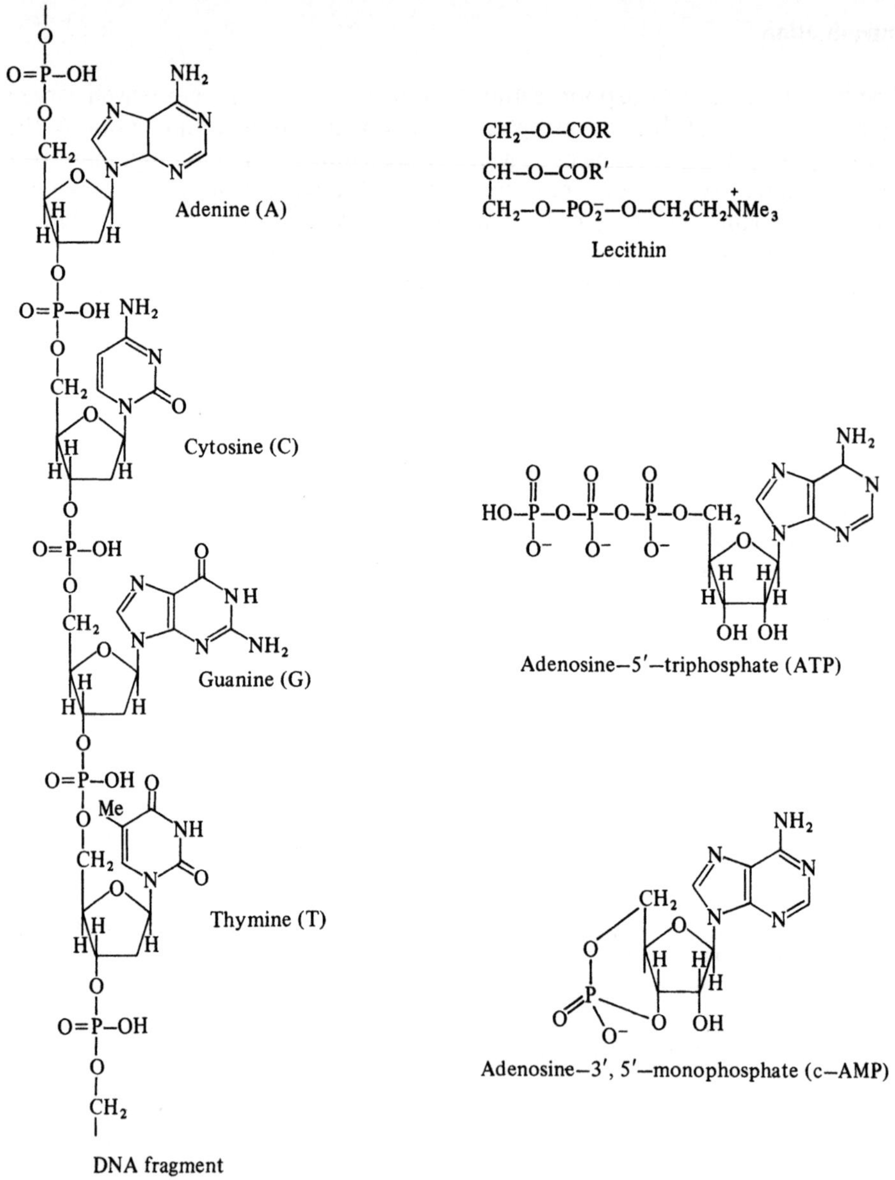

Fig. 1. Structures of key biological organic phosphates

and di-esters are the organic phosphates produced by all living cells, and the diversity of the organic group is the essential feature here. In several cases the phosphate will be a combination of acid, salt and ester, i.e., $(RO)P(O)(OH)O^- M^+$, but so long as the phosphorus has at least one ester group attached to it then it is referred to as an "organic phosphate".

Phosphorus is a vital and irreplaceable element in all living systems [4]. As DNA, RNA, phospholipids, ATP, and c-AMP it plays a key part in almost all the essential functions of a living cell such as creation, structure, operation and reproduction. The nucleic acids of the cell are polymers based on phosphate esters; a DNA fragment is illustrated in Fig. 1 and RNA is similar. Phospholipids such as lecithin (Fig. 1) can form bilayer-lipid membranes, the R and R′ groups being derived from a variety of long chain acids such as palmitic, ($C_{15}H_{31}CO_2H$), linoleic ($C_{17}H_{29}CO_2H$), and others. ATP (Fig. 1) is used in cells in the synthesis of fats, carbohydrates and proteins. Outside cells the related molecule cyclic-AMP, adenosine-3′,5′-monophosphate, acts as a messenger in conjunction with hormones. Not all essential phosphates are organic phosphates. Calcium phosphates have a crucial contribution to make as the basic component of bone and other solid phase structures in organisms.

Unlike the other non-metal elements that are essential to life, phosphorus has only the one environmentally stable form, phosphate. And whether the phosphate be an inorganic salt or an inorganic ester it is non-volatile, so that on the surface of the Earth phosphorus is restricted to the lithosphere and hydrosphere. Moreover it is unevenly distributed and this leads to local shortages, especially in regions where the land has been above sea level for a long period. Although restricted to the condensed phases of the planet, phosphate is still quite mobile and a distinguishing feature of phosphate is its tendency to move from the lithosphere to the hydrosphere and there to concentrate in places inaccessible to living things such as at the bottom of lakes or the sea. Because of the downward movement of phosphate there is a chronic shortage of this nutrient in ecological systems undisturbed by Man.

The Primary Inorganic Cycle

In order of natural abundance in the Earth's crust phosphorus comes 10th with an average concentration of 0.1% by weight [5]. The movement of this phosphate can be resolved into three cycles: two biological cycles superimposed on an inorganic cycle. The inorganic cycle turns very slowly, its rate of revolution being measured in Gy (10^9 yr). The two biocycles turn rapidly in comparison – the land based cycle is made up of several components and there may be an annual turnover of some of the phosphate, although there are reservoirs of phosphate in the soil that can store it for centuries. In the water based biocycle phosphate turnover is measured in days and months.

It is difficult to quantify the amount of phosphate moving through these three cycles although attempts have been made to do this [6, 7] and a useful summary based on Pierrou's paper [7] is shown in Fig. 2. The picture is incomplete because there are areas in which little work has been done as yet,

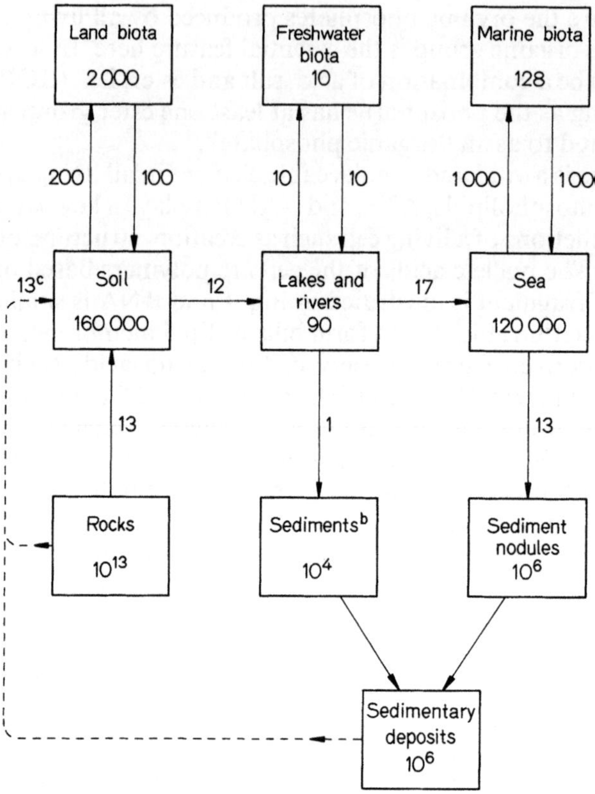

Fig. 2. The amounts of phosphorus in the various compartments of the environment and the annual natural movements between them (Mtonnes P)[a]

[a] 1 tonne P = ca. 5 tonnes fluoroapatite rock

[b] Lake sediments are the least easy to quantify

[c] Deposits mined by Man for fertilizer use

e.g. the phosphorus in wind borne dust and sea spray, the uptake of soil phosphorus by microbes, and the utilization of phosphate by bacteria in water. These parts of the natural cycle are likely to be relatively small however and would not significantly alter the data in Fig. 2.

Although we speak of the primary inorganic cycle of phosphate a better analogy would be a spiral – Fig. 3. Phosphate comes originally from igneous rocks and passes through the soil to the rivers and on to the sea, where it eventually ends up as sediment. If the sediment is part of the deep ocean then there is little chance of that phosphate going through the cycle again, although as we shall see some ocean phosphate does return to the land via ocean currents, fish and the birds that feed on them. If the sediment is part of an inland sea or continental shelf there is a good chance it will once again be part of dry land as a result of geological uplift. The phosphate in the sea is present largely as phosphate nodules and there is an estimated 300 Gtonnes of these [8].

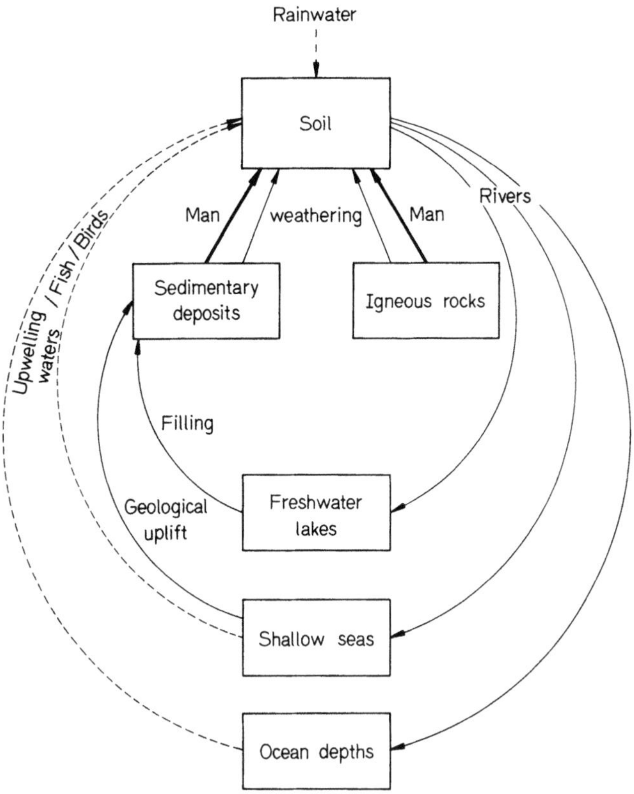

Fig. 3. The primary inorganic phosphate cycle – spiral

The overall natural movement tends to concentrate phosphate in large deposits in a few areas of the globe, chiefly along the western coasts of the continents [9]. There the spiral would end since such deposits are inaccessible to plants, lying as they do several metres below the surface. Phosphate thus has a natural tendency to move in a downward spiral and only the intervention of a highly organized species, such as industrialized Man, can release this locked-up phosphate to the inorganic cycle again.

There are rocks that have still not been through the inorganic cycle once and some of these igneous rocks are so rich in phosphate that they are mined. Most igneous rocks contain about 0.2% phosphate and this is the origin of the vast majority of the phosphate that is moving or has moved through the inorganic cycle. Meteorites have contributes about 100 Gtonnes to the total and still add about 35 tonnes per year [10].

Phosphate Minerals

There are over 200 known phosphate minerals [11] but in terms of quantity only the calcium phosphates, and especially the so-called apatites, are important. The alkaline earth metal, calcium, has a dominant effect on the inorganic chemistry of the Earth's surface since it locks-up three key non-metal elements – carbon, sulfur and phosphorus – as their insoluble calcium salts $CaCO_3$ (limestone etc.), $CaSO_4 2H_2O$ (gypsum), and $Ca_5(PO_4)_3(OH,F)$ (apatites).

Calcium and phosphate ions may crystallise in several forms, only one of which is soluble in water, Table 1. Although examples are found in nature of all the insoluble salts, only one is important and that is fluoroapatite. Hydroxyapatite is the salt that precipitates from an alkaline solution of calcium when phosphate is added (conditions that pertain in sea water), and then the hydroxide radical is displaced wholly or partly by fluoride.

Table 1. Calcium phosphates (Mineral name)

		Ratio Ca:P	
Calcium dihydrogen phosphate hydrate	$Ca(H_2PO_4)_2 \cdot H_2O$	0.5	Soluble
Calcium hydrogen phosphate dihydrate (brushite)[a]	$CaHPO_4 \cdot 2H_2O$	1.0	Insoluble
Octacalcium phosphate pentahydrate	$Ca_8H_2(PO_4)_6 \cdot 5H_2O$	1.33	Insoluble
Calcium phosphate (whitlockite)[b]	$Ca_3(PO_4)_2$	1.5	Insoluble
Hydroxyapatite	$Ca_5(PO_4)_3OH$	1.67	Insoluble
Fluoroapatite	$Ca_5(PO_4)_3F$	1.67	Insoluble

[a] A dehydrated form, $CaHPO_4$, is found as the mineral monetite
[b] Two polymorphs are known, the α- and β-forms, but only the β-form is stable in contact with water and the mineral whitlockite is β-$Ca_3(PO_4)_2$

The solubility product of $Ca_3(PO_4)_2$ is 2.0×10^{-29}, but that of hydroxyapatite may be as low as 10^{-51} and fluoroapatite even smaller. This is the least soluble form of all, but it does not necessarily follow that it will be the one to precipitate even in the presence of fluoride. Sometimes the crystal form of naturally occurring hydroxyapatite may be like that of octacalcium phosphate, showing that this was the form which precipitated. It then reverted to the chemical composition, but not the structure, of hydroxyapatite on standing [12]. In neutral solution $CaHPO_4 \cdot 2H_2O$ and $Ca_8H_2(PO_4)_6 \cdot 5H_2O$ appear to precipitate most rapidly; in acid solution $CaHPO_4$ may precipitate first.

Nevertheless, whichever calcium phosphate comes out of solution as the solid phase, it slowly reverts to hydroxyapatite and/or fluoroapatite and these are the chief forms of the large deposits of phosphate rock. Deposits of over 10^9 tonnes are listed in Table 2. These deposits are not the same as phosphate "reserves", a term that is used to describe those that are mineable with present day technology. For instance the USA "reserves" are quoted as 3 Gtonnes representing mainly the Florida deposits which are mined on a large scale. World "reserves" are ca 60 Gtonnes [8].

The deposits in Table 2 are all fluoroapatite – indeed 95% of the world's solid phosphate is fluoroapatite [13]. In addition to these major deposits there are smaller but workable deposits of igneous origin in South Africa, Rhodesia and Brazil, of sedimentary type in Israel, Jordan, Senegal, Togo, China, and Peru, and of guano origin in Nauru, Christmas Island and Ocean Island.

Table 2. Phosphate deposits

Location	Size Gtonnes	Type of rock
Phosphoria deposit, Western USA[a]	100	Sedimentary
Oulab-Abdoun, Morocco	50	Sedimentary
Spanish Sahara[b]	25	Sedimentary
North Carolina	10	Sedimentary
Tunisia	6	Sedimentary
Australia[c]	3	Sedimentary
Algeria	3	Sedimentary
Bone Valley, Florida, USA	2	Secondary sedimentary[d]
Kola Peninsula, USSR	1.5	Igneous
Vyata-Kama, USSR	1.5	Sedimentary
Kara Tau, S. Kazakhstan, USSR	1.5	Sedimentary

[a] Not economic to mine at present
[b] Now under Moroccan control
[c] Only discovered in 1966
[d] Produced by weathering of phosphoric limestone

The phosphate rock of guano origin is found in the Pacific and derives, as its name implies, from bird droppings. Where fish is plentiful sea birds flourish and over the centuries sizeable deposits of guano accumulate. Birds are very inefficient at metabolising phosphate, yet take in large amounts of phosphate in their diet. The excreted phosphate is leached by rainwater and if the underlying rock is calcium carbonate – as it was on these Pacific Islands – then a chemical reaction occurs in which calcium phosphate is formed. Although small in size relative to the deposits in Table 2 (the largest, on Christmas Island, is only 0.2 Gtonnes) the guano deposits are economically profitable to mine.

The phosphate deposits of marine origin, such as the Florida and Morocco ones, cover vast areas. The Oulab-Abdoun basin is 5,000 km^2 and the Florida deposit 6,400 km^2. The deposits are several metres thick. The deposits around the Mediterranean region (Spanish Sahara, Morocco, Tunisia, Algeria, Egypt, Israel, Jordan, and Syria) were laid down in the late Cretaceous-Eocene age, i.e. about 60,000,000 yr ago. The USA phosphoria deposit is around 200,000,000 yr old. Most of the sedimentary deposits formed in regions of upwelling ocean at a time when sedimentation, other than apatite, was slight. The process of converting sediment to rock is called diagenesis and takes millions of years [14].

The Land-based Phosphate Biocycle

Soil Phosphate

Soil formation is known technically as pedogenesis and is an on-going process that is partly mechanical (weathering) and partly as a consequence of supporting living things. Pedogenesis slowly releases the inorganic phosphate from the apatite fractions of the original rocks to the water in the soil, when it becomes available for plant nutrition. However, this source of phosphate is released too slowly to represent a significant contribution to the phosphate required for a particular growing season, although it represents of course the original source of all the phosphate in the soil.

The water in soil generally has a concentration of 10^{-5}–$10^{-6}M$ in phosphate, but may be as low as $10^{-8}M$ in some tropical soils. The top soil, down to 30 cm depth, has around 6 cm of soil water and at $10^{-5}M$ this represents 0.04 kg ha^{-1} of immediately available phosphorus, less than 1% of that required to sustain normal plant growth for a season. Moreover, when plants grow in such soil it has been observed that the concentration of phosphate in the soil water remains the same [15]. In other words there are insoluble sources or "pools" of phosphate in the soil that can supply the soil water with phosphate to maintain an equilibrium concentration.

Previously it was stated that there were no volatile compounds of phosphorus in the environment, but this does not exclude some phosphate being supplied to the soil in rainwater, which contains 10–100 μg P l^{-1} [16]. The total atmospheric deposition of phosphorus has been estimated at between 6–13 Mtonnes per year coming from a variety of sources such as dust, sea-spray [8], meteorite burn-up [10], forest fires, volcanoes and coal-burning [17]. Though the amount supplied in rainwater is tiny it is thought to be sufficient to compensate for the loss of phosphate from soil by leaching. In certain areas the influx of phosphate from the atmosphere has been measured throughout the year [18].

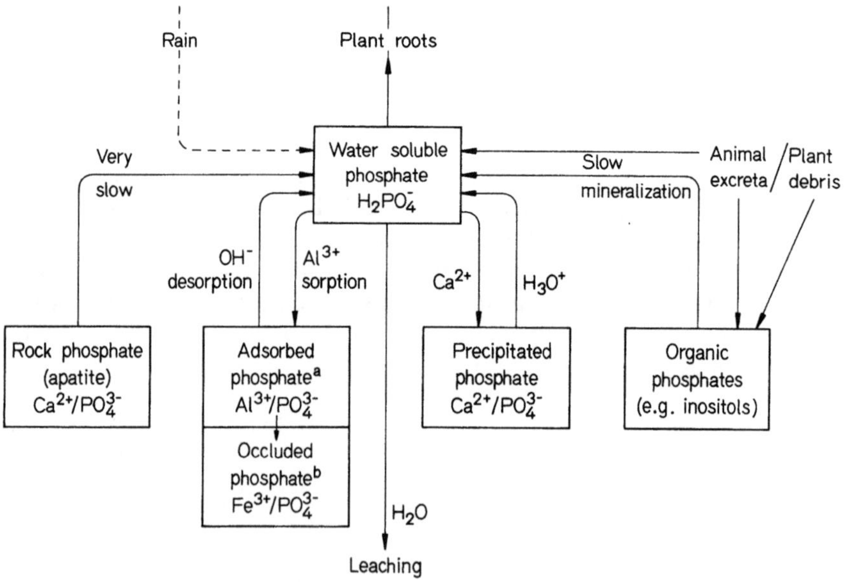

Fig. 4. Phosphate in the soil
[a] Extractable with NH_4F and dilute NaOH solution
[b] Extractable only with concentrated NaOH solution

In addition to the supply of phosphate from the original rocks and rainwater, it is possible to identify three "pools" of phosphate that can supply it to the soil water and these are shown in Fig. 4. We can describe them as either non-equilibrium pools, such as the original apatite and the organic phosphate, from which phosphate is slowly released, or equilibrium pools such as the precipitated and adsorbed phosphate pools. These labile pools can release phosphate to soil water or remove phosphate from soil water as conditions permit.

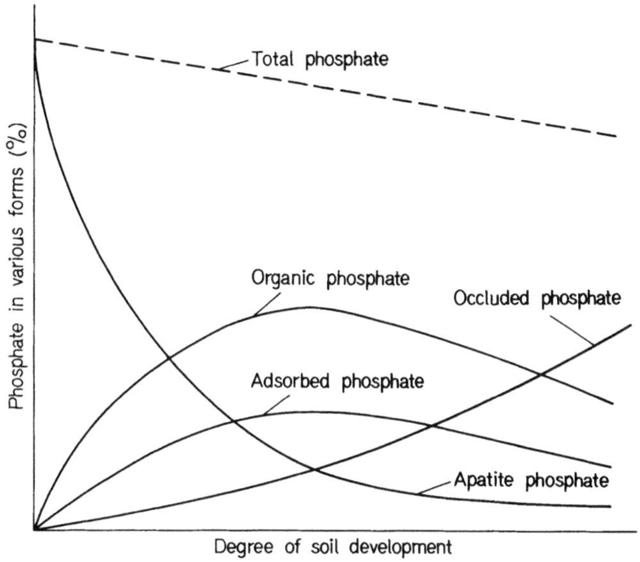

Fig. 5. Phosphates in the soil vary with soil development

As pedogenesis continues, the relative sizes of the sources and pools of phosphate change as shown in Fig. 5 [19]. The build-up of organic phosphates in the soil is the most dramatic change and as time progresses this becomes the chief reservoir of reserve phosphate in the soil. In tropical soils it may be the only one. The reasons for this are their insolubility and chemical stability. Being the salts or metal complexes of phosphate esters they release their phosphate by hydrolysis, but only very slowly [Eq. (1)]. Phosphate esters can have.

$$(RO)PO_3M + H_2O \longrightarrow ROH + HPO_4^{2-} + M^{2+} \quad (1)$$

half lives of hydrolysis of hundreds of years. The process can be greatly speeded up by the action of phosphatase enzymes in the soil whose job is to facilitate reaction (1) by catalysing it.

Several components of the organic phosphate fraction have been identified [20]. These are (i) inositol phosphates, (ii) nucleotides, (iii) phospholipids, and (iv) others.

(i) *Inositol phosphates* can account for half the total of organic phosphate in the soil. In plants and their debris the main component is *myo*-inositol hexaphosphate (II). Other inositol

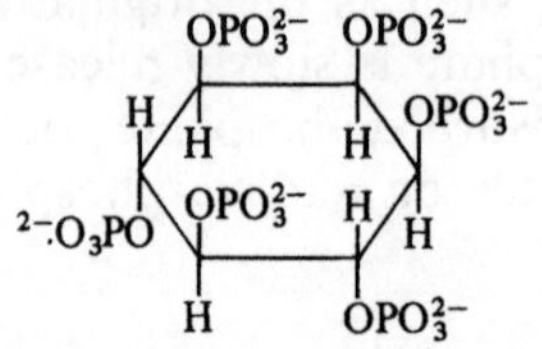

(II)

phosphates have been found in soil [21] and these come from microbes. The inositol phosphates are present as their magnesium, calcium, iron and aluminium salts which are very insoluble. Under acid conditions the rate of hydrolysis is increased and inositol hexaphosphate releases its phosphate stepwise and in soil the penta-, tetra-, tri-, di- and monophosphates have been detected.

(ii) *Nucleotides* derived from DNA are found in soil but these represent only about 2% of the total organic phosphate being relatively easily hydrolysed. They are mainly microbial in origin rather than from plants or animals.

(iii) The *phospholipids* are a motley collection of cell membrane residues etc. and account for about 2% of the organic phosphate [22]. A few have been identified such as lecithin (Fig. 1) and phosphatidyl serine (III).

CH_2OCOR
|
$CHOCOR'$ (III)
|
CH_2-O-PO_2H-$CH_2CH(CO_2^-)\ NH_3^+$

(iv) A large fraction of the organic phosphate in soils remains to be identified. Ribitol and glycerol phosphate have been shown to be present in small concentrations [23].

The two other "pools" of phosphate, precipitated and adsorbed, are both inorganic. When $^{32}PO_4$ is added to a soil it can be shown that at first this is loosely held by the soil and all is freely available for plant use. After a month, however, only 70% can be extracted by plants, the remaining 30% being permantly held by the soil. This latter percentage increases month by month until after a few years all the phosphate is tightly bound to the soil.

To explain this behaviour it is necessary to consider the three metals with which phosphate has a special relationship in soils: calcium, iron and aluminium. The first of these will form an insoluble calcium phosphate but this does not remove the phosphate permanently and must exist in equilibrium with a certain concentration of phosphate dissolved in the soil solution.

Aluminium hydroxide on the soil particle surface is the principal holding agent for inorganic phosphate which acts as a ligand to it in the $H_2PO_4^-$ form [24]. It holds the HPO_4^{2-} ion less securely and for this reason an increase in soil pH will release phosphate bound to aluminium. Thus liming soil (adding Ca $[OH]_2$) has the effect of releasing more aluminium bound phosphate than it precipitates as calcium phosphate. How the amounts of $H_2PO_4^-$ and HPO_4^{2-} vary with pH are shown in Table 3.

Table 3. Composition (%) of inorganic phosphate in water at various pHs[a]

pH =	5	6	7	8	9
H_3PO_4			Negligible		
$H_2PO_4^-$	99	94	61	14	2
HPO_4^{2-}	1	6	39	86	98
PO_4^{3-}			Negligible		

[a] Bates, R. G., Acree, S. F.: J. Res. Nat. Bur. Stand. *30*, 129 (1943)

Phosphate attached to aluminium is labile and crystals of definite composition may form, such as taranakite, $K_3H_6Al_5(PO_4)_8 \cdot 18H_2O$. Phosphate attached to iron is immobile and described as occluded. Ferric hydroxide sites on the soil surface can thus hold phosphate permanently and the gradual build up of occluded phosphate is due to its migration to and retention by these sites. Phosphate attached to iron is not labile nor does it crystallise [25].

The Land-based Biocycle

Roots are very good at scavenging phosphate from the soil water and it then begins a journey through the biocycle shown in Fig. 6. The ratio of plants to animals is the determining factor in the rate of phosphate flow through the system. In systems with little animal activity, such as woodland, the uptake of

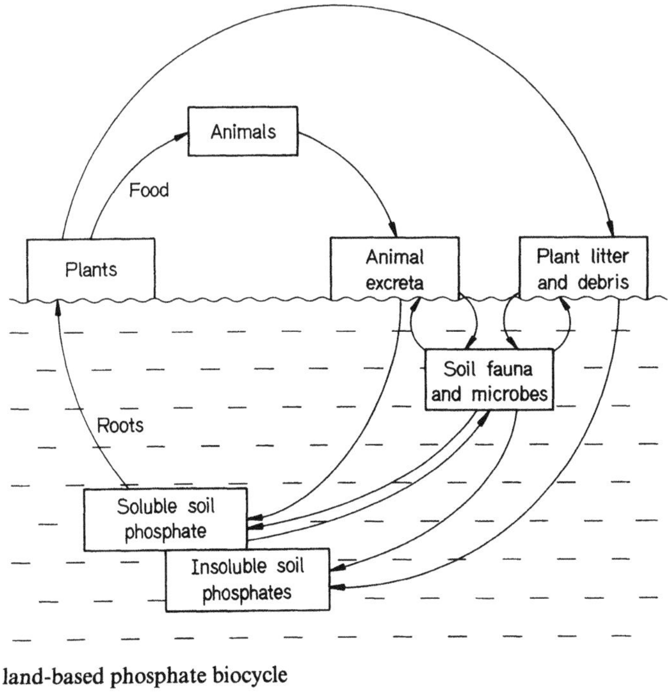

Fig. 6. The land-based phosphate biocycle

inorganic phosphate from the soil each year may be only half that of grassland supporting a grazing flock of animals [26].

How phosphate from the soil reaches the cells of a plant leaf is very complex and not entirely understood as yet. Basically there are three chemical steps: (i) esterification when the phosphate is absorbed by the roots; (ii) de-esterification to inorganic phosphate when it is transferred to the xylem for movement up the stem; and (iii) re-esterification on being absorbed by the leaf cells.

Within 5 s of a root being immersed in 10^{-5}M KH_2PO_4 solution phosphate is absorbed and incorporated into a variety of organic forms such as nucleoside di- and tri-phosphates, hexose phosphates etc. [27]. This has been demonstrated using $^{32}PO_4^{3-}$.

To move around the plant the phosphate needs to be soluble $H_2PO_4^-$ and hence the necessity of a dephosphorylation step, which itself must be fairly rapid, since within 10 min labelled phosphate from solution can pass through the roots and be moving up the stem of young barley plants. On arrival at leaf cells the phosphate is again esterified by both oxidative and photosynthetic phosphorylation processes.

In the cell only 1% of the phosphate is inorganic phosphate – the rest is organic [28]. Cells can absorb excess phosphate even though the concentration of important metabolic compounds, such as ATP, remain virtually unchanged. The surplus phosphate is transferred to vacuoles for storage in the forms of precipitated phosphates, polyphosphates with $n \cong 200$ [29], and as inositol hexaphosphate which is the chief form. Enzymes have been discovered that can transfer phosphate directly from inositol hexaphosphate to ADP [30].

At grass roots level the transfer of phosphate from soil water to root cell is still somewhat obscure. Microorganisms living on the root surface and within the cells are involved. The widespread root mycorrhizas (fungi) live in beneficial symbiosis with roots and this species is able to pass phosphate from its own store to root cells.

Other soil fungi and bacteria are able to utilise insoluble phosphates such as apatites, thus bringing this phosphate into circulation.

Organic phosphates in the soil can be broken down by phosphatase enzymes which may originate in the root itself or be released by the associated microorganisms. This has been demonstrated for maize roots [31].

Recently another factor has emerged, which influences the uptake of phosphate by roots, and that is the need for the non-metal boron in the process [32].

Whatever the mechanism, roots can transfer phosphate against a concentration gradient from soil water to plant. By so doing they create in turn a concentration gradient for the precipitated and adsorbed phosphate in the soil, causing them to release phosphate to the soil solution. In addition roots can cause release of phosphate from the organic phosphate "pool" by stimulating phosphatase enzyme development, and this they do in soils of low inorganic content.

The role of soil fauna and microbes is a key part of the phosphate cycle.

These species of decomposer organisms require phosphate for their own needs and a simulation model analysis has shown that they take up 4–5 times more soil phosphate than do the plants living in the same soil [33]. They too are responsible for adding to the organic phosphate reservoir, and it has been estimated that 75% of organic phosphate in the soil comes from microbial sources as shown by ^{14}C and ^{32}P labelling experiments [34]. This can also be seen in the analysis of the inositol phosphates in the soil. Plants produce only the *myo* isomer, whereas soils contain sizeable fractions of the other isomers which can only have come from microbial sources [23].

Decomposer organisms, working on animal excreta which is a rich source of soluble inorganic phosphate, may serve to convert some of this to organic forms thereby adding it to the organic phosphate pool.

Figure 6 applies to an undisturbed ecosystem, and does not apply to land from which there is large scale crop removal. How much phosphate is removed depends very much on the crop, Table 4, but human vegetable crops

Table 4. Phosphate in human crops[a]

Crop	kg P removed $ha^{-1}\ yr^{-1}$	Crop	Global yield Mtonnes	Global P removed Mtonnes
Beans	11	Milk	424	0.37
Turnips	11	Wheat	360	1.23
Potatoes	10.5	Rice	323	1.00
Wheat	7	Potatoes	293	0.16
Trees[b]	0.5	Maize	293	0.32
Sheep[c]	1			

[a] Sawyer, C. N.: Ref. [1], p. 633
[b] Depends on type of tree. Whole tree harvesting, which includes root removal, may remove 1 kg P $ha^{-1}year^{-1}$, assuming a 30 yr growth period
[c] Wool and meat

and milk yields remove amounts of phosphate that must be replaced by fertilizers. On a global scale the top five human food crops remove 10 Mtonnes of phosphate per year from the land. To replace this requires several times as much calcium phosphate fertilizer since only about 25% of fertilizer phosphate is useable by plants – the rest is lost to the occluded soil fraction [35].

In certain areas of the globe the soils of whole continents, such as Australia, may be depleted of phosphate through millions of years exposure to weathering. The inorganic apatite in such soils is almost exhausted and the labile pools dried up[2]. Tropical soils may have all their phosphate in the

2 In certain Australian soils an excessive amount of phosphate fertilizer has to be added to refill the phosphate "pools" in the soil before there is any noticeable effect on crop yield. Amounts of 450 kg phosphate ha^{-1} were required on some soils before phosphate ceased to be the limiting factor to plant growth. (Anderson A.J., Maclachlen, K.D.: Aust. J. Agric. Res. *2*, 377 (1951). Piper, C.S., de Vries, M.P.C.: ibid *15*, 234 (1964)

organic reservoir and this is in the uppermost layer of the topsoil. Enough phosphate can be released by burning the forest to support crops for a few years but this cannot be done on a large scale without exposing the soil to irreversible weathing.

New hope of increased yields from tropical soils has come with the introduction of the leguminous tree *Leucaena leucocephala*. Because its main root is a deep tap root it reaches for water and phosphate at depths well below the exhausted top soil. Moreover it can fix nitrogen via *Rhizobium* bacteria in its root nodules. The tree grows rapidly and can be harvested as fuel, timber, animal fodder, fertilizer, etc. Once grown to shield other crops from wind or sun, it is now being developed in its own right as the ideal tropical soil crop. It can yield 20 tonnes $ha^{-1}year^{-1}$ of organic matter [36].

Table 5. Leaching of phosphate from the land[a]

Type of land	Global area/m^2	Phosphate loss mg P $m^{-2}yr^{-1}$
Agricultural	1.47×10^{13}	156
Meadow	2.99×10^{13}	34
Forest	4.03×10^{13}	105
Other	4.90×10^{13}	< 50

[a] See Ref. [7]

The loss of phosphate from the land by leaching is relatively small: Table 5. The downward loss by leaching is probably compensated for by the phosphate arriving in rainwater. Even heavily fertilized agricultural land loses very little phosphate each year [37], although flooding may remove quite a lot.

The routes for conveying phosphate from the land cycle to the water based cycles are the rivers. It has been estimated [7] that the world's rivers carry 17 Mtonnes P to the seas each year, half from natural processes, half due to Man's activities. About 3 Mtonnes is conveyed as soluble phosphate and the rest is held in sediment particles.

The Water-based Phosphate Biocycle

This is shown in Fig. 7. There is a rapid throughput of phosphorus in the water-based system [38]. Two factors determine the extent of the biomass in most lakes and seas: (i) the availability of phosphate [39–41]; and (ii) the amount of solar energy. Again we speak of a phosphate cycle when we really mean a downward spiral, since the twin processes of precipitation and sedimentation are responsible for the rapid loss of phosphate from the aqueous phase to the solid phase at the bottom.

When $^{32}PO_4^{3-}$ was added to an aquatic system the results showed its uptake into the biomass was extremely rapid [42]. Within one minute of its addition

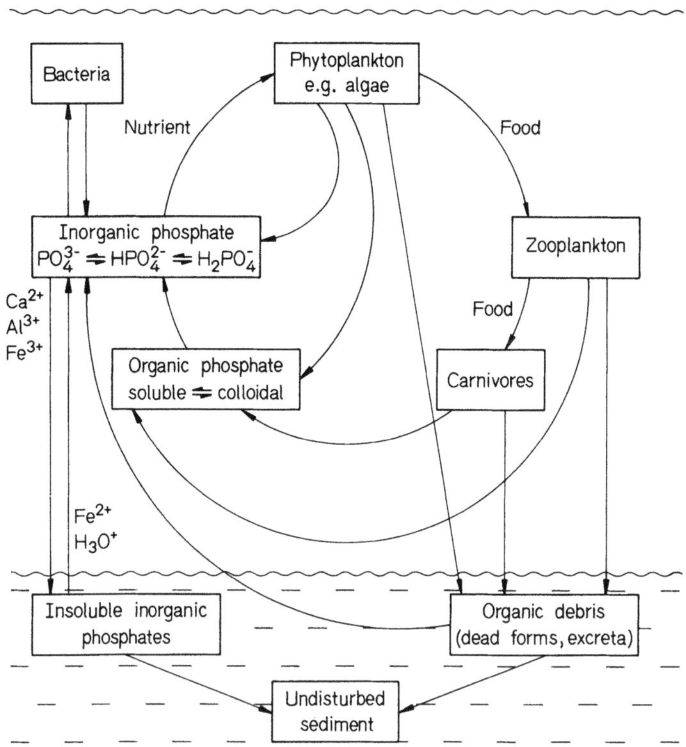

Fig. 7. The water-based phosphate biocycle

to lake water 50% of the added phosphate had been absorbed into organisms such as bacteria and algae. Within three minutes 80% was absorbed. Algae, like land plants, can take in more phosphate than they actually need and can store enough surplus phosphate to support them through three cell divisions without outside help.

Once absorbed, the labelled phosphate can be followed as it progresses through the various species [43]. Studies show how turnover of phosphate is rapid with algae and becomes slower as it moves along the food chain, and enters the plants which grow at the bottom last of all. Using $^{32}PO_4^{3-}$ it has been shown that there is a daily loss of 1–2% of the phosphate in the aqueous phase to the sediment. This observation has been demonstrated not only in the microenvironment of an aquarium [43] but for a natural lake in North America [44].

Two parts of the aquatic system have particular relevence to the phosphate cycle – the algae and the sediment.

The Algae

Algae are best able to utilise the inorganic phosphate in water, and given the right combination of a plentiful supply and enough sunlight they can multiply to such an extent as to colour the water red or green, mostly the latter. When they do this they are said to bloom and blooming may be a twice yearly natural phenomenon in certain lakes.

In Spring it is sunlight that promotes such a bloom, the algae feeding on the accumulated phosphate brought to the lake by the rivers throughout the winter months. In Autumn it is phosphate from the deeper layers, called the hypolimnion, that suddenly come to the surface, that enables a second blooming to happen[3].

Perpetual blooming of the algae is caused if an endless supply of phosphate-rich river water feeds them. The result is eutrophication – a word that literally means good feeding but has come to mean excess breeding of the algae, since this species alone can exploit the inorganic phosphate. By sheer numbers they blot out the sunlight at the top and their decaying debris consumes all the oxygen at the bottom. Conditions become intolerable for other aquatic species and they leave or die. Several studies of this widely publicised effect have been made [45–47].

Blooms may look substantial but the actual biomass may be quite small [48]. It is their unpleasantness to human beings, who want to use the water for drinking, swimming or fishing, that causes blooms to be condemned as a form of environmental pollution.

Several lakes have suffered this fate. Those most at risk are ones that are fed by rivers passing through large cities whose sewage works discharge their waste water into them. Until recently phosphate removal from sewage was not considered necessary, but with the entrophication of some of the Great Lakes of North America the problem demanded a solution [49]. Part of the blame was laid on sodium triphosphate used in detergents from 1945 onwards. The growth in the use of these seemed linked to the growth in the eutrophication of the lakes in the 1950's and 1960's.

A detergent phosphate ban came into force in many States of the USA and in Canada in the 1970's, but more important was the introduction of a tertiary phase of sewage treatment. This was the use of aluminium salts to precipitate the phosphate before the water was allowed to flow into the rivers. Given time, algal blooms disappear and several lakes are now slowly returning to a more balanced state. The dramatic reclamation of Lake Washington in the mid-1960's showed that reversal of eutrophication was possible [50]. Algae cannot grow and multiply if the concentration falls below a certain value, which for single cell species has been measured at less than 0.06 μg P per 10^6 cells [51]. However, algae can absorb up to 0.4 μg P per 10^6 cells and this is sufficient for 3 cell divisions without the need to take in outside phosphate.

Many years ago it was observed that the concentration of inorganic phosphate in a body of water suffering an algal bloom could be extremely low.

3 In Summer the colder layers at the bottom of a lake pick up phosphate released from the sediment, but because of the temperature-density stratification of the lake these lower layers are held at the bottom by the warmer, less dense surface layers. Only when the surface waters cool in Autumn can the lower layers well up to the surface bringing their phosphate

This seemed to be at odds with the concept of phosphate as the limiting factor to growth. Concentrations of as little as ca. 5×10^{-6}M were found sufficient to sustain a bloom: a paradox of low concentration of limiting factor supporting a high population. However it is not the concentration of inorganic phosphate which is important – it is the total phosphate in the system that matters.

Phosphate in algae can be taken up and released rapidly. Within hours of algae dying their phosphate is released again to the water. Within a day 90% of the phosphate from within the cells has been released [52]. Nor does this have to be inorganic phosphate for it to be reused. Other algae can utilise soluble organic phosphates such as glucose phosphates and phospholipids [53]. Indeed algae and their associated microbes are well adapted to scavenge phosphate from almost all forms including polyphosphates, which they can break down with the help of an enzyme in their cell walls.

Some phosphate is lost to the bottom of a lake as algal debris, but a lot is used as food since algae are the producer organisms of a lake, converting inorganic nutrients such as phosphate, nitrogen and carbon dioxide into chemically more complex molecules that can sustain the rest of the biomass. Their role in the aquatic environment is the same as that of plants on land.

Sediment

The final resting place for phosphate is at the bottom of a lake. Here it may remain for millions of years, once the layer of sediment has reached a certain thickness and the phosphate is permanently trapped. Phosphate in the top layer of sediment can return to the waters of the lake by various means: decomposer organisms, enzymes, worms and other bottom creatures may assist in this. Certain forms of organic phosphate can form stable complexes with metals and this phosphate is not released [54].

Feeding on the phytoplankton are the zooplankton and these in turn are food for the carvinores such as fish. The debris and faecal pellets from these species are responsible for the downward movement of phosphate to the sediment. Some species of zooplankton excrete an amount of phosphate equivalent to 100% of their total body phosphate per day. About half the excreted phosphate is inorganic, the rest organic. Although most reaches the bottom a lot is broken up into colloidal particles before then and its phosphates released and reused by algae.

When the water of a lake is stratified the lowest layer may become anaerobic – its oxygen depleted in degrading the organic debris falling from the upper layers. The lower layer may also become slightly acidic. Both these conditions of a reducing and acidic solution will facilitate the return of phosphate from the sediment back into solution. Phosphate which is held in the sediment as insoluble $FePO_4$ will be reduced to soluble $Fe_3(PO_4)_2$, and insoluble $CaHPO_4$ will be acidified to soluble $Ca(H_2PO_4)_2$. Under these conditions the release of phosphate from the sediment can sometimes exceed the downward flow of phosphate to it [55, 56]. Overall the mechanisms for the return of phosphate to solution can only convert a fraction of it to solution[3]. There is a net loss of phosphate to the bottom of a lake.

The top layer of sediment of a eutrophic lake may hold enough phosphate to keep the lake in a eutrophic state despite steps taken to curb the inflow of phosphate to the lake. Thus Lake Trummen in Sweden did not respond to controls of the phosphate inflow and it was necessary to dredge the top layer of sediment before the lake recovered [57].

The fate of all lakes is to fill up with sediment to become marshes and finally dry land. The rate at which a lake disappears depends upon the rate at which it is supplied with nutrients. Eutrophication can prematurely age a lake and Lake Erie, one of the North American lakes most polluted with algae, has been estimated to have aged 15,000 yr this century [58].

Although many rivers flow into lakes, most flow into the sea and there the fate of the phosphate depends very much on local conditions.

The Sea

In theory the sea can absorb all the phosphate poured into it simply by precipitating it as calcium phosphates. The concentration of Ca^{2+} in the sea is high (0.400 g litre$^{-1} \cong 0.01$ M), and the sea itself is alkaline (pH $\cong 8$). Both these factors would contribute to the precipitation of $Ca_5(PO_4)_3OH$ at concentrations of inorganic phosphate as low as 10^{-12} *M*.

Below 1,000 m the concentration of phosphates in the sea is ca. 2.7×10^{-6} *M* which in theory should cause precipitation. However the phosphate is not present solely as inorganic phosphate but also as organic and colloidal phosphate. Nevertheless, it is surprising that the sea can support such relatively large concentrations of phosphate.

The profile of the phosphate concentrations of the major oceans is drawn in Fig. 8, which shows how similar they are below 1,000 m. At such concentration the sea should be capable of sustaining an enormous biomass. What is lacking however is sunlight. Contrary to popular opinion the population of fish in 99.9% of the sea is very small. Half the fish in the sea live in 0.1% of its surface area [59], and these particular regions are very fertile due to the upwelling of the deeper waters rich in phosphate. Where they come to the surface they can sustain a large phytoplankton and zooplankton population upon which the fish feed.

Upwelling regions are to be found in the mid-Pacific (where a lot of the phosphate has found its way into guano rock), the Pacific coast of Peru (where islands rich in guano are found), Antarctia (where there is sufficient fish to support large colonies of penguins in the least-hospitable of climates) and Arabian seas.

The enigma of the sea is that all the sunlight is at the top and all the phosphate at the bottom [59]. The phosphate concentration at the surface is low because of the steady downward drift of organic debris. Because of the enormous pressures in the sea most of this debris is crushed and its phosphate released before it reaches the bottom. Indeed it is thought that only certain skeletal parts, such as teeth, can survive the journey to the deep ocean floor.

The total amount of phosphorus in the sea is 120,000 Mtonnes which is equivalent to 600,000 Mtonnes hydroxyapatite. This represents about 0.01%

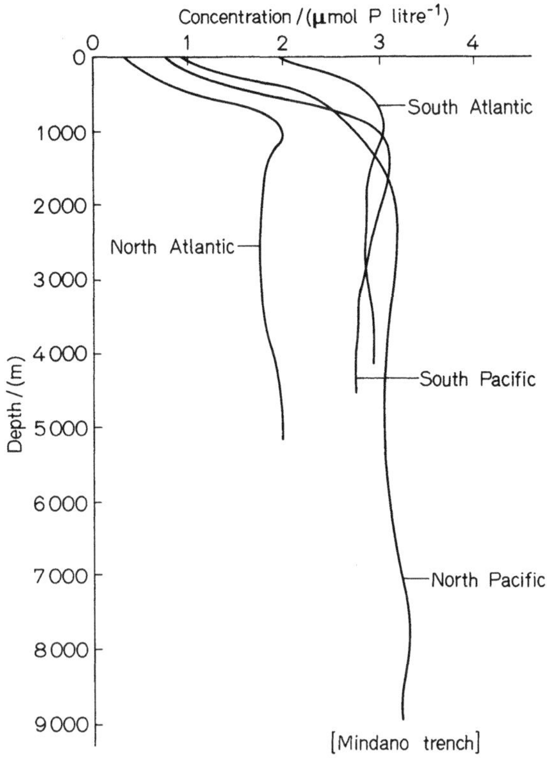

Fig. 8. Phosphate profile of the major oceans [60]

of the phosphate in the primary inorganic cycle. Rivers supply about 8 Mtonnes of naturally derived phosphorus (as opposed to that from human activity) per year which gives an average residence time of phosphate in the sea of 15,000 yr. This is relatively short for a non-volatile element. Sodium for example has a residence time of 260,000,000 yr, but unlike phosphate, there is no slow but relentless loss of sodium to the bottom.

Where the sea is shallow, such as near coasts and river estuaries, there may be algal blooming, This is caused by the phosphate rich lower waters being carried shore-wards by counter currents produced by the faster flowing river water at the surface.

Counter currents can also carry the phosphate-rich lower waters away, and this is what happens to the phosphate in the Mediterranean. This is the largest body of impoverished sea water in the world, despite its being fed by the nutrient rich Nile. Evaporation from the Mediterranean exceeds the flow of its surrounding rivers so that water flows in through the Gibraltar Straits and the Bosporus. This surface inflow creates a lower counter-flow which carries the phosphate out into the Atlantic and Black Sea. This is shown in Table 6, based on the paper of Gulbrandsen and Robertson [60].

Table 6. The phosphate of the Mediterranean and adjacent seas

Depth (m)	North Atlantic phosphate concentration[a]		Mediterranean phosphate concentration[a]		Black Sea phosphate concentration[a]
10	0.26	→	0.05	←	0.12
100	0.74		0.11		1.03
			0.34		
1000	1.14	PO_4 ←		PO_4 →	7.00
2000	1.09	out	0.32	out	7.46

[a] μ mol P litre^{-1}

The Atlantic is unaffected by the flow of phosphate from the Mediterranean but the lower depths of the Black Sea is greatly enriched. There we find phosphate concentrations three times that of the major oceans, and twenty times that of the Mediterranean at the same depth. Not that eutrophication threatens this body of water since the concentration at the surface is still low.

References

1. Griffiths, E.J. et al. (eds.): Environmental phosphorus handbook. New York: Wiley 1973
2. Phosphorus in the environment: Ciba Foundation Symposium 57 (new series). Amsterdam: Elsevier – Excerpta Medica – North Holland 1978
3. Emsley, J., Hall, C.D.: The chemistry of phosphorus. London: Harper and Row 1976
4. Williams, R.J.P.: Phosphorus biochemistry, p. 95 in [2]
5. Brinch, J.W.: World resources of phosphorus, p. 23 in [2]
6. Stumm, W.: Water Research *7*, 131 (1973)
7. Pierrou, U.: Nitrogen, phosphorus and sulphur global cycles. SCOPE Report 7. Svensson, B.H. and Söderlund, R. (eds.). Ecol. Bull. (Stockholm) *22*, 75 (1976)
8. Emigh, G.D.: Eng. & Mining *90*, 173 (1972)
9. Gulbrandsen, R.A.: Econ. Geol. *64*, 365 (1969)
10. Moore, C.B.: Phosphorus in meteorites and lunar samples. Chap. 1 in [1]
11. Peck, D.R.: The history and occurrence of phosphorus, in Mellor's Comprehensive Treatise on Inorganic and Theoretical Chemistry. Vol. VIII, Supplement III (Phosphorus), Sect. I. London: Longman 1971
12. Brown, W.E.: Solubilities of phosphates. Chap. 10 in [1]
13. Rankama, K., Sahama, T.G.: Geochemistry. Chicago: University of Chicago Press 1949
14. Broecker, W.S.: Chemical oceanography. New York, Chicago, San Francisco: Harcourt-Brace-Jovanovich Inc. 1974
15. Fried, M.: Soil Sci. *84*, 427 (1957)
16. Rigler, F.H.: Fundamentals of limnology. Ruttner F. (ed.), p. 263. Toronto: University of Toronto Press 1974
17. Bertine, K.K., Goldberg, E.D.: Science *173*, 233 (1971)
18. White, E.J., Turner, F.: J. Appl. Ecol. *7*, 441 (1970)
19. Williams, J.D.H., Walker, T.W.: Soil Sci. *107*, 213 (1969)
20. Anderson, G.: Soil Biochem. *1*, 67 (1967)

21. Cosgrave, D.J.: Aust. J. Soil Res. *1*, 203 (1963); Nature *200*, 568 (1963); Soil Sci. *102*, 42 (1960)
22. Hance, R.J., Anderson, G.: Soil Sci. *96*, 94 and 157 (1963)
23. Halstead, R.O., Anderson, G.: Canad. J. Soil Sci. *50*, 111 (1970)
24. Bromfield, S.M.: Nature *201* (1964)
25. Tandon, H.L.S., Kurtz, L.T.: Proc. Soil Sci. Soc. Amer. *32*, 799 (1968)
26. Harrison A.F.: Phosphorus cycles of forest and upland grassland ecosystems and some effects of land management practices, p. 175 in [2]
27. Loughman, B.C.: Metabolic factors and the utilization of phosphorus by plants, p. 155 in [2]
28. Loughman, B.C.: Plant Physiol. *35*, 418 (1960)
29. Tinker, P.B.: [2], p. 170
30. Biswas, S., Burman, S., Biswas, B.B.: Phytochem. (Oxford) *14*, 373 (1975)
31. Clark, R.B., Brown, J.C.: Crop Science *14*, 505 (1974)
32. Pollard, A.S., Parr, A.J., Loughman, B.C.: J. Expt. Biol. *28*, 831 (1977)
33. Cole, C.V., Innis, G.S., Stewart, J.W.B.: Ecology *58*, 1 (1977)
34. Fuller, W.H., Fox, R.H.: Soil Sci. *97*, 421 (1964)
35. Russell, E.W.: Soil Conditions and Plant Growth, 10th edn., Chap. 23, p. 555. London: Longman 1973
36. Vietmeyer, N.D.: 1979 Yearbook of Science and the Future p. 239, Encyclopaedia Britannica of Chicago 1978
37. Nicholls, K.H., MacCrimmon, H.R.: J. Env. Qual. *3*, 31 (1974)
38. Rigler, F.H.: Ecology *37*, 550 (1956)
39. Schindler, D.W.: Science (New York) *195*, 260 (1977)
40. Vollenweider, R.A.: Schweiz. Z. Hydrologie *37*, 53 (1975)
41. Dillon, P.J. Rigler, F.H.: Limonol. & Oceanog. *19*, 767 (1974)
42. Lean, D.R.S.: Science *179*, 678 (1973)
43. Whittaker, R.H., Ecological monographs *31*, 157 (1961)
44. Chamberlain, W.M.: Ph. D. thesis University of Toronto, Canada 1968. See Rigler F.H. chap. 30, p. 544 in [1]
45. Rohlich, G.A.: Eutrophication: Causes, Consequences, Correctives. Nat. Acad. Sci. Washington D.C. 1969
46. Lund, J.W.: Proc. Roy. Soc. B (Biol. Sci.) *180, 371* (1972)
47. Porter, K.S.: Nitrogen and Phosphorus: Food Production, Waste and the Environment. Ann. Arbor: Ann. Arbor Sciences 1975
48. Reynolds, C.S., Walsby, A.E.: Biol. Revs. Camb. Phil. Soc. *50*, 437 (1975)
49. Report of the International Lake Eire and Lake Ontario – St. Lawrence River Water Pollution Board. Vol. I. Queen's Printer, Ottawa, Ontario, Canada 1969
50. Edmondson, W.T.: Int. Verein für Theor. und Angew. Limnol. *18*, 284 (1972)
51. Mackereth, F.J.H.: J. Exp. Bot. *4*, 296 (1953)
52. Marshall, S.M., Orr, A.P.: J. Marine Biol. Assn. U.K. *34*, 495 (1955)
53. Hooper, F.F.: Chap. 9 in [1]
54. Jackson, J.A., Schindler, D.W.: Int. Verein. für Theor, und Angew. Limnol. *19*, 211 (1975)
55. Lean, D.R.S., Charlton, M.N.: Env. Biochem. *1*, 283 (1976)
56. Sonzogni, W.C., Larsen, D.P., Malueg, K.W.: Water Res. *11*, 461 (1977)
57. Björk, S.: Ambio *1*, 153 (1972)
58. Beeton, A.M.: Limnol. Oceanog. *10*, 240 (1965)
59. Morris, I.: Chem. in Brit. *10*, 198 (1974)
60. Gulbrandsen, R.A., Robertson, C.E.: Chap. 5 in [1]

Metal Cycles and Biological Methylation

*P. J. Craig**

School of Chemistry, Leicester Polytechnic
Leicester LEI 9BH, England

Biogeochemical Cycles and Methylation

Introduction

The equilibrium between the various forms, species and locations of each heavy metal is a dynamic not a static one. For this reason the importance of heavy metals in the biosphere has to be considered in the context of a cycle between these forms and locations. For many years the idea of a cycle for elements or compounds such as carbon, nitrogen or water has proved invaluable in a discussion of their role in the biosphere. In more recent years the idea of a cyclic system for the various heavy metals has also proved important in an assessment of their environmental importance and this approach will be taken in this section. In most cases the details of heavy metal cycles are less well known than are the cycles for the elements oxygen, carbon, nitrogen, sulfur and phosphorus and for water which are considered in other sections of this volume. Quite recently much information has been accumulated for certain metals (e.g. mercury) and the very use of the cyclic approach in some other cases points out areas where our knowledge is inadequate and calls for more research.

The approach taken will be to compare and contrast the natural pre-man mobilization of various heavy metal elements to the present day mobilization which includes anthropogenic activity. It will be seen that even in cases where man's contribution to heavy metal mobilization is small compared to the natural flux on a global scale, in local environments it can be in considerable excess of natural movements.The heavy metals that appear of special interest in view of their toxicity and industrial importance are

---

* This chapter was written while the author was on sabbatical leave at the Gray Freshwater Biological Institute, University of Minnesota, Navarre, Minnesota 55392, USA

mercury,
lead,
tin and
cadmium.

Roughly speaking the above order parallels our knowledge of the environmentally important properties of these elements. Other heavy metals will also be considered (e.g. thallium, platinum, palladium and tellurium). Transformations of other less toxic or less well researched metals (e.g. iron, manganese, molybdenum, and copper) will be discussed briefly and suitable references to papers and reviews will be given. Environmental cycles for the metalloidal elements arsenic and selenium have been well researched in recent years. The important conclusions will be discussed and references to other reviews will be given.

In the case of mercury one particular stage in the biogeochemical cycle has been discovered that has been of much general importance. This is the pathway

$$Hg^{2+} \rightarrow CH_3Hg^+,$$

i.e. biomethylation

In view of the well known importance of this step for mercury, emphasis will be placed on the possible occurrence of biomethylation for the other elements. In addition processes responsible for demethylation of methyl metal species will also be covered.

Direct steady state determinations of heavy metal concentrations in various parts of the biota, water, sediment, rocks, etc. will not be reported here in detail except when they have increased importantly due to man. Such determinations are innumerable in recent years and tell us little about the rates of cycling of heavy metals in the geosphere. Steady state or ambient concentration data can be found in many of the references in the reviews cited here. This discussion will concentrate mainly on cycling rates and quantities, reservoirs and residence times, and on chemical speciation.

Biological redox conversions of metals have been reviewed recently [1] and will be only briefly covered here. Similarly, toxicity properties of metals have been much discussed in recent years [2,3] and will be covered only incidentally.

Biogeochemical Cycles

A number of materials are naturally circulated through the atmosphere, lithosphere, hydrosphere and the biosphere in balanced integrated systems. Each of these entities may be considered as a reservoir or sink for the material at a particular time. The concentration and distribution of the material within the reservoir, the rate of transfer between reservoirs and the type and rates of chemical modification within a reservoir constitute the biogeochemical cycle for that element. Where possible mass balance models should be used to describe quantitatively the dynamics of these cycles. Feedback loops which modify or exagerate the importance of an input signal may also occur in the cycle (i.e. negative or positive feedback).

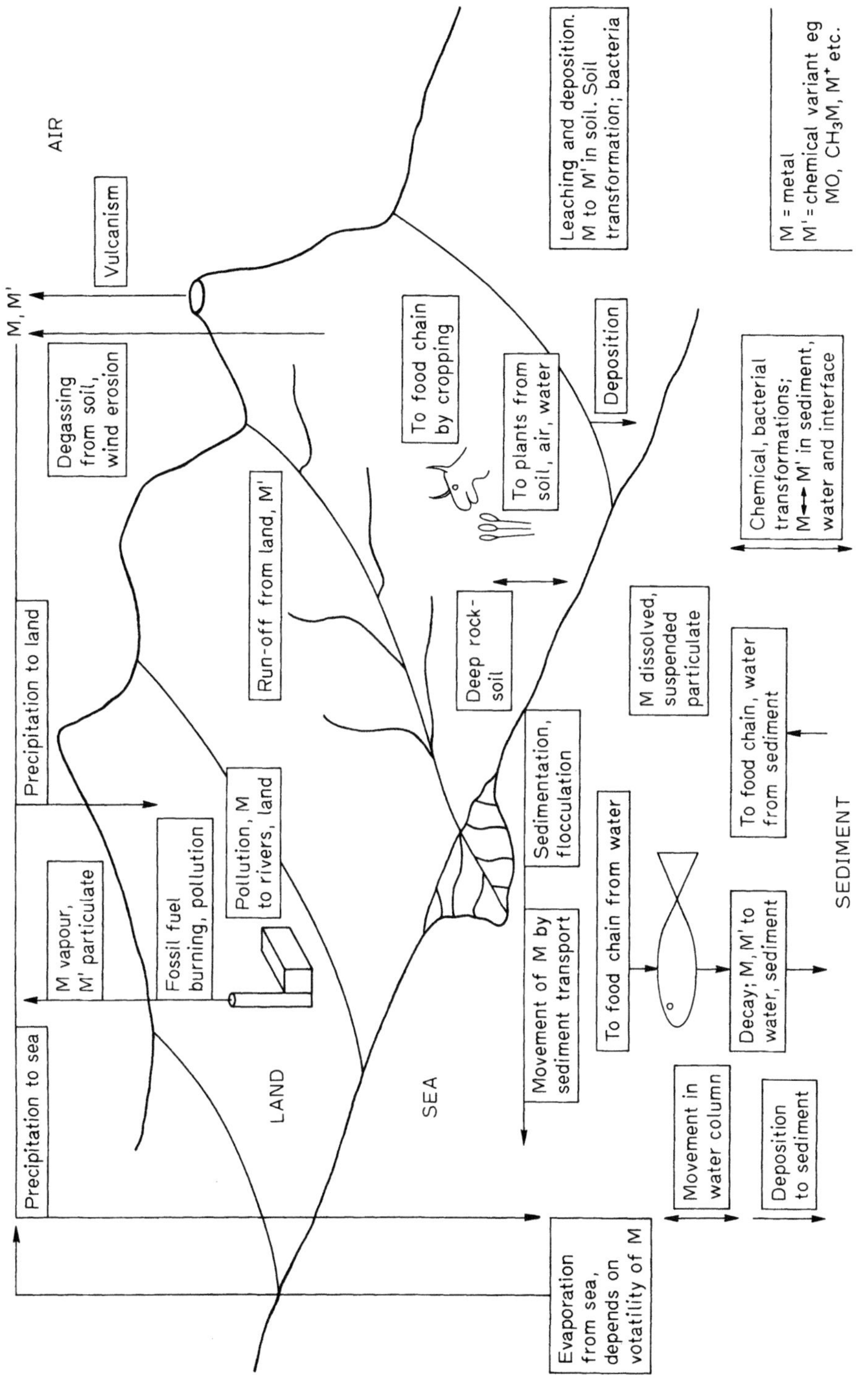

Fig. 1. Biogeochemical cycle for metal M

In order to properly describe the cycling of a metal in nature it is necessary to know the chemical speciation of the metal in the differing phases of the cycle. The state of the metal regarding charge and chemical combination needs to be known for the vapor, the suspended particulate, the flocculated, the solution and the lithospheric phases. Sometimes the chemical form of the element may change considerably during its encounter with the biota – biological methylation being an extreme example.

In addition, the oxidation-reduction characteristics (Eo) and often the Eh, pH, and pCl of the immediate environment of the metal are of great importance. The element will be distributed between the various phases as a result of its volatility, diffusibility, solubility, the ionic strength of surrounding aqueous media, its adsorption-desorption chemistry, various ion exchange processes, and it's precipitation characteristics. In addition to describing the various environmental reservoirs then, knowledge of the rates of exchanges between them is perhaps even more important.

The properties of a metal that determine its importance in the biogeochemical cycle are discussed in detail in a later section.

A generalized biogeochemical cycling system is shown in Fig. 1.

Methylcobalamin and Biomethylation Mechanism

In view of the intimate involvement of mercury in our knowledge of metal biomethylation, this section will also constitute the mechanistic discussion on mercury methylation. It should be read in conjunction with the section on mercury.

Three observations originated the discovery that inorganic mercury could be methylated under environmental conditions. First it was realized that mercury present in fish was present mostly or entirely in the methyl form [4]. Then it was discovered that inorganic mercury added to non-sterilized aquarium sediments generated methyl mercury detectable in the aquarium sediments [5]. Also it was found that methyl cobalamin containing methanogenic bacteria could methylate inorganic mercury in sediments [6]. There is some direct evidence that methyl cobalamin is the in vivo methylating agent in fish [7].

The role of methyl cobalamin (CH_3CoB_{12} – see Fig. 2) as the biological source of methyl groups for metal methylation has been central to discussions on this topic. If it is assumed that methylation takes place essentially to an oxidized form of the metal (M^{n+}), then biological methylating agents transferring methyl groups as, for example, the carbonium ion, CH_3^+ would be unlikely to play a role in metal methylation. Methylation by CH_3^- transfer seems much more likely.

Of the three known biological methylating species only methyl cobalamin can transfer its methyl group as CH_3^-. The others, s-adenosyl-methionine and N^5-methyltetrahydrofolate derivatives transfer the group as CH_3^+. Not surprisingly then, debate has focused strongly on methyl cobalamin as the responsible agent for metal methylation. Recent work on cobalamin species has demonstrated a remarkable versatility in the number of ways the methyl

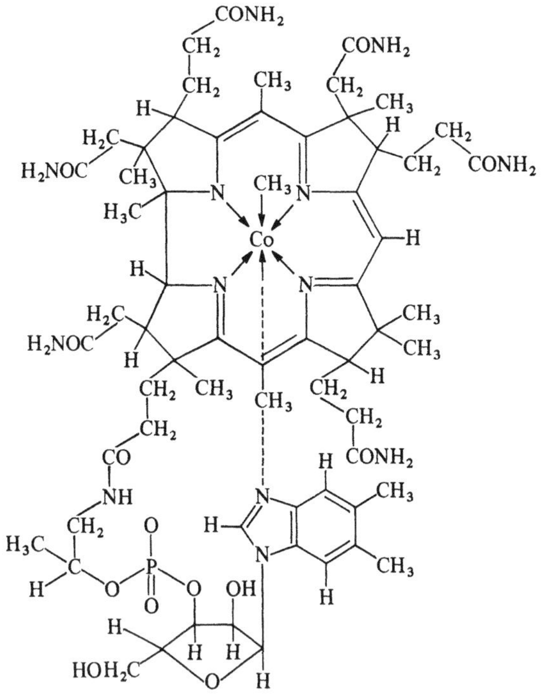

Fig. 2. The structure of methyl cobalamin, CH_3CoB_{12} (distribution of charges not shown)

group may be transferred [8] and it is now too simplistic to assume the carbanion route is the only one, even perhaps in the case of mercury.

The role of cobalamin as a methyl carbanion doner to an oxidized state of mercury and various other metals has been exemplified by Wood and co-workers in a number of publications [8–17]. Methylation of mercury (II) compounds e.g.

$$CH_3CoB_{12} + Hg^{2+} \rightarrow H_2OCoB_{12}^{+} + CH_3Hg^{+}$$

in aqueous solution under abiotic conditions by CH_3B_{12}, thereby demonstrating the feasibility of the role of CH_3B_{12} in nature, has been studied by a number of workers, who have in general agreed that methylation under these conditions occurs by electrophilic attack of Hg^{2+} and formal transfer of CH_3^- from CH_3B_{12} [18–23]. There is still uncertainty about whether CH_3Hg^+ or $(CH_3)_2Hg$ is the dominant environmental product [18] [analytical methods used usually convert $(CH_3)_2Hg$ to CH_3Hg^+]. The second methyl group can be added in a further and analogous methylation step. However in abiotic model systems using CH_3B_{12} the monomethyl product is generated at a rate 6,000 times faster than the $(CH_3)_2Hg$ product [19]. In some cases there is no evidence for $(CH_3)_2Hg$ formation in these model systems [22]. Although in model studies the reaction is not catalytic, under natural conditions the cobalamin species is remethylated and can subsequently donate another methyl group to a new mercury species i.e. the process is enzymatic [24]. No attempts have been made yet to render a model abiotic system catalytic.

Interestingly it has been suggested that mercury methylation can be a transmethylation process by methyl tin derivatives [25], i.e.

$$(CH_3)_nSn^{4-n} + Hg^{2+} \rightarrow CH_3Hg^+ + (CH_3)_{n-1}Sn^{5-n}.$$

This idea will be discussed further in the section on tin. Such transmethylation reactions are common in abiotic organometallic systems. In the above case the involvement of cobalamin is uncertain.

Although the discussion above is chiefly concerned with transfer of CH_3^-, for certain metallic elements transfer is more likely to take place by a free radical mechanism. The route may be controlled by the Standard Electrode Potential (E^0) of the metal; metals that are easily reduced (accept electrons) tend to react with electron rich CH_3^- groups as M^{n+}. This route seems to occur when $E^0 > 0.8$ V (e.g. Hg/Hg^{2+}). Metals which are not so easily reduced tend to react with methyl radicals in a formal oxidative addition reaction with an already reduced state of the metal [5]. For this type of reaction E^0 is usually <0.6 V. This route is discussed further for each relevant metal, but formally we can write for the example of a M_2^+/M_3^+ couple of <0.6:

$$CH_3Co^{III}B_{12} + M^{II} \rightarrow CH_3M^{III} + Co^{II}B_{12}.$$

A detailed discussion of the various ways by which CH_3CoB_{12} may transfer methyl groups has very recently been presented [8]. These routes are shown in Fig. 3 and are discussed briefly here.

Route 1: This pathway involves a single electron oxidation of methyl cobalamin by an electron acceptor. The oxidized methyl cobalamin product is extremely labile and produces a methyl radical and aquocobalamin.

Route 2: This reaction involves heterolytic cleavage of the Co–C bond with the transfer of a carbanion to the attacking metal ion. This is the way mercury (II) is methylated, certainly in abiotic systems.

Route 3: This reaction has been described as a "Redox-Switch" mechanism. The metal ion in its lower oxidation state, first forms a complex with the corrin macrocycle, followed by a two electron transfer from the complexed metal ion to a two electron acceptor (often a higher oxidation state of the same metal). Oxidation of the complexed metal ion facilitates carbanion transfer from the cobalt to the complexed metal.

Route 4: This reaction pathway involves homolytic cleavage of the Co–C bond by free radical attack. This route appears to exist for tin methylation.

Route 5: In this mechanism the reagent transfers a single electron to methyl cobalamin. The product of this reaction is very labile, producing a methyl radical and cobalamin$^-$.

Route 6: This reaction involves heterolytic cleavage of the Co–C bond by an attacking nucleophile which displaces a carbonium ion to produce cobalamin$^-$.

Routes (2) and (4) seem to be the predominant reaction mechanisms for a number of metals and metalloids under environmental conditions. Methylco-

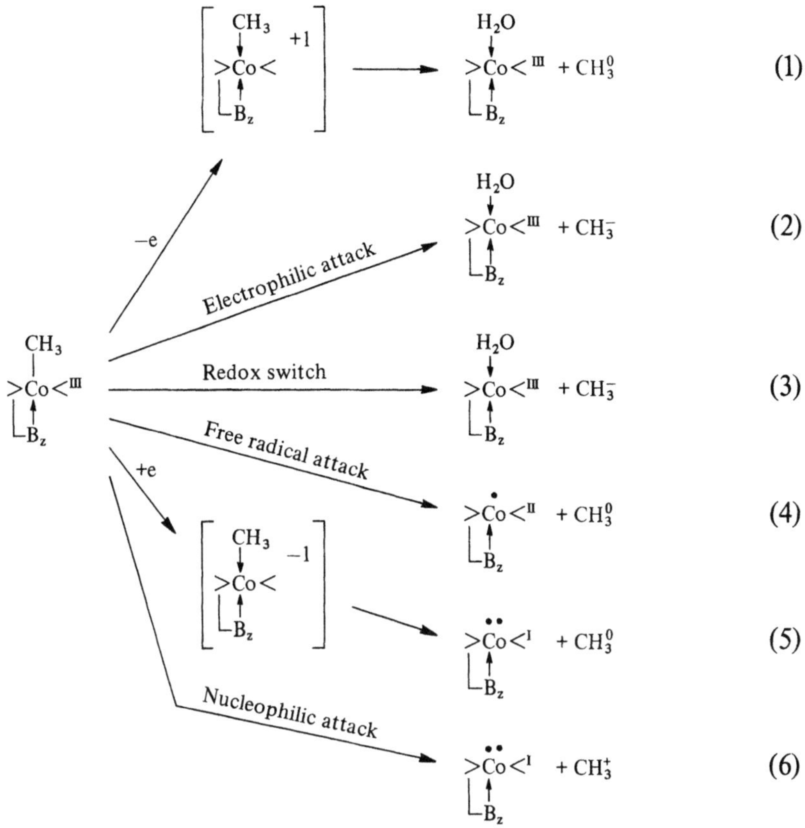

Fig. 3. Reaction mechanisms for the cleavage of the cobalt carbon σ-bond of CH_3CoB_{12} [8]. (After: Wood, J.M., Fanchiang, Y.T.: Proc. 3rd European Sympos. Vitamin B_{12} and Intrinsic Factor, Zürich. Walter de Gruyter, Berlin 1979)

balamin has been found to be very stable to nucleophilic attack, but very susceptible to electrophilic attack and free radical attack.

Route (1) might be possible under certain conditions but the oxidizing potential has not yet been measured. Route (5) seems a possible route for the methylation of lead(II) and thallium(I) by radical transfer under very anaerobic conditions. (See appropriate sections)

Non-Cobalamin Dependent Methylation Routes

E. coli which does not produce (though it can utilize) methyl cobalamin does methylate mercury (II) [26, 27]. This represents either use of trace amounts of CH_3B_{12} in experimental nutrient media (i.e. really a B_{12} route) or a route not involving B_{12}. Biosynthesis of methyl mercury occurs in Neurospora, an organism whose metabolism is not known to involve B_{12}. Methylation seems to occur by transfer of a methyl group to mercury bound to homocysteine [28].

The origin of the methyl group and its electronic state are unknown, but if this process is regarded as an "incorrect" synthesis of methionine the methyl group could be approximately regarded as CH_3^+. Certainly there was no evidence that the methyl group came from methionine [$CH_3S(CH_2)_2$-$CH(NH_2)COOH$] itself or from s-adenosylmethionine. Although it seems well established that non-enzymatic direct transfer of CH_3 groups to free mercury(II) cannot occur from s-adenosylmethionine or N^5-methyltetrahydrofolate, it is not known whether or not enzymatic transfer can occur to mercury complexed to a natural organic ligand system.

Another, perhaps more prosaic, route to mercury methylation is by the totally abiotic photochemical methylation of mercuric acetate [29, 30]. Up to 3% of mercuric acetate was converted to methyl mercury per day by UV irradiation, a much more rapid process than occurs through microbial methylation. Chloride ion inhibits this process suggesting that it is not a route that occurs in sea water. This route does suggest that extra care should be taken in interpreting kinetic or microbial methylation experiments using mercuric acetate as substrate. Similarly, photolysis of aliphatic α-amino acids in abiotic systems in the presence of mercury (II) results in the formation of methyl mercury [31]. As intense UV light is required in these processes it is doubtful if they could occur in most natural sediments.

Non cobalamin based methylation routes may also exist for arsenic. Coenzyme M (2,2′-dithiodiethane sulfonic acid) is involved in the methylation of arsenate [32]; s-adenosylmethionine is involved in arsenic methylation by fungi [33]. However more recent work has suggested that two independent-biomethylation pathways for arsenic may exist and that one of them is cobalamin dependent [34]. The situation here is still not completely clear and more research is needed.

Hydrogen sulfide may be involved in a methylation route with methyl mercury as substrate, as follows [35]:

$$2\,CH_3HgCl + H_2S \rightarrow (CH_3Hg)_2S + 2HCl \rightarrow (CH_3)_2Hg + HgS.$$

This recent work will be discussed later. Significantly it is a route which could lead to complete methylation from unsaturated methyl metal systems e.g. $(CH_3)_3Pb^+$ to $(CH_3)_4Pb$.

As mentioned, many transmethylation reactions for abiotic systems are known, i.e.

$$RM + M^1 \rightarrow M + RM^1$$

and some of these may be significant in environmental methylations, particularly in grossly polluted situations. The main requirement seems to be that a water stable organometallic species (or one stabilized by complexation) occurs in the vicinity of a suitable substrate metal. Certain organo tin and lead species seem suitable in this context [25].

A rather more exotic transmethylation process to mercury has also been observed – certain water soluble trimethylsilyl salts will methylate mercury in aqueous solution [36].

Demethylation Processes

Though these will be considered here in general terms, knowledge of the process is mainly confined to methyl mercury where the great bulk of the research has been carried out. We also point out the possibly trivial case of demethylation by hydrolysis of unstable biogenerated methyl derivatives of heavy metals e.g. monomethyl lead derivatives.

Bacterial degradation of methyl mercury to methane and mercury in lake sediments has been demonstrated by Spangler et al. [37]. The species involved were identified as pseudomonads. Similar growth and decay patterns for methyl mercury have been observed both for sediment doping experiments with $HgCl_2$ and in cases where mercury was already present in the sediments and was disturbed on sampling [5, 24]. This suggests a balance of methyl mercury production and degradation in sediments and perhaps accounts for a general trend in many cases for the ratio of methyl to total mercury in sediments to be about 0.4‰ [29]. The balance between production, decay and loss of methyl mercury in sediments is not well understood although it seems to involve physical, chemical and biological features [38].

In the atmosphere dimethyl mercury is decomposed by photolysis to methane, ethane and mercury (0) and is therefore present only to a small extent [~ 1% (see the second part of the following section)].

Certain di and tri methyl tin products used commercially are demethylated by hydrolysis to produce ultimately tin oxide species. (See section on tin.)

Biogeochemical Cycles for Mercury

Introduction

Mercury compounds are not known to be involved in normal metabolic function and their presence in the cells of living organisms must therefore be due to contamination of natural or anthropogenic origin. In this discussion aspects of mercury transfer between the various reservoirs will be considered primarily; toxicological considerations are treated briefly in another volume of this Handbook and elsewhere [2, 3]. It should be noted that there is some disagreement on estimates of flow rates and pool sizes. Average values are shown in the figures and uncertainties are discussed in the text.

The role of mercury in the environment has been the subject of a recent comprehensive report [39], and in view of the historic importance of this topic a large number of detailed works on the chemistry and toxicity of mercury exist. Where relevant these are referred to in the text.

The Natural Cycle

It is not easy to accurately assess mercury flux rates between ocean or land and the atmosphere, or between deep ocean and continental shelf water masses. Little is known about the rate of flux of alkyl mercury derivatives under

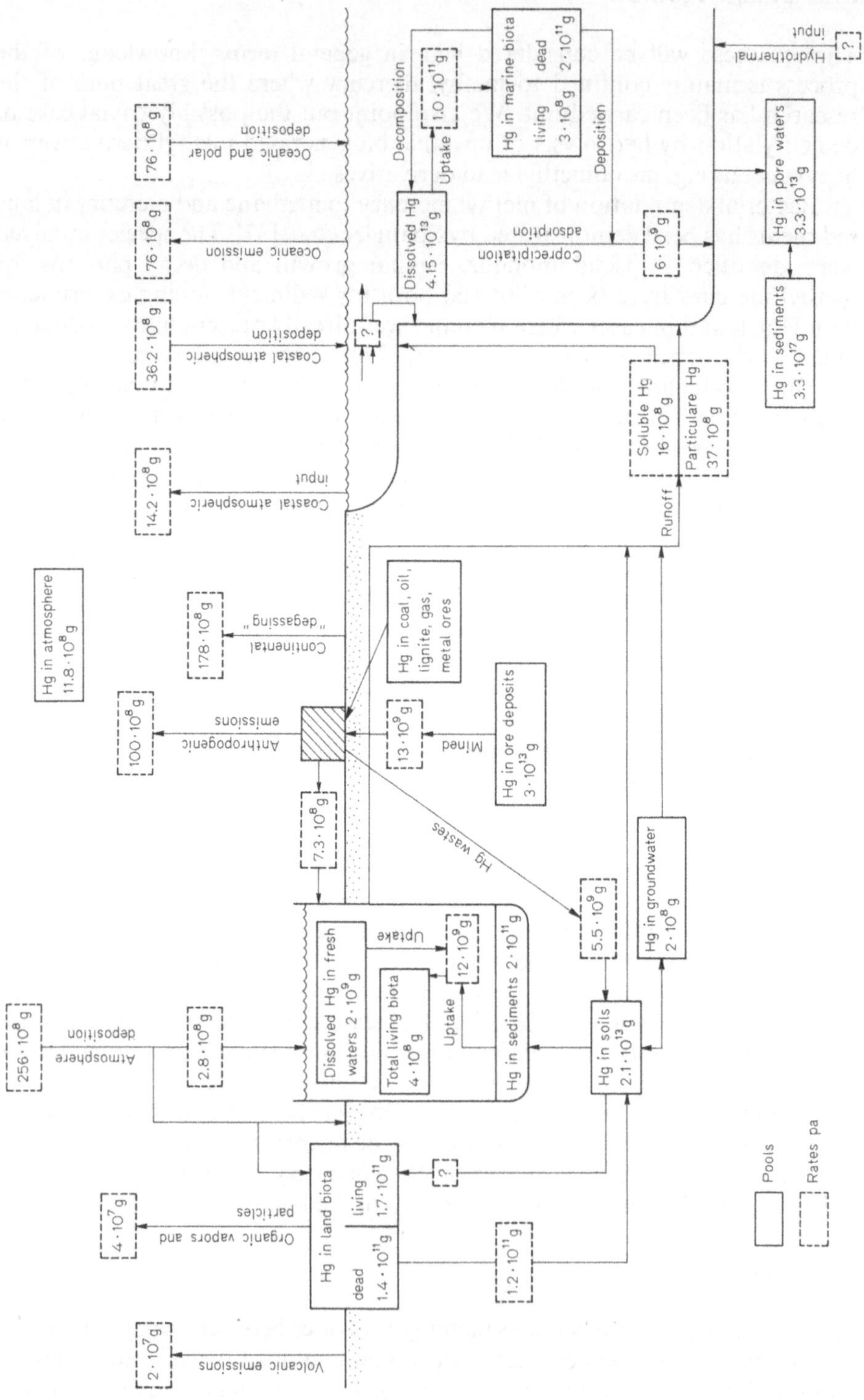

Hg in atmosphere $11.8 \cdot 10^8$ g
Volcanic emissions
$2 \cdot 10^7$ g
Organic vapors and particles
$4 \cdot 10^7$ g
Atmosphere deposition
$256 \cdot 10^8$ g
$2.8 \cdot 10^8$ g
Anthropogenic emissions
$100 \cdot 10^8$ g
Continental "degassing"
$178 \cdot 10^8$ g
Coastal atmospheric input
$14.2 \cdot 10^8$ g
Coastal atmospheric deposition
$36.2 \cdot 10^8$ g
Oceanic emission
$76 \cdot 10^8$ g
Oceanic and polar deposition
$76 \cdot 10^8$ g
Hg in land biota
dead $1.4 \cdot 10^{11}$ g
living $1.7 \cdot 10^{11}$ g
?
$1.2 \cdot 10^{11}$ g
Dissolved Hg in fresh waters $2 \cdot 10^9$ g
Total living biota $4 \cdot 10^8$ g
Uptake
Uptake
$12 \cdot 10^9$ g
Hg in sediments $2 \cdot 10^{11}$ g
$7.3 \cdot 10^8$ g
Hg wastes
$5.5 \cdot 10^9$ g
Hg in soils $2.1 \cdot 10^{13}$ g
Hg in groundwater $2 \cdot 10^8$ g
Mined
$13 \cdot 10^9$ g
Hg in ore deposits $3 \cdot 10^{13}$ g
Hg in coal, oil, lignite, gas, metal ores
Runoff
Soluble Hg $16 \cdot 10^8$ g
Particulare Hg $37 \cdot 10^8$ g
?
Dissolved Hg $4.15 \cdot 10^{13}$ g
Uptake
$1.0 \cdot 10^{11}$ g
Decomposition
Hg in marine biota
living $3 \cdot 10^8$ g
dead $2 \cdot 10^{11}$ g
Deposition
Coprecipitation adsorption
$6 \cdot 10^9$ g
Hg in pore waters $3.3 \cdot 10^{13}$ g
Hg in sediments $3.3 \cdot 10^{17}$ g
Hydrothermal input
?
Pools
Rates pa

natural conditions. However a number of authors have now devised global mercury models [39–48]. Differences in detail regarding calculation exist between some of these reports, and even differences in basic assumptions also occur. A good deal of work has been done on biogeochemical cycles in recent years however and the following general conclusions may be stated about the natural mercury cycle. Overall mercury cycles are shwon in Figs. 4 and 5.

1. Transfer of mercury between land and atmosphere (usually as mercury vapor) [49–52] is much greater than direct transfer between land and oceans (usually as divalent mercury associated with organic matter).

2. Movement of alkyl mercury derivatives is a very small proportion of the total quantity of mercury cycled. In some circumstances though it can be of dominant importance.

3. Degassing of mercury from land accounts for about twice as much mercury movement as degassing from the oceans.

4. Loss of mercury to the atmosphere from the oceans is more important than deposition in sediments. The latter however is a better storage reservoir for mercury.

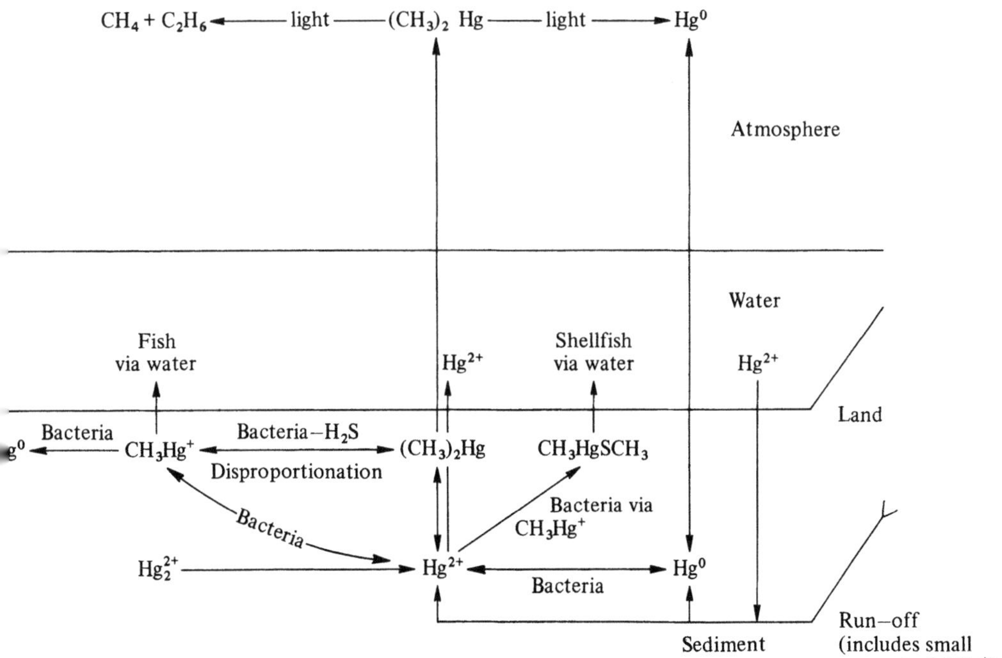

Fig. 5. Aquatic mercury cycle

◀ **Fig. 4.** Present day global mercury cycle (speciation not shown) [39] (After: An assessment of mercury in the environment. Report prepared by the panel on mercury; National Academy of Sciences Washington, 1978. Reproduced with premission)

5. The average residence time of mercury in the various reservoirs has been calculated to be as follows: atmosphere 11 days, ocean 3,200 yr, ocean sediments 2.5×10^8 yr, soils 1,000 yr [53].

A number of uncertainties in the pool burdens, flux rates and species in Figs. 4 and 5 exist. Estimates of the natural land – atmosphere flux are largely based on Californian measurements, and degassing rates used by various authors range from 0.0014 to 10 $mgm^{-2}d^{-1}$ [44]. Also some authors assume degassing takes place over land only, but Wollast 1975 [54] has argued convincingly that it must also take place from oceans, though at a lower rate.

Removal of mercury from the atmosphere has usually been assumed to occur mainly by precipitation although removal by direct impact of vapor is now thought to be significant also. This could account for the findings of greater mercury concentrations in Greenland snow cores nearer the surface [55], an observation which may have been mistakenly attributed solely to increased pollution in recent years.

It is now believed that mercury concentrations are not constant with increasing height in air [42]; there seems to be an exponential decrease with height. This suggests earlier estimates of the atmospheric mercury pool may have been too high. The main species present is mercury (0) vapor; little dimethyl mercury is found. Vapor species average as follows (%): Hg(0) 49, Hg(II) 25, CH_3Hg 21, $(CH_3)_2Hg$ 1. Particulates constitute 4% [50]. These results should be regarded as tentative in view of the great complexity of the analytical work involved and the dilution of the species. The chemical species of mercury involved in precipitation are not well known; concentration in rural areas varies from 0.01 to about 0.5 μg dm^3 [51]. Atmospheric concentrations of mercury in remote areas have a mean value of 4.0 (land) and 0.7 (sea) (6) ng m^{-3} [56]. It is not yet known how efficiently mercury is removed by heavy rain from the atmosphere [40, 50]. Mercury concentrations in non-mineralized unpolluted soils average at 71 ng g^{-1} and can vary by two orders of magnitude. A number of reviews of mercury levels in the lithosphere generally have appeared recently [16, 17, 46, 57, 58]. In many rocks and mineralized areas or under anoxic conditions much of the mercury present is in the form of mercuric sulfide (cinnabar). Under these conditions it is relatively immobile although in certain sediments it may not be. In mineralized areas the mercury content of soils can exceed 500 μg g^{-1} and cinnebar ore deposits usually contain from 0.5 to 1.2% mercury. Non mineralized, non polluted freshwater sediments have a mean of about 0.3 μg g^{-1}. Oceanic sediments have not been fully studied but similar concentrations seem likely [59].

Certain bacteria of the pseudomonas species can carry out the reduction Hg(II)→Hg(0) in soils and sediments and this process probably forms the basis of soil degassing [60, 61]. Methylation may also take place in soils and sediments – see Vol. 2, Part A.

The residence time of mercury in soils and sediments has been calculated at about 1000 yr, but perhaps the chief importance of mercury in these matrices (particularly for sediments) is the ease in which the methylation and demethylation reactions may take place [62, 63]. Previously sediment bound mercury was considered to be an inert sink; however, the occurrence of methyl mercury in fish, and work showing that even mercuric sulfide is not inert, has demonstrated that sediment bound mercury may be the most important matrix regarding pollution and food chain effects. This is considered in detail later. In rock weathering reactions mercury may be released as mercury(0) or (II) or in solid particulate form. In natural waters this mercury becomes mostly associated with suspended particles present in the water and then becomes deposited in bed sediments [64]. Ericksson has estimated 10^{10} tons of rock are weathered each year, suggesting a loss of 800 tons of mercury each year [65]. This 800 tons may be deposited in soils, sediments or transported to the sea. Some may evaporate. Most of the mercury in waters will eventually be deposited in bed sediments (up to 97%) [64].

The chemical forms of mercury present in natural waters are not fully understood. A number of authors have predicted on thermodynamic grounds the species of mercury expected under various conditions [22–24, 45, 47, 54, 66–70]. Obviously redox, pH and ligand conditions are very important. Hem [68] and Wollast [54] showed that in well oxygenated waters (Eh $>$ 0.4 v) dissolved mercury should exist as mercury (II) compounds; in moderately oxidizing to mildly reducing conditions (Eh -0.2 to $+0.4$) as mercury (0) or (II); and in reducing conditions (Eh < -0.2) as mercury (0) or HgS_2^{2-}. Anfault [66] has discussed the importance of the presence of chloride species; mercury in freshwaters should exist as $Hg(OH)_2$, $HgCl_2$ or Hg(OH)Cl depending on pH and pCL. Dyrssen and Weetborg [67] calculate that the most important species in sea water are $HgCl_4^{2-}$, $HgCl_3Br^{2-}$, $HgCl_3^-$, $HgCl_2Br^-$, and $HgCl_2$ at relative concentrations of 65, 12, 12, 4, and 3 respectively. Wollast [54] showed that Hg_2^{2+} is unlikely to exist in natural waters.

However it is very difficult to apply these thermodynamic calculations directly to real environmental conditions. A number of kinetic factors may considerably modify the behavior of mercury in natural waters. Much of the so-called dissolved mercury exists in association with suspended organic particulate matter (e.g. clays, hydrous oxides or detrital organic matter). In addition, in sediments and at the sediment-water interface, methylation and demethylation reactions may occur; in view of their importance these are treated separately. In fresh and marine waters dissolved methyl mercury species have been detected only in polluted systems. Non polluted waters must contain less than 0.2–1.0 mg dm^{-3} of dissolved methyl mercury [71, 72]; up to 50 or 60% of mercury present in seawater may be present as organic compounds or in association with organic matter [73].

It appears then that the majority of mercury in natural waters exists in combined or adsorbed form [72, 74–77]. Lindberg has pointed out the importance of organic coordination to mercury through functional groups such as SH in preventing the precipitation of mercuric sulfide in reducing sediments [76]. The role of hydrous metal oxides in removing and changing the behavior of mercury in natural waters has also been discussed [54, 78]. In sea water it appears by contrast that most of the mercury is present in dissolved form [79], albeit complexed to various organic and other species.

Most observations suggest that dissolved mercury in fresh water varies between 0.02 and 0.06 μg dm^{-3} and in the oceans averages from about 0.01–0.03 μg dm^{-3} [74, 80]. The main reactions of mercury(II) (formed in well oxygenated waters) may be deposition and reduction in anaerobic sediments to mercuric sulfide, reduction to mercury(0) and subsequent degassing, and conversion to alkyl mercury derivatives.

Much mercury appears to be deposited from water to sediment by flocculation processes in the transition zone between fresh and saltwater [72].

The Influence of Man

It has been suggested that 25%–30% of the present atmospheric mercury load is derived from man's activities and that certain river and lake burdens (water, and bottom and suspended sediments) have risen generally by two to four fold. Total increases in oceanic mercury concentrations due to man are negligible; similarly the total soil content has increased by only 0.02%. Freshwater and estuarine sediments appear on average to have a burden 2–5 times pre-man levels [53].

The industrial, agricultural and other activities that led to these changes have been discussed in another Volume of this handbook; here they will only be considered in passing where they are relevant to the global cycles and movements of mercury [81]. Figures 4 and 5 show the main aspects of the mercury cycle today.

It is likely that previous estimates of the mercury burden of the atmosphere may have been excessive; this has been caused by incorrect assumptions about how mercury concentrations vary with height [42] and by dubious assumptions concerning increases in the mercury content of Greenland ice cores in recent times [55].

Present data is insufficient for us to determine with confidence if the mercury content of freshwater and land biota has increased generally recently. There is evidence for an increase in fish and in sediments from freshwater lakes, estuaries and coastal waters (where a magnification of two to five fold may have occurred) [46, 73, 74, 76, 82]. Of course, very considerable increases have been recorded in localized environments. Transfer of mercury from soil to plants does not in general seem to be a critical problem.

Amongst the chief ways by which mercury reaches the atmosphere from man derived activities are smelting and fossil fuel burning. The amount from smelting is unknown; estimates from coal burning (3,000 tons/a) and oil fuel (1,800 tons/a) have been made [45]. Compared to natural degassing these figures are small.

Another assessment of the proportion of mercury in the atmosphere has also been presented recently. Using the concept of Global Interference Factors, defined as:

$$\frac{\text{Particulate} + \text{Fossil fuel heavy element fluxes}}{\text{Natural flux}} \; 100\%.$$

MacKenzie et al. have assessed the input of mercury to the atmosphere compared with pre-man conditions as 8,000 [83, 84]. This suggests that perhaps

80 times as much mercury is being released to the atmosphere as would be under pristine conditions. Similarly the Interference Index [85] – the ratio of emission rates to atmosphere compared to rain out from clean air – for mercury is 1.25. For a fuller discussion of these criteria see Section on other metals.

Gavis and Ferguson have suggested that, based on estimated 1970 rates, the release of mercury from the lithosphere by man is about 18 times more rapid than release by weathering. They have also estimated that at 1970 rates of release, and with certain assumptions, it will take about 3,450 yr to double the concentration of mercury in the oceans. Therefore present day annual anthropogenic release to atmosphere and oceans is much greater than natural release; however these have not occurred for sufficient time yet to effect greatly the concentrations in the various reservoirs. It should again be pointed out that important exceptions occur in localized areas.

The other main route of anthropogenic mercury loss arises to water from the uses of mercury as an industrial catalyst or biocide and in agriculture. Much work has been carried out on the fate of anthropogenic mercury in natural waters. The increased burden due to man's activities usually leads to increased particulate and sediment loadings rather than to drastically increased dissolved mercury levels. Although losses to water due to man is greater than total estimated natural losses to water (5,500 versus approximately 800 tons), it is particularly important that these losses take place in localized areas, often at high concentrations.

Much work has been carried out in recent years on the estimated impact of man on various river, estuary and lake systems e.g. Thomas et al. in the Great Lakes [86], the Ottawa River Project [87], the U.K. Mersey Survey [46, 88]. Since this and other work is discussed in more detail in another Volume only certain overall conclusions will be noted here.

(1) Anthropogenic losses of mercury to waters are not generally reflected in proportionate increases in dissolved mercury or methylmercury levels in these waters. Most of the mercury is lost to the sediments. Methylmercury has been detected in water only rarely.

(2) Where loss has occurred the concentration of mercury in sediments and fish can be greatly magnified over natural levels in those areas, e.g., mercury and methyl mercury concentrations in typical polluted river sediments – not cases of mineralization – are typically of the order of 1–20 $\mu g\ g^{-1}$ and 5–100 $ng\ g^{-1}$ respectively. This compares to the natural levels of the order of 0.3 $ng\ g^{-1}$ for mercury in freshwater sediments.

Biomethylation of Mercury

The mechanistic aspects have been covered in detail earlier (Methycobalamin and Biomethylation Mechanisms). Some of the environmental questions are discussed here.

Following on the original discoveries suggesting methylation of inorganic mercury, it has since been demonstrated on many occasions that mercury added to bottom sediments systems may be converted to methyl mercury.

Work with pure strains of various microorganisms has also produced many examples of bacterial methylation of mercury (e.g. by bottom dwelling bacteria, intestinal bacteria, soil anaerobes and yeast) [63, 89–101].

The question of methyl cobalamin involvement is discussed elsewhere but it seems likely that methylation from methylcobalamin is likely to be the primary route for mercury methylation and mobilization from bottom sediments in polluted waters. There is evidence that maximum methylation rates take place at oxidative anaerobic sediment interfaces (Eh between +0.4 and −0.2 V). However this does not completely explain the high levels of methyl mercury found in some deep ocean fish (e.g. tuna, swordfish). Moreover these levels do not seem to have changed much in the last 100 yr [102, 103]. The ability of tuna fish liver to methylate mercury(II) has been clearly demonstrated as also has the likely involvement of cobalamin [7]. However, the great speed of diffusion of methyl mercury across cell membranes has also been demonstrated [104, 105], and this, taken with the throughput of water through the gills, suggests a bioconcentration of methyl mercury rather than a biomethylation. The question of bioconcentration versus biomethylation is not clearly resolved. Certainly methylation by anaerobic or oxidative anaerobic sediment microorganisms should not be considered the only ultimate source of methylmercury in fish; mercury may be methylated by the normal bacterial flora of the gills and guts of the fish; even human intestinal bacteria have been shown to methylate mercury [106, 107]. Certainly very recent work suggests that diffusion rates may be more important than lipid solubility in mercury accumulation in various species. There are few reports of methyl mercury being analyzed in natural waters [108].

Various estimates have been made for the rates of methylation for mercury under environmental conditions. However the evidence suggests an equilibrium process leading to stable methyl mercury levels exists. Interestingly the ratio of methyl to total mercury present in various polluted sediments in different locations seems to vary between 0.1% and 1.5%. If mercury was really being removed from the sediments by methylation then in many locations no more mercury would be left in situ. However, breakdown processes of methylmercury to mercury exist and these have been demonstrated in lake sediments [92]. These will return mercury to the sediment. There is then a dynamic equilibrium between methylation and demethylation in sediments.

The occurrence of methylmercurythiomethyl (CH_3HgSCH_3) in shellfish has been noted [109, 110]. The possible routes leading to this compound have been also discussed [14]. Interestingly the toxicity of methyl mercury appears to decrease in the presence of normally toxic selenium salts [111, 112]. A mechanism has been discussed [14].

A potentially important cycling process for organo mercury compounds is shown by the role of hydrogen sulfide in the mobilization of methyl mercury as a volatile organo mercury derivative. This was originally thought to have been a methyl mercury sulfur compound [113]; the volatile material is now known to be dimethylmercury, originating as follows [114]:

$$2CH_3Hg^+ + H_2S \rightarrow (CH_3Hg)_2S \rightarrow (CH_3)_2Hg + HgS.$$

This process has been demonstrated for methyl mercuric chloride in water and added to sediments; it has not so far been demonstrated for "natural" mercury-containing sediments. Thus its effects in cycling mercury in the biosphere cannot be easily assessed. However as monomethyl mercury does not usually appear to build up in sediments beyond the 1.5% level compared to total mercury, then it may be adjudged as one of the processes for cycling and transforming methyl mercury in the environment (the others being dissolution, demethylation and absorption in the biota). Release rates of mercury from sediments have been studied by a number of authors, particularly by Krenkel et al. [115].

The observation of a "growth and decay" effect for methyl mercury in sediments is also interesting. Characteristically it is often observed that addition of inorganic mercury salts to sediments leads to a production and then a decline of methyl mercury in the sediments (demonstrating methylating and demethylating or removal processes). Disturbance of mercury-containing sediment samples on collection may also show a growth and then a decay of methyl mercury without any experimental addition of mercury. The methyl mercury arises by methylation of the reservoir of inorganic mercury present. This effect appears to occur mainly in anaerobic-oxidizing sediment zones. These observations also provide circumstantial evidence for the role of methyl cobalamin in methylation.

A particularly important situation arises when the rainfall in a district is acidic owing to sulfur and nitrogen oxide air pollution originating elsewhere. Under acid conditions monomethyl rather than dimethyl mercury is the stable species, and very high (methyl) mercury levels have been observed from fish from acid lakes [116].

Abiological methylation of mercury in soil has also been reported recently [117, 118].

Biogeochemical Cycles for Lead

Introduction

There is an interesting historical precedent in discussing the generalized health and pollution problems arising from anthropogenic lead. The massive use of lead in Roman times for aqueducts and piping has led to suggestions that the fall of the Empire was largely caused by endemic lead poisoning [119]. Roman per capita lead consumption at times reached 4 kg/a [120], amazingly comparable with the USA figure of 6 kg/a for 1965 [121]. Interesting accounts of the historic use of lead exist in many general journals and history of science texts.

A comprehensive work edited by J. R. Nriagu (The Biogeochemistry of Lead in the Environment) has recently appeared [122]. This author (PJC) is glad to acknowledge his debt to this work for part of the discussion presented here.

The Natural Cycle

Lead occurs in nature at an average crustal abundance of 16 μg g^{-1} and is found in trace amounts in rocks, soil, water, biota and the atmosphere. Like mercury it is not known to be an essential trace element. Total amounts of lead in the earth's crust and soil are estimated to be 3.8×10^{20} and 4.8×10^{15} g respectively. About 1.9×10^{10} g of lead pa is introduced by natural routes into the atmosphere (4% of the amount now introduced by man). The main sources of natural lead emission are suspended dust particles (85%), plant exudates (10%) and forest fires (3%). The mean residence time of lead in the unpolluted atmosphere has been calculated to be 14 days, but there is some variation in this view [123]. The natural pre-man lead concentration of the atmosphere was about 0.6 ng m^{-3}, about one sixth of the average concentration now [124]. The chemical composition of airborne lead particles under pre-man conditions probably corresponded to the originating mineral forms of lead. Present day compositions from rural sites now mainly reflect anthrogenic lead output. Most of the lead aerosols present in the atmosphere are eventually removed by precipitation or dry deposition. Lead concentrations in rain and snowfall from remote locations (the nearest we can get to estimating pre-man conditions) have been measured at 7.5 [125] and 0.02 μg dm^{-3} [126] respectively, the former an order of magnitude lower than for rainfall from polluted areas. Greenland ice cores dated 800 BC contain 0.001 μg kg^{-1} of lead [127]. In some areas dry deposition may be more important than rainfall as a mechanism of removal of lead from the atmosphere. Nevertheless anything other than core data measured today is likely to be severely affected by anthropogenic inputs.

99% of the lead entering the seas with the suspended load of rivers settles on the continental shelves. The major fraction of dissolved lead probably reaches the open seas [128]. Residence time in the ocean is calculated as 200 yr. In the past lead entered the oceans from surface waters at a natural rate of 1.7×10^{10} g pa [128]. An approximate world average natural lead level in surface water has been calculated; 0.5 μg dm^{-1} [129] varying considerably with location. This is one tenth the present level. In deep ocean waters the lead concentrations have been measured at 0.01–0.05 μg dm^{-3} level [130], presumably unchanged in recent times. More recent work suggests that deep ocean lead levels may be as low as 0.002 μg dm^{-3} [131]. Much of our knowledge in this area is tentative. There is still uncertainty about residence times both in oceans and freshwater lakes [132–134].

The prehistoric natural lead cycle in the oceans seems to have been as follows [128]. Lead entered the oceans from rivers either dissolved (complexed by water or organics) or on suspended particles. Most of the particulate lead settled into the coastal sediments. Much of the dissolved lead became incorporated in various organisms, eventually ending as organic detritus on ocean floors where some might be redissolved into the sediments. About 1.7×10^{10} g of dissolved lead pa was removed from the oceans in prehistoric times by settling [135]. 75% of the dissolved lead passes beyond the continental shelves and slopes into the deep ocean [136]. Chemical speciation of lead in waters under natural conditions is not well known; about 75% seems to be in

suspension and 25% in solution in rivers. Humic and fulvic acids may play a major role in environmental lead binding. In sea water lead may be equally divided between particulate and dissolved forms [137, 138], but there is disagreement on the extent. Lead(II) oxide seems to be the solid form of lead in stable equilibrium in oxygenated sea water [139], and Goldberg has suggested that dissolved lead species in seawater may be mainly lead(II) [140]. There is uncertainty as to the amount of lead in natural waters as soluble organic complexes or associated with suspended or colloidal matter. One study suggested that lead in sea water was 66% combined with colloidal organic species, 24% bound to inorganic colloids and 10% in solution as free (1%) or complexed ions [137]. Dissolved lead in freshwater between pH 6 and 8 and in contact with the atmosphere will exist mainly as carbonate species, especially $Pb(CO_3)_2^{2-}$ and in more acid conditions as lead carbonate [141–145]. The same situation appears to exist in sea waters [141, 146]. A minority view suggests lead chloride species are more important in sea waters [147]. There is some lack of agreement between the various models. In interstitial waters as in soils lead complexation by humic and fulvic acids may be important. Chow has summarized [148] that lead levels in natural waters are controlled by:
(a) formation of lead-inorganic complexes;
(b) formation of lead-organic complexes;
(c) dissolution and precipitation of inorganic and organic lead complexes;
(d) metal sorption onto hydrous and clay colloids;

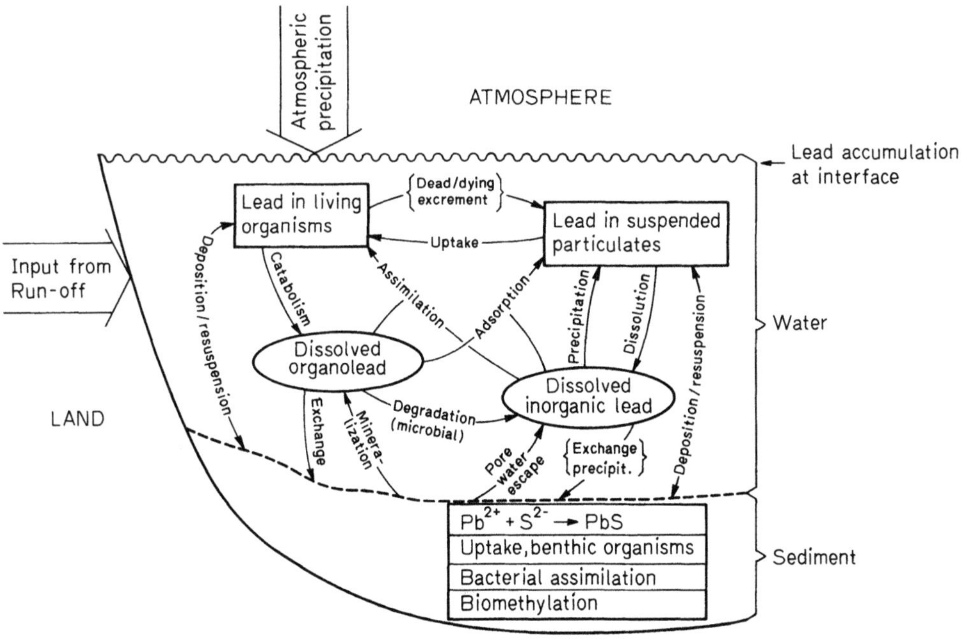

Fig. 6. Cycling of lead in an aquatic Eco-system [122]. [After: Richard, D.T., Nriagu, J.O. In: The biogeochemistry of lead in the environment (Nriagu, J.O. ed.). Elsevier, North Holland 1978, p. 276]

(e) lead sorption onto organic colloids and particulates;
(f) decomposition of organic biomass to liberate lead and
(g) coprecipitation of lead with other inorganic mineral phases.
It has been suggested that the major lead containing solids in freshwater systems are probably lead phosphates [149].

The cycling of lead in aquatic ecosystems is shown in Fig. 6.

Lead moves in aquatic systems as follows. Lead may enter a natural system via run off or streams as ionic lead(II), as inorganic or organic complexes or bound to suspended or bed sediments. The major source of lead in natural water systems is atmospheric precipitation (dry or rainfall depositions, 60%).

Within the water body the lead partitions into storage pools – organic biomass, suspended particulates, and sediment pools; and active pools-dissolved organoleads and dissolved inorganic lead. Another active lead pool occurs at the water-air interface [150]. The rates and amounts of exchange between the pools are determined by the physicochemical conditions and the biological activity of the system. Lead may enter the biomass as soluble ions, organolead molecules (from pollution) or with particulate matter. There then may occur a progressive concentration up the food chain (as with mercury).

The average crustal abundance for lead has been noted (16 μg g^{-1}). In uncontaminated soils lead ranges from less than 1 μg g^{-1} to over 10% in ores. It is better to state that it varies between 2 and 200 μg g^{-1}. Natural sources of lead in soils include weathering of rocks, vulcanism, fires and blowing dust. The lead content of most coals is 15 μg g^{-1}, the average being 9.6. Peat contains slightly more, oil much less (average about 0.3 μg g^{-1}).

Like soils in the atmospheric system, sediments are the primary sink for lead in the aquatic environment. For deep ocean sediments the natural average value is about 47 μg g^{-1}. Of course there are wide fluctuations. Chow and Patterson believe that two thirds of lead in marine sediments arises from lead previously dissolved in the sea water [128]. Shallow sea sediments average at 16 μg g^{-1}, the crustal abundance. Lead values in bay, estuarine and other coastal sediments have been much altered by man's activities. Pre industrial sediments have values which vary considerably and generally approximate to those of the surroundings. For unpolluted lake sediments (pre industrial strata) the value again is 16 μg g^{-1}, as it would be for unpolluted river sediments. Lead in suspended matter is, as for mercury, higher than in the corresponding bed sediments [151, 152]. The extent to which clay minerals versus dissolved and suspended organic materials are responsible for suspended sediment transportion of lead is unclear [153].

We can briefly summarize the transportation routes of lead within and between the various reservoirs as follows. The precipitation of sparingly soluble species, the formation of stable organic complexes and the adsorption of lead to particulate matter are the main routes for the environmental cycling of lead in natural waters. Similarly lead is adsorbed largely by organic matter and clays in soils and sediments. Some reports suggest in soils that adsorption by organic matter is greater than that by clays [154].

The chemistry of lead in soils is largely controlled by (a) adsorption at soil mineral surfaces, (b) by the formation of stable organolead complex ions and insoluble organolead chelates, particulates and residues and (c) by the precipitation of sparingly soluble lead compounds and precipitation of lead with the common soil minerals [155].

In anaerobic sediments most of the lead will be present as lead sulfide. Lead cysteine complexes may also be important [156]. Within the sediment medium lead may be methylated to alkyl lead derivatives. This has not been observed under natural lead conditions; it is discussed in the sections concerned with man's influence and the biomethylation of lead. The escape of alkyl lead compounds may be another pathway for returning the lead in the sediment pool to the overlying waters and atmosphere.

The Influence of Man

For lead this topic has been discussed, reviewed and investigated more perhaps than any other heavy metal pollution question. Though aspects of the man related lead cycle are still mysterious (e.g. biomethylation) much is now known in general of the speciation and transport of anthropogenic lead. This topic has very recently been the subject of a single comprehensive work wherein much detail is given concerning the concentrations of lead in atmosphere, biota, waters and the lithosphere. For such detail the reader is referred to this work edited by Nriagu [122]; in this section the overall conclusions about transport routes, species and averaged concentrations for lead will be presented. Biomethylation possibilities are not covered in detail in Nriagu's work and are reviewed in the following section.

Release of lead to the atmosphere is heavily man oriented. Only 4% of the lead flux now comes from natural sources. Of the approximately 4.4×10^5 tons of lead released annually, about 61% originates from leaded gasoline, 23% from production of steel and other metals, 8% from lead mining and smelting and 5% from other fossil fuel burning [157, 159]. MacKenzie has calculated an atmospheric global interference factor of 7,000% for lead (i.e. a 70 fold ratio of man derived to natural lead flux) [83, 84]. There are some different opinions about the importance of gasoline derived lead in the atmosphere, i.e., between 55% and 90% of lead release to air is attributed to gasoline use by various workers [160]. The background concentration of lead in the present day atmosphere is probably about 3.7 ng m^{-3} (compared to 0.6 pre-man). So unlike the mercury case man has drastically modified the lead levels now existing in the atmosphere [161]. In urban atmospheres lead ranges from 0.5 to 10 μg m^{-3}. In road tunnels, near highways and in traffic higher levels are usually found, e.g., 32.2 μg m^{-3} in the road tunnels at Heathrow Airport, London, U.K. [162]. Atmospheric lead levels decrease in general as follows:
central districts > high density residential > shopping commercial > heavy industrial > inner city parks > medium density residential > light density residential > open fields.

Much has been said about traffic and airborne lead and the main conclusions are noted here. Lead concentrations relate to traffic density and driving conditions (highway release > city release). Climatic conditions are also important as is uptake of lead by vegetation and soil. Lead concentrations in the atmosphere sharply decrease with increasing distances away from the road. Usually the decrease is about 50% between 10 and 50 m from the road. Also the heavier particles drop out of the atmosphere close to the road (50% of the particles > 6.5 μm diameter had done so by 600 m away) [163]. Why so much deposition of lead should occur so close to the highway is still not clear.

The composition of airborne lead particles from gasoline has been investigated. The principle products are lead halides (e.g. $PbCl_2$, $PbBr_2$ and principally PbBrCl), mixed hydroxo-halides [e.g. Pb(OH)Cl], double salts (e.g. PbO $PbSO_4$) and phosphates [e.g. $Pb_3(PO_4)_2$]. For the USA it has been calculated that 10% of lead released from automobiles to the atmosphere is in the organic form [164]. Mining activities release lead carbonates, oxides and sulfides. Elemental lead is emitted from smelting activities. Lead products (e.g. lead arsenate) are also released. In rural airborne lead determinations it was found that $PbCO_3$ (30%), $(PbO)_2PbCO_3$ (28%), PbO (20%), $PbCl_2$ (5.4%), $PbOPbSO_4$ (5%), Pb(OH)Cl (4%), $PbSO_4$ (3.2%), PbBrCl (1.6%), $(PbO)_2PbCl_2$ (1.5%), $(PbO)_2PbBrCl$ (1.0%), and $PbBr_2$ (0.1%) were present [165]. Lead halides become photochemically decomposed in the atmosphere releasing free bromine and chlorine [166]. Atmospheric lead is diminished away from the zone of emission by dilution and by precipitation.

Of the total airborne lead, about 1.0%–15% in cities is in the organic from (up to 62% has been reported [167]). Harrison and Laxon predicted typical rural and urban ratios of about 1.0%–4.0% organic lead. Surprisingly in some rural areas much higher ratios were found. They related this to environmental methylation of lead [168].

The mean residence time for atmospheric lead appears to be about 6–12 days. Faster turnover occurs in polluted areas (e.g. 10 h over US cities, 0.6–5 h within 100 m of roadways). Deposition is to soil and plants. Not surprisingly plants near highways have more lead than those in remote areas. Mosses and lichens have been used as sensors for aerial lead and other metals in (a) estimating atmospheric concentrations and (b) estimating flux rates. Average annual lead deposition from the atmosphere is 0.8 mg m^{-2} and 0.4 mg m^{-2} (northern and southern hemispheres respectively) [169]. It can of course be much higher in certain locations e.g. 6,000 mg m^{-2} near lead smelters and 400 mg m^{-2} in cities. About 10 mg m^{-2} is the US average. Dry deposition of lead tends to be more important in polluted areas (and of course in dry areas).

Extrapolations as to future lead emission into the atmosphere are difficult. Reduction of lead in gasoline will probably depend largely on geopolitical events concerned with gasoline availability and economic prosperity generally. There is still "no epidemiological data showing unambiguous relationships between ambient air lead concentrations, elevated blood levels and frequency of lead related diseases" [170]. However it would be reminiscent of 19th century practice to advocate waiting until positive epidemiological correlations exist with certainty before doing anything about the problem. There is much strong debate on this topic and it is discussed in detail by several authors [171, 172].

The US average level of lead in rainfall is about 34 μg dm^{-3}; there is considerable variation depending on location e.g. from 4.3 μg dm^{-3} in rural

Nebraska to over 300 μg dm^{-3} in San Diego [173]. The drinking water limit in the US is 50 μg dm^{-3}. Present day lead concentrations in Greenland snows are about 500 times the estimated natural level [127].

Lead in surface waters is now on average about 10 times greater than the natural levels (5 μg dm^{-3} versus 0.5 μg dm^{-3}) [124]. Lake and river waters usually vary from 1 to 10 μg dm^{-3} [174]. A vast amount of work has been carried out in estimating lead levels in various waters. Most of the lead originates from automobiles via the atmosphere. In waters lead halides from autos become hydrolyzed to insoluble oxohydroxides or form surface complexes with ferromanganese oxohydroxides. Much has also been written about lead in drinking water and the problem of conveyance through lead piping. The WHO drinking water limit of 100 μg dm^{-3} compares with the US limit quoted above. In some soft water regions with slightly acid pH the water can possess sufficient solvency for lead to result in appreciable concentrations of dissolved lead after overnight standing in lead pipes, i.e., levels greater than the drinking water limits. Drinking waters are not treated for lead removal as initial lead content is usually less than 50 μg dm^{-3} anyway; lead contamination arises by standing in lead pipes during supply.

Lead is transported from cities in storm water run off and sewage (100–500 μg dm^{-3} in industrial areas). Lead emission from mining, smelting and refining operations may also cause pollution of surface waters. The fact that some surface ocean waters contain much greater amounts of lead than deeper waters is cited as evidence for increased pollution loading [130].

The present day input of lead in the oceans appears to be as shown in Fig. 7. It is estimated that more than 90% of the lead now discharged to US streams and lakes is from man made sources [175]. This is probably typical for

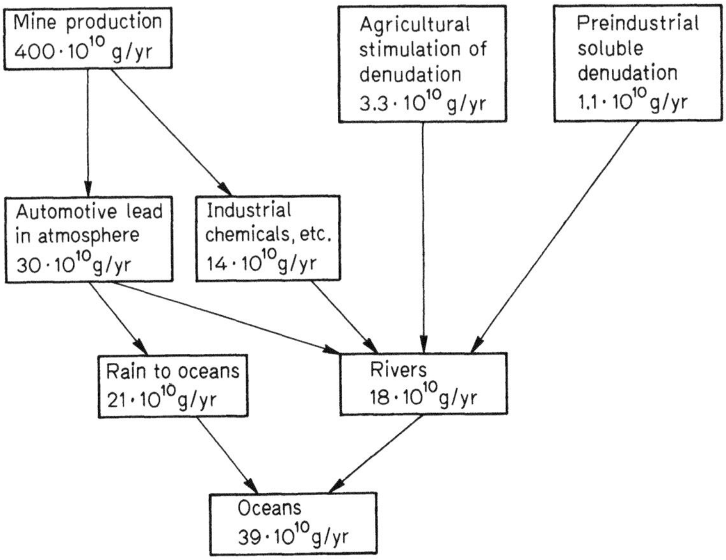

Fig. 7. Lead input to oceans (Northern hemisphere) [122]. [After: Nriagu, J. O. (ed.): The biogeochemistry of lead in the environment. Elsevier, North Holland 1978, p. 25]

industrialized countries. Some recycling of lead from ocean sediments may occur through biomethylation and desorption processes. The rates and extent are unknown.

Soils and sediments are the major sinks for pollutant lead (Fig. 8). Automobile sources are also the main sources of lead pollution in soils. Lead concentrations are related to traffic density and road type – there is more lead in soils near highways than in soils near city roads – climatic conditions and vegetation type. Foliage can trap out a significant proportion of atmospheric lead so that soils in wooded areas have less lead than those in open fields. Lead levels in soils fall steeply with distance from the highway and depth from the surface. In most cases lead decreases by 50% between 10 and 20 m from the highway. Lead concentrations in roadside soils decrease in the order

central business/high density residential > commercial > industrial > open ground.

In general lead in roadside soils is probably between 10–100 times elevated above natural levels. The increased lead burden for roadside crops, grass and trees may be 5–20, 50–200, and 100–200-fold respectively over non contaminated levels.

Lead in street dusts in cities may vary from natural levels to 6.5% by weight [176]. Nowadays this is due to leaded gasoline; previously coal burning helped elevate these levels. This means that water run off from city streets is an important transport route for pollution lead. Lead in various city environments has been involved in a number of cases of lead poisoning in children.

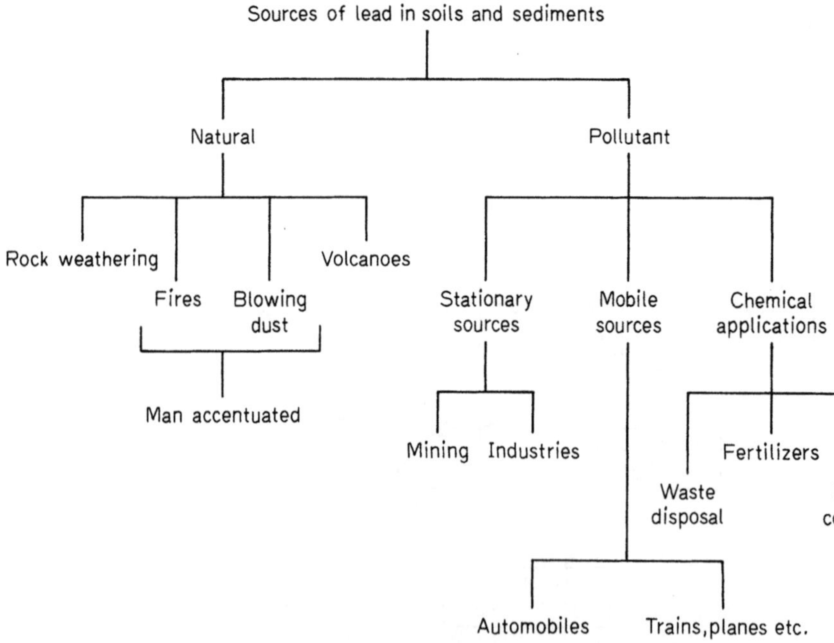

Fig. 8. Sources of lead in soils and sediments [122]. [After: Nriagu, J. O. (ed.): The biogeochemistry of lead in the environment. Elsevier, North Holland 1978, p. 25]

Lead emitted from stationary sources into the atmosphere may finally be deposited in the soil. Soil concentrations are obviously dependent on rate of lead release from the source, on wind, rainfall and other atmospheric phenomena, on vegetation around the source, on erodability of the soil and on depth in the soil (there is a near exponential decrease with depth). Unlike in the case of roadside soils, elevated lead levels in soils related to fixed sources may be found more than 25 km from the source. The half life of pollutant lead in soils is less than 20 yr; the turnover time in top soils is probably a few decades [177, 178].

Use of industrial or domestic waste as land fill or fertilizer adds a burden of lead to the soil. These wastes are usually high in lead (e.g. 100–450 μg g^{-1} for coal ash [179, 180], 100 μg g^{-1} for domestic sludge [181]), and may increase the concentration of the top 10 cm of soil by about 25 μg g^{-1} in the area used [182]. Lead also enters agricultural soils with lime and artificial fertilizer. Limestone may contain 50–600 μg g^{-1} lead [183] and fertilizers may contain 10–450 μg g^{-1} [184]. The use of lead insecticides has been shown to be mainly responsible for elevated concentrations of lead in orchard soils (perhaps up to a 20 fold elevation). The ubiquity of lead in and around households is well known – urban housing may be as important in contributing to elevated lead in nearby dust and garden soils as automobile emission [185].

Sediments are the primary sink for lead in aquatic environments. Natural deep sea pelagic sediments have an average lead content of 47 μg g^{-1} (q.v. crustal abundance of 16 μg g^{-1}). Chow and Patterson [128] believe two thirds of the lead content is derived from lead originally dissolved in sea water. Shallow sea sediments approximate to the crustal value of 16 μg g^{-1}. Loading of lead into sediments in bay, estuarine and marsh environments has been much affected by man. Lead contents of near shore sediments have been measured between 1 and 11,000 μg g^{-1}. Lead profiles in most undisturbed polluted sediments tend to decrease exponentially with depth, acting as markers for human influence. There is evidence that lead from grossly polluted sediments is relatively more easily released than lead than from less polluted sediments [186, 187]. Lead in lake sediments has been measured at background to over 5,000 μg g^{-1} in polluted sources – the average for lake sediment is six fold elevated over pre industrial concentrations. There is usually a sharp decrease of lead concentrations with depth.

Lead accumulates in lake sediments from stream sediments and waters, from ground water, erosion of lake banks, fall out from the atmosphere and effluent discharge. Lead in solution in the lake may become incorporated in sediments by precipitation of mineral phases, settling out of organic matter and sorption by organic matter and inorganics. The loading rate to sediments varies considerably from less than 0.1 μg cm^{-2} pa to more than 130 μg cm^{-2} pa [188]. Atmospheric input is the main source (60–80%) of lead in recent sediments.

Average lead levels in river sediments which show some evidence of pollution is estimated to be 98 μg g^{-1} – reported ranges vary from 3.9 to 3,700 [189]. Taken overall an average 30% increase in river sediment lead values has taken place owing to man's activities.

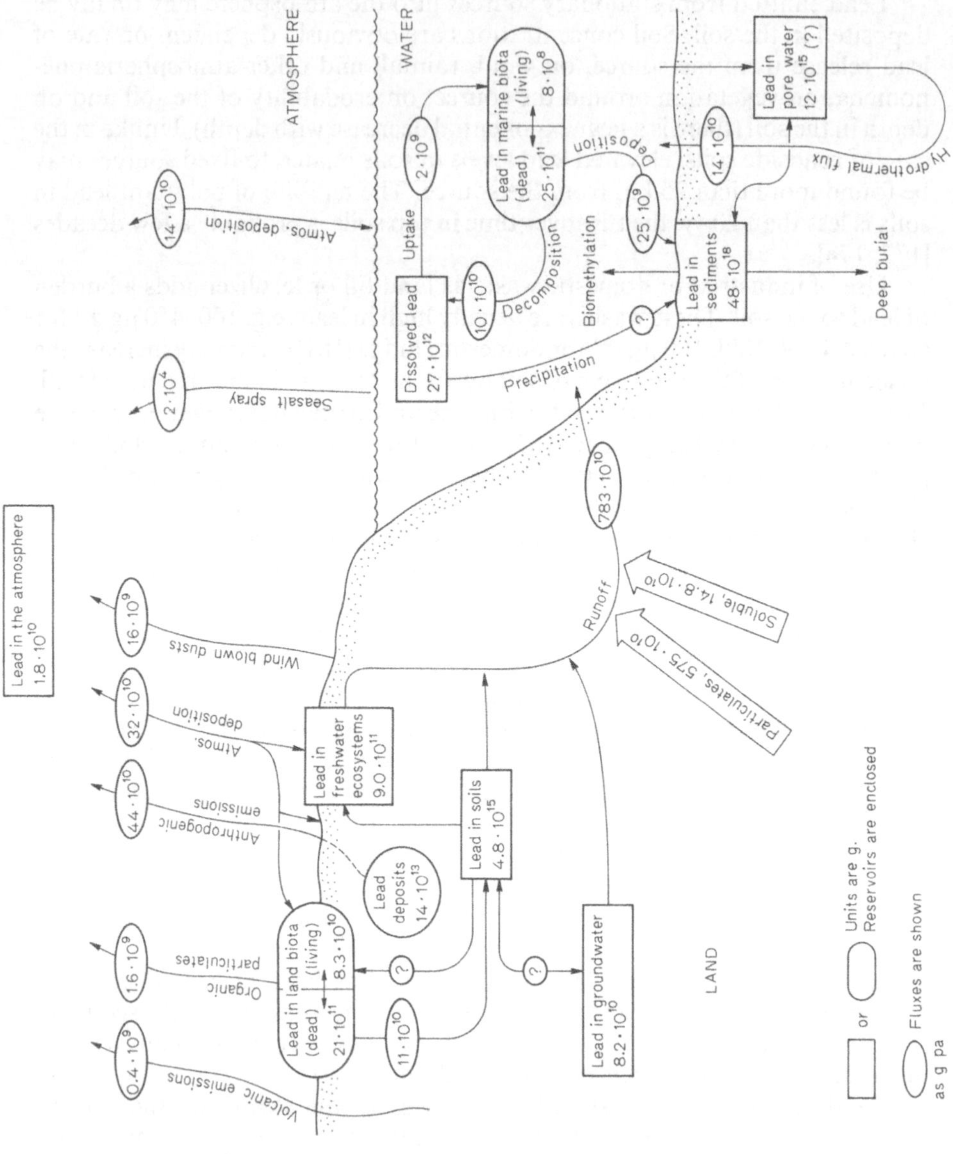

Fig. 9. Present day global lead cycle [122]. [After: Nriagu, J.O. In: The biogeochemistry of lead in the environment (Nriagu, J.O. ed.). Elsevier, North Holland 1978, p. 11]

The correlation between lead concentration in river water and river sediment is not established clearly, and neither has the relationship between lead in suspended matter and solution been fully evaluated. As for mercury, lead concentration in suspended material is higher than for bed sediments owing to the smaller average size of the suspended particles.

Suspended matter is chiefly responsible for retention and transport of lead in rivers. The importance, within this medium, of clay minerals, colloidal fractions, ferromanganese oxide coatings and dissolved and suspended organic material for lead transport varies. Humic and fulvic acids have a large binding capacity for lead, and both organic and inorganic components are responsible for transport of lead in rivers [190].

It is within the sediment zone that recent work has suggested that biomethylation of lead occurs (see appropriate section).

An overview of the present day global lead cycle is presented in Fig. 9.

Biomethylation of Lead

Although the concept of a biological conversion of inorganic lead to tetramethyl lead (TML) has been much discussed and is of obvious practical significance, to date only four papers have appeared where this topic is investigated under realistic environmental conditions. There are significant disagreements in the conclusions of the four papers. There have been a number of studies on model inorganic organometallic systems but the relevance of these to real environmental events is not conclusive. The question of lead methylation has recently been reviewed [191].

The first report that lead compounds could be biomethylated by microorganisms appeared in 1975 [192]. A mixture of Great Lakes water and sediments together with nutrients generated TML without any laboratory addition of lead compounds, i.e., the sediment already contained lead. Addition of $(CH_3)_3PbOAc$ greatly enhanced the amounts of TML produced, the authors assumed by a biomethylation process. The possibility that excess TML arose by disproportionation reactions of the $(CH_3)_3Pb^+$ salt or from the acetate moiety was considered at the time although later work showed that both biomethylation and disproportionation reactions were occurring [193]. It was also reported that in some cases inorganic lead(II) salts also caused enhanced production of TML in this system. In addition several species of pure bacteria were able to convert $(CH_3)_3Pb^+$ salts into TML; this did not occur with lead(II) salts. The conclusion that addition of lead(II) salts generated excess TML is very important, but it was observed in not all of the total number of experiments and could possibly be caused by factors other than biomethylation, e.g. displacement of pre-existing, weakly bound and previously unanalyzable TML from coordination sites in the sediment by incoming lead(II) salts. There was no demonstration that the lead added was the same as the lead later detected as TML.

A later paper shows that $(CH_3)_3Pb^+$ salts could be methylated to TML by a chemical disproportionation mechanism. This involved conversion by sulfide present in the sediments to $[(CH_3)_3Pb]_2S$ and subsequent disproportionation to TML [194]. Significantly addition of $(C_2H_5)_3PbCl$ to the sediment produced $Pb(C_2H_5)_4$ not $Pb(C_2H_5)_3CH_3$, suggesting the substrate had been disproportionated not methylated. Such a mechanism has been demonstrated to exist in the analogous mercury case, i.e., CH_3Hg^+ is converted by H_2S to $(CH_3Hg)_2S$ which disproportionates to $(CH_3)_2Hg$ and HgS [35]. These observations might suggest that biomethylation had not taken place for $(CH_3)_3Pb^+$ salts. However the sulfide does have a biological origin. These workers observed no methylation of lead(II) salts in sediments or using methyl cobalamin in model systems.

While bearing the above observations in mind, a further paper postulated the occurrence of a biological methylation for lead in aqueous solutions containing microorganisms on the basis of excess amounts of TML present above those expected to occur from a disproportionation alone [193]. However the biological proportion was deduced to be only a maximum of 20% of the whole methylation [196]. This result is not easy to interpret clearly as the different experimental conditions between sterile and biological samples could have effected the rate of the disproportionation route on which the calculation depended. However, this work also noted that certain lead(II) salts could be converted to TML in water containing microorganisms. The lead(II) salt used was lead acetate – in certain circumstances the methyl group could arise from the acetate grouping [197]. However even if this occurred it still demonstrates production of TML from lead(II) in an aqueous environment – a very significant result.

A more recent paper [198] has suggested that both sulfides and biomethylation are both causes of lead methylation from $(CH_3)_3Pb^+$ salt (as acetate). However as the sulfides are biologically generated, lead methylation therefore occurs by an indirect biologically dependent route and also by direct biological attachment of a methyl group to the trimethyl lead moiety. The biological route may have been 10 times faster than the sulfide route. Even more important, these authors found that lead(II), salts added to St. Lawrence River sediments produced TML, in disagreement with the earlier results (from different sediments) of Jarvie, Markell, and Potter. Two out of three sediment sites produced TML from lead(II) nitrate.

The chemical alkylation of lead(II) salts in aqueous media to TML is not inherently impossible. It has been accomplished using boron alkyls, e.g., BR_3 or $NaBR_4$ and lead(II) salts [199–202]. TML and lead metal were produced by disproportionation of transient $Pb^{II}R_2$. The alkylating agent was assumed to be R^-, analogous to the methyl carbanion generated from methyl cobalamin (CH_3CoB_{12}), the commonly assumed environmental alkylating agent. Therefore there is no theoretical reason why lead(II) cannot be biomethylated in the aqueous environment. Although the initial monomethyl lead cations produced are unstable in water, if the rate of further attachment of methyl groups (producing more stable di- and tri-methyl lead cations) is faster than the rate of decomposition, then TML may finally result. Furthermore, the unstable mono-methyl lead species may well be stabilized in the environment by coordination to naturally occurring ligand species.

The role of methyl cobalamin has also been suggested by its reaction with lead(IV) salts. CH_3CoB_{12} has been observed to be demethylated by $Pb(OAc)_4$, PbO_2, and Pb_3O_4 [203]. Using ^{14}C labeled CH_3CoB_{12}, loss of radioactivity, presumably as a volatile product, was observed in parallel with aquocobalamin ($H_2OCoB_{12}^+$) formation, i.e. loss of the ^{14}C labeled methyl group from B_{12}. This could be caused by attachment to lead. No attempt was made to identify the volatile products. CH_3CoB_{12} reacts with dialkyl lead(IV) halides to give volatile tetraalkyl lead compounds in non aqueous media [9]. However a number of workers have reported no reaction of lead(II) salts in abiotic aqueous solutions with methyl cobalamin [205, 206]. If CH_3CoB_{12} is involved in the biomethylation of lead then stabilization by natural organic complexing agents of the lead substrate must occur. In this respect the situation differs from mercury where model experiments in aqueous solution easily demonstrate methylation by CH_3CoB_{12}. Dimethyl cobaloxime a model compound for CH_3CoB_{12} reacts with lead(II) in isopropanol to give a stable insoluble methyl lead product of unknown structure.

An alternative route to TML through CH_3CoB_{12} may arise from radical generation under anaerobic conditions and subsequent reaction with lead metal (Fig. 3, route 5). This has been observed under abiotic conditions with methyl radicals generated in other ways.

Further evidence suggesting that biomethylation of lead may be occurring has also recently emerged. On the basis of reverse air movement projections and an enhancement of the alkyl to total lead ratios normally found in airborne particulate materials, Harrison and Luxon concluded that in certain Cumbrian UK intertidal sediments a conversion of lead salts, (present through natural or anthropogenic sources) to organic lead was taking place [168]. Direct confirmation by measurement of TML evolution from these sediments has not yet been reported.

Also, paralleling the mercury case, it has been reported that 10%–24% of total lead in some cod samples, and 39% in mackerel muscle, is in the alkyl form [204]. This suggests either a biological methylation is occurring or that strong selective concentration of the alkyl forms are taking place in fish.

There is then strong but not overwhelming evidence that a biological methylation of lead salts may occur under environmental conditions in polluted sediments. This question has been reviewed in more detail recently [191].

Biogeochemical Cycles for Tin

Natural Cycles and the Influence of Man

Tin occurs widely in nature as oxides (e.g. cassiterite, SnO_2) and as organic complexes in peats and coals. Like lead and mercury, tin has been of importance to man since historic times, giving the name of one of it's alloys to the Bronze age. Tin rarely occurs as the free metal. It may occur in rocks moderately susceptible to weathering, e.g. feldspars, biotite and can therefore be mobilized as these rocks are weathered. Among igneous rocks it is found at the level of 2 $\mu g\ g^{-1}$, in shales at 6 $\mu g\ g^{-1}$ and sandstones and limestones at 0.5 $\mu g\ g^{-1}$. In freshwater it occurs at 4×10^{-2} ng g^{-1}, in seawater at 3 ng g^{-1} and in air at less than 10 ng m^{-3}. It occurs in soils at an average level of 10 $\mu g\ g^{-1}$, varying between 2 and 200 $\mu g\ g^{-1}$ [207].

Variations of tin levels in soils are usually related to the bedrock from which the soils are derived – from traces, to 30–300 μg g^{-1} in the ashes of some Finnish peats [208]. Of nearly 900 US soil samples only 1% contained more than 10 μg g^{-1} and the maximum found was 20 μg g^{-1}. Soils of tin districts of course differ; e.g. 1,500 μg g^{-1} from soils of the Lost River region of Alaska [209].

US public water has been found to contain 1–2 ng g^{-1} tin. Arizona sources varied from 0.8 to 30 ng g^{-1}. Some dusts from industrial areas may contain up to 10,000 μg g^{-1} of tin [210]. Marine plants contain tin at an average level of 1 μg g^{-1}; land plants have levels of less than 0.3 μg g^{-1} except for bryophytes and lichens. Marine animals have levels of tin between 0.2 and 20 μg g^{-1}; land animals generally contain less than 0.15 μg g^{-1}. The residence time in seawater is 100,000 yr [207].

Although commonly found in plants and animals tin has not been thought to be an essential trace element. However in 1970 Schwarz et al. demonstrated that various tin compounds are necessary for the growth of rats – interestingly alkyl tins were amongst the necessary compounds [211].

Plants do appear to be able to concentrate tin present in soils e.g., there was a 3–10 fold elevation of tin in plants from a tin ore region over those from outside the region. However there is not unanimity on the question of bioconcentration; several bioaccumulation coefficients of less than one have been found [209]. Bioconcentration of commercial organo tin compounds seems small, as is half life in animals. Bioconcentration of tin from sea water is large; 2,900 for plankton and 92 for brown algae [207].

The most obvious source of tin in foods would appear to be from tin plated cans and tin additives in plastic food wrappings rather than from soil. Tin concentrations in canned food and drinks are usually less than 100 μg g^{-1} although standing can produce higher levels [212]. These levels are not directly toxic although they do raise the question of whether such tin compounds may be biomethylated. A tin compound (stannous chloride) is added to soft drinks as an antioxidant and the same question is raised. The use of stannous fluoride in toothpastes is also interesting in this context, as is the use of tin in the tubes themselves.

Tin(II) compounds are considered very toxic to fungi and angiosperms (i.e. produce toxic effects at concentrations below 1 μg g^{-1} in the nutrient solution). For small mammals tin is considered moderately toxic (i.e. LD_{50} between 10 and 100 mg kg^{-1} body weight) when taken by mouth [207]. In general inorganic tin has been considered to be relatively nontoxic by some authors [213]. Perhaps a chief interest here will be to ascertain the feasibility of biomethylation for tin compounds. Recent analytical work on tin(IV) organic compounds in (1) rain, (2) freshwater (Tampa Bay, Fla.), (3) estuarine (Tampa Bay), (4) saline water, and (5) tap water (Fla.) gave results as follows [214].

Total	(μg tr^{-1})	% Tin (IV)	% Mono-	% Di-	% Trimethyl
1	25	44	24	30	0.88
2	9.1	46	22	15	16
3	12	63	19	14	3.7
4	4.2	40	15	33	12
5	9.2	24	47	14	15

As these total tin levels are less than the world wide ocean averages the effects of man on tin mobilization cannot be judged from these figures. The presence of organo tin compounds does not imply biomethylation; organo tin compounds are important commercial products.

Tin is widely mined (0.2×10^{12} g pa) and used as the metal in plating steel food cans, bearing alloys, solder, pewter, type metal, bell metal, bronze and phosphor bronze. Uses of tin in the USA (%) are plate (50), solder (25), bronze (7), babbit metal (3), tinning (3), and other uses (12). Organo tin(IV) compounds are also widely used.

In view of the widespread use and production of organo tin compounds attention will be paid to the environmental transport and cycling of these materials. 25,000 tons of organo tin compounds are produced each year (4.26% of annual tin metal production).

Two thirds of this production is for thermal- and light-stabilizer additives for polyvinyl chloride (PVC) plastics – a small amount of degradation of the tin-free polymer may lead to diminished optical clarity and darkening. The mode of action of the organo tin stabilizers seems to be to prevent excessive dehydrochlorination of the polymer. Typical tin stabilizers are dialkyltin compounds containing thio or ester anions. Dibutyl and dioctyl tin compounds are the chief derivatives. In recent years methyl tin compounds have also been used.

8.5% of organo tin production is devoted to various forms of biocide including disinfectants, pest control chemicals and marine antifouling paints. The compounds are trialkyl tin derivatives particularly the tributyl and triphenyl species. For use as antifungal or antibacterial agents only triorgano tin agents are successful amongst the organo tins; diorgano tin compounds are used as anthelmintic (antiparasitic) agents. Hexamethylditin ($[CH_3]_6Sn_2$) has been introduced as an agricultural insecticide – biomethylation possibilities for this compound do not seem to have been investigated yet.

Unlike the lead and mercury methyl derivatives widespread in the environment, the organo tin compounds used commercially appear to be a less important problem. Their direct toxicity to mammals is not high – that property is partly responsible for the choice of commercial organo tin compounds – and also they have low half lifes (of a few days generally) under environmental conditions. The final degradation products are inorganic tin oxides. Only a few organo tin compounds are very toxic, e.g., triethyl tin salts. Tetra alkyl tin compounds are toxic to mammals and this makes the question of biomethylation of di- or triorgano tin compounds important. Generally trimethyl and triethyl tin compounds tend to be the most toxic in the alkyl series, acting upon the central nervous system. In the dialkyl chloro series the butyl compounds are the most toxic with ethyl and methyl derivatives being somewhat less toxic. About 0.5% of manufactured alkyl tins are lost to the environment at that point of manufacture. Most final disposal is from the consumer, principally through municipal sewers and waste disposal for the stabilizers. For the biocidal organo tin derivatives environmental input will be through normal weathering or leaching of the consumer product during use.

The industrially important organo tin compounds have low volatility. Although the vapor pressure of some derivatives at 25 °C can exceed standards for tin in workplace air (e.g. $[(nBu)_3Sn]_2O$) it does not seem to be sufficiently high to rapidly mobilize large quantities of organo tin compounds to the atmosphere.

Leaching of organo tin derivatives from soil has been investigated using ^{14}C-labeled compounds. Even with quite drastic treatment (hot methanol) complete removal from soil did not occur and it was concluded that soil mobilization is a very slow process [217]. This does not mean that it is insignificant in terms of its importance environmentally, but it does suggest that organo tin compounds disposed on landfills will be largely immobilized until decomposition to oxides. However it is possible that experiments in this area have not been carried out over a long enough time scale to fully evaluate environmental impact.

The situation is different in the aqueous phase. It is estimated that about 0.5×10^3 tons of tin biocides are released to the environment annually via sewers, rivers, harbors and oceans. Organo tin compounds are susceptible to attack at the carbon-tin bond by electrophilic, nucleophilic and free radical processes. The significant reactions in this context result in stepwise removal of organo groups from tin. The relative ease of removal is $R_4 > R_3 > R_2 \cong R_1$, decreasing with increasing size of the group and increasing with aromatic and unsaturated groups. Under environmentally realistic conditions solvolysis of R_4Sn proceeds 10–100 times more rapidly than for R_3SnX. Eventually hydrated inorganic tin oxides are produced.

Microbiological degradation is also important (producing tin oxides) with half lifes of seconds to days. From this it seems that organo tin derivatives are not persistent enough in aqueous solution to constitute an environmental problem in the same sense as methyl mercury. In addition the tin organics used commercially are not those most toxic to mammals. Finally bioconcentration in mammals does not seem to be a crucially important factor for organo tin compounds; where tin compounds are taken up by animals they are in general rapidly eliminated.

Unlike for the lead and mercury cases there is by comparison sparse data on the effects of man on the natural cycling processes. Except for some immediate industrial environments it is not easy to compare the impact of man-derived tin fluxes with those arising naturally. As tin and its compounds are of generally low volatility it is likely that atmospheric concentrations are not much elevated over natural levels. Harbor, river and lake levels are likely to be higher in inorganic tin content than the natural levels owing to the widespread use of tin compounds in agriculture and marine paints. This would also lead to elevated sediment levels in the immediate environment. It is unlikely that ocean concentrations of tin have been much modified. Owing to the ease of degradation the level of organo tin compounds in the aqueous environment is likely to be small. In general there is not sufficient data in the tin case to come to firm conclusions – in particular biomethylation of tin species could be responsible for much environmental cycling of tin. This possibility is little investigated yet.

The applications and biological effects of organo tin compounds have been discussed by J. G. A. Luijten in Volume 3 of the comprehensive work edited by A. K. Sawyer [216]. The environmental impact of organo tin compounds is discussed by Monaghan et al. and by Zuckerman et al. in the American Chemical Society Symposium series [217]. Aspects of the geochemistry of tin are reviewed in Volume II of Geochemistry and the Environment, the US National Academy of Sciences publication [215] and by Stumm in a Dahlem Konferenzen report [218]. The toxicity, bioconcentration and occurrence of inorganic tin is discussed in detail in Bowen's "Trace Elements in Biochemistry" [207].

Biomethylation of Tin

Evidence has been obtained for the methylation of tin (IV) by a tin and mercury tolerant strain of pseudomonas isolated from Chesaspeake Bay, USA [219]. The strain was grown on a sterile medium consisting of casamino acids, yeast extract, glucose, agar and various salts (NaCl, KCl, $MgCl_2$) at pH 7.2. On incubation with a tin(IV) salt a volatile tin species was detected in the atmosphere above the agar media. This was identified by mass spectrometry and fluorescence spectroscopy as $(CH_3)_3SnCl$ or $(CH_3)_2SnCl_2$. Interestingly the former tin species only was able to transfer a methyl group to mercury(II) salts to produce monomethyl mercury [220]. The authors of this paper speculated that the principal methyl tin metabolite produced in the agar may be other than the compounds noted above. These authors suggested that methyl tin compounds in the environment may undergo transmethylation reactions with mercury(II) to produce methyl mercury.

Tin(IV) salts have not been observed to react with methyl cobalamin. However tin (II) in the presence of iron(III) or cobalt(III) reacts with cobalamin in a free radical mechanism to produce monomethyl tin (IV) species [221]. The oxidizing species iron(III) or cobalt(III) were necessary for the reaction to occur – tin(II) alone would not react. Use of $^{14}CH_3CoB_{12}$ showed no evidence for production of $^{14}CH_4$, $^{14}CH_3OH$ and $^{14}HCHO$ suggesting production of a water-stable $^{14}CH_3$-Sn bond. A mechanism for these observations has been presented, viz:

$$\text{Sn (II)} + \text{Fe (III)} \rightleftharpoons \text{Sn (III)} + \text{Fe (II)}$$

$$\text{Sn (III)} + CH_3Co^{III}B_{12} \longrightarrow CH_3\text{Sn (IV)} + Co^{II}B_{12}.$$

The methyl tin species was identified by 270 MHz NMR.

The extensive use of dialkyl tin (IV) compounds as stabilizers for PVC and other plastics and the use of trialkyl tin compounds as biocides in agriculture and forestry prompts a question concerning their ability to undergo biomethylation. The R_2Sn^{2+} and R_3Sn^+ cations are relatively stable in water and therefore seem suitable substrates for methyl carbanion transfer reactions from methyl cobalamin in model abiotic systems or with cobalamin dependent species in sediments in the real environments. The products might be expected to be $R_2Sn(CH_3)_2$ or $R_3Sn(CH_3)$ respectively. These species would be more volatile than the di- or tri-organo species from which they were formed and might be more effective in the transport of tin in the environment. As ethyl and methyl tin species are more toxic than the higher alkyl group tin species used commercially, tin methylation would also represent an increase in toxicity of the tin compounds. Rather surprisingly biomethylation possibilities for R_2Sn^{2+} and R_3Sn^+ species appear not yet to have been investigated either in vitro or in aqueous sediment systems. Incomplete methylation of lower tin species to (say) the trialkyl stage might also be important. Generally $R_3Sn^+ > R_2Sn^{2+} > RSn^{3+}$ in terms of toxicity. The R_4Sn derivatives seem of the same order of toxicity as the R_2Sn^{2+} compounds. Tin methylation is by no means an academic question only; in 1975 more than, 25,000 tons of organo tin compounds were manufactured throughout the world.

Use of dialkyl tin compounds as stabilizers in food wrapping materials prompts questions about methylation possibilities for R_2Sn^{2+} derivatives in the human intestine. Similarly alkyl tin compounds have been detected in humans, possibly arising from tin anti-oxidants in soft drinks [222].

Other Metals – Natural Cycles and the Influence of Man

Introduction

Predictably most research and discussion on global cycles has been concerned with those elements that interact strongly with man through ubiquity, biological role, or toxicity. It is impossible to cover individually cycles for all metallic elements in a work of this length, and the remaining elements will be covered in a more general way considering the main principles determining global transport and flux. Many of these principles will have emerged from a study of the cycles for mercury, lead and tin discussed previously. This section will therefore constitute an overview and summary based on these cycles. Mercury, lead and tin are the most researched heavy metals – in this section another well researched element, cadmium, will also appear as an illustrative example of general principles. From the previous cases we can see that whether or not the biogeochemical cycling of a metal is "important" to man will be determined by the following criteria, grouped into Chemical, Biological and Physical properties for that metal:

Chemical Properties

1. Ease of formation, transportation and stability of it's alkyls. This relates to toxicity.
2. Redox properties, determining important precipitation and solubility characteristics.
3. Related to [2], the stability and solubilization properties of it's sulfides in an environmental "sink" (usually sediment).
4. Ionic association and dissociation in waters also determining transport routes. Ion exchange reactions are also important in this respect.
5. Chemical complexing properties with organic and inorganic ligands (measured by stability constant values). These are important, particularly in competitive situations. Their values determine the outcome of competition by metals for naturally occurring scarce ligands and vice versa. The outcome of competitive complexation will determine for example which elements are transported in water, which elements are retained in situ in soils or sediments and which are transported.
6. Stability and persistence under environmental conditions. This is usually applied to organic compounds but is also relevant to heavy metal compounds.

Biological Properties

1. The toxicity of the element and its compounds to man.
2. The rate of accumulation in food chains and biota; concentration factors e.g. for cadmium the bioaccumulation over water is 900 for marine plankton and brown algae and 1,600 for freshwater plants. Rice and wheat accumulate significant quantities of cadmium from soil – consumption of cadmium accumulated in rice has led to itai itai disease in Japan (cadmium poisoning).
3. The retention time in an organism for the element or its compounds. For cadmium in man this is 10–40 yr.
4. Microbial oxidation and reduction reactions.
5. Biological methylation.

Physical Properties

1. Natural occurrence, availability and extent of industrial use of the metal.
2. Volatility of the metal and its compounds.
3. Adsorption and desorption properties an particulate materials. These properties determine aqueous transportation capabilities for the metal e.g. distribution equilibria between biota, sediment and mineral phases.
4. Transport by diffusion through biological membranes and solubility of metal species in lipid material.

Jernelöv [223] has also discussed the impact of man's additional contribution of heavy metals to that occurring in the natural flux. Taking into account man's activities we should look at the following parameters in order to evaluate the importance of the metal and its cycle:

1. Production and emission of the metal in relation to the natural flux.
2. Residence times in the various sinks.
3. Bioaccumulation, both active (harmful) and passive (not known to be directly harmful).
4. Physico-chemical properties of the metal under environmental conditions relating to dispersion, volatility, adsorption characteristics, colloidal chemistry and precipitate formation.
5. Toxicity to aquatic species.
6. Toxicity to mammalian species (including man).
7. Long term effects on ecosystem metabolism.
8. Biotransformation reactions mediated by organisms.

These latter transformations are in general of two kinds: (a) redox conversions between inorganic forms, and (b) conversions between inorganic and organic forms (typically methylation and demethylation). Type (a) transformation have been reviewed recently (1) and will not be covered in detail here.

Discussion

Perhaps the basic factor of importance for heavy metals is their availability and accessibility to biota. Wood [10] has drawn up a table of elements under these characteristics.

Table 1. After: Wood, J. M.: Science *183*, 1049 (1974). Copyright 1974 by the American Association for the Advancement of Science

Non critical		Toxic and accessible		Toxic but insoluble or rare	
Na	S	Be	Pd	Ti	Ir
K	Cl	Co	Ag	Hf	Ru
Mg	Br	Ni	Cd	Zr	Ba
Ca	F	Cu	Pt	W	
H	Li	Zn	Au	Nb	
O	Rb	Sn	Hg	Ta	
N	Sr	As	Tl	Re	
C	Al	Se	Pb	Ga	
P	Si	Te	Sb	La	
Fe			Bi	Rh	

The crustal abundance of some key elements is also given [as the Clarke or $\mu g\ g^{-1}$ (ppm)]. This is an important background consideration but must be taken in conjunction with human industrial activities, physico-chemical properties and toxicities.

An immediate and vivid measure of the extent of man's interference with natural cycles is obtained by comparison of the rate of the man – derived flux of an element into the atmosphere with the natural flux. This has been done

Table 2. Some crustal abundances (as μg g^{-1}). (Except where shown as %)

H	1400	Pd	0.01
Be	2	Ag	0.05
C	250	Cd	0.2
N	20	Sn	3
O	46.6%	Sb	0.2
Si	27.7%	Te	0.01
P	700	I	0.5
S	260	W	1
Cl	130	Ir	0.0001
V	140	Re	0.01
Cr	100	Au	0.005
Mn	950	Hg	0.08
Fe	5%	Tl	0.5
Co	25	Pb	16
Ni	70		
Cu	50		
Zn	70		
As	1.8		
Se	0.1		
Br	2.5		
Mo	1.5		

using the Global Interference Factors [83, 84] discussed previously – restated here in slightly different form (omitting the percentage factor).

Table 3. Interference factors for elements (as ratio of anthropogenic to natural flux). From [84]. [Adapted from: Global chemical cycles and their alterations by man (Stumm, W., Pytklowicz, R. M. eds.) Dahlem Konferenzen, Berlin, 1976]

V	1.3	Mo	29
Ni	0.9	Cd	5.2
Cu	2.3	Sn	3.5
Zn	4.6	Sb	28
As	3.3	Hg	80
Se	14	Pb	70

A variation on this method is the use of the Interference Index [85], the ratio of emission rates to the atmosphere owing to man's activities compared to the annual quantity of "rain out" of the element from clean air. Implicit in the use of this index is the assumption that rainout quantity from clean air equals annual natural emission of that element from earth to atmosphere. The index therefore compares man-made to natural fluxes (percentage factor omitted).

Table 4. Interference indices for elements. From [83]. Adapted from MacKenzie, F.T. and Wollast, R. in: The Sea (Goldberg E.D. et al., eds.) Wiley Interscience, New York, Vol. 6, p. 763, 1977

Pb	1.29	As	0.26
Cu	1.11	Hg	1.25 - 10.00
V	4.50	Zn	0.70
Ni	0.42	Se	0.30
Cr	0.71	Sb	1.00

That the Interference Indices are so different from Interference Factors for the same element (i.e. smaller) appears due to rainout of the element even from clean air now being much greater than in pre man conditions, i.e., the atmosphere is now more polluted. This has the effect of giving excess weight to natural fluxes. What the Indices do show however when compared to the Factors is that

a) present day atmospheres contain higher loadings of metals than in the past and that

b) comparing current emissions to metal concentrations present in the atmosphere now gives a measure of the present day growth of an element's concentration in the atmosphere.

Similarly the volatility of an element is an important factor in its availability. Some trace elements are more concentrated in particulate form in the atmosphere with respect to their content in sedimentary rocks than are elements with higher boiling points (measured as oxides) [225]. Elements with high Enrichment Factors seem to fall into two groups:

a) elements that are emitted to the atmosphere in vapour form and later removed as dissolved gases in rain (e.g. mercury, arsenic and selenium) and

b) elements which are in part released in vapour form but which condense and are removed from the atmosphere principally as solid particles in rain or dry fallout (e.g. lead, zinc and cadmium).

Some element Enrichment Factors are presented in Table 5, using their concentrations in comparison with aluminum as a base line [225, 226].

Table 5. Some enrichment factors. Defined as $\log \dfrac{[M]_{atm}}{[Al]_{atm}} \Big/ \dfrac{[M]_{rocks}}{[Al]_{rocks}}$. Adapted from Mackenzie, F. T. and Wollast, R. in: The Sea (Goldberg E. D. et al., eds.). Wiley Interscience, New York, Vol. 6, p. 764, 1977.

Cd	3.3	V	1.7
Sb	3.0	M	1.0
Se	3.0	B	0.9
As	2.8	Cr	0.6
Zn	2.3	Ti	0.1
Hg	2.0	Sc	0.1
Pb	2.0	Al	0.0 (by definition)
Ni	2.0		

Significant amounts of some high volatility elements are present in atmospheric *particulates* at lower concentrations than they would be if oxide boiling points alone are a reasonable index of availability.These elements are likely to be present in the atmosphere in gaseous or chemically combined states (e.g. methyl mercury – mercury in particulates is less than one tenth of total atmospheric mercury).

The combining properties of an element are of obvious importance in its environmental impact. The broad preference of a metal cation either for large easily polarizable, low electronegativity anions (e.g. sulfide) or smaller, more electronegative anions (e.g. oxides, nitrogen) has been generalized into a chemical theory (the Class A or B element theory or the Hard and Soft Acid-Base Theory) [227–230]. The broad ligand preferences of important environmental metals may be listed using the Hard-Soft approach – whose essential principle is that Hard Acids (metals) are small, of higher positive charge and not easily polarizable, and prefer coordination with hard bases (ligands). Soft acids are of larger size, smaller positive charge and are easily polarizable – they preferentially combine with soft bases. As can be seen from

Table 6. Classification of metals [227], [Adapted from: Pearson, R.G.: Chem. in Brit. *3*, 103 (1967)]

Hard acids (prefer oxide, nitrogen complexation)	Borderline	Soft acids (Prefer sulfide complexation)
Mn^{2+}	Fe^{2+}	Cu^{+}
La^{3+}	Co^{2+}	Ag^{+}
Cr^{3+}	Ni^{2+}	Au^{+}
Co^{3+}	Cu^{2+}	Tl^{+}
Fe^{3+}	Zn^{2+}	Hg^{+}
As^{3+}	Pb^{2+}	Pd^{2+}
$(CH_3)_2Sn^{2+}$	Sn^{2+}	Cd^{2+}
VO^{2+}	Sb^{3+}	Pt^{2+}
MoO^{3+}	Bi^{3+}	Hg^{2+}
$(CH_3)_2Be$		CH_3Hg^{+}
Cr^{6+}		Pt^{4+}
		Te^{4+}
		Tl^{3+}
		$(CH_3)_3Tl$
		M^{o}

Table 6 for metals of environmental interest there is a preference for combination with soft bases (e.g. sulfide) although a number of borderline cases exist. Those elements or species (e.g. CH_3Hg^+, Hg^{2+}) which form sulfides in anaerobic sediments may at first site appear to be permanently unmobilized in non soluble form in the sediments. However re-oxidation to sulfate may occur by oxidizing bacteria (e.g. HgS by thiobacilli) and mobilization through disproportionation may also occur. (e.g. CH_3Hg^+)

For example the approximate order of increasing stability of cadmium (II) complexes with the various doner ligands is

$$F < Cl < Br < I < S4 < NH_3 < PO_4 < OH < CO_3 < P_2O_7 < P_3O_{10} < CN < SH.$$

Each of these ligand types is known to be present in different environmental situations from oxygenated surface waters to anaerobic sludges [231]. Such an affinity sequence must also be coupled with mass balance availability constraints in each case in order to find the distributions of the metals in each system. A suitable value for a stability constant does not imply the complex will be present in the environment – competition for ligands and redox variation may be very important for speciation e.g. The mercury(II) – SeH interaction is stronger than the equivalent-SH interaction, but as sulfur is more common than selenium in the earth's crust, mercury sulfur complexes are more often found. [Mercury(II) is a very strong acid and complexes with both strong and weak bases occur: Se > S > O > N.] For similar reasons of availability mercury-chloride complexes are found in sea water and mercury-fatty acid complexes are found in the presence of certain anaerobic bacteria in freshwater systems although comparitive stability constants are not strongly favorable to their formation.

Competition between metal cations for available ligands also determines the species found environmentally. Comparitive stability constant data, for example for cadmium, led to some relevant environmental conclusions [232, 233].

1. Cadmium(II) should not displace the essential metals (Mn, Fe, Co, Cu, and Zn) or the toxic metals (Hg, Pb) from oxalate ligands.
2. Cadmium(II) should displace Mn and Fe from ethylenediamine but not Zn, Co, Ni, Cu or Hg.
3. Cadmium(II) is bound more strongly to sulfur than all the metals except Cu, Hg and Pb.

The following affinity sequence for cadmium(II) binding to some bi dentate organic ligands has been produced:

oxalate < glycine < ethylenediamine < mercaptoethylamine (strongest).

Similarly for oxygen donating ligands:

acetate < oxalate < phthalate < citrate < salicylate < nitrilotriacetate [233].

Stability sequences for complexes of a number of metals have recently been discussed. Metals forming more stable complexes may displace other metals previously (and less strongly) complexed; hence a metal other than that added to the waterway as a pollutant may be mobilized. Similarly the competition between pollutant organic ligands in an effluent may be important in determining which metals are mobilized. Some stability sequences are shown in Table 7.

The consequences of the trends shown in Table 6 are that, for example with salicylate ligands, the calcium complex is the most stable and the copper the least. If calcium is added to a system in which available salicylate is complexed to other metals, these metals will be displaced and released to the water

Table 7. Some stability sequences [234–236]. All metals in the (II) state except where indicated. [Adapted from: Fates of pollutants, research and development needs, US National Research Council, Washington, pp. 66 and 67; 1977]

Ligand	Metals								
Salicylate	Cu <	Zn <	Fe <	Mn <	Cd <	Ca			
Glycine	Ca <	Mg <	Mn <	Cd ~	Fe <	Zn ~	Pb <	Cu <	Hg
Cysteine	Mn <	Zn <	Fe <	Cd <	Pb <	Hg			
Hydroxide	Ca <	Mg <	Cd ~	Mn <	Zn <	Co <	Fe <	Cu <	Pb < Hg
Polygalacturonic acid – $[(RCOOH)_2]n$	Ni <	Zn <	Cd <	Cu					
Silicia (adsorption)	Mg <	Ca <	Co <	Cd <	Zn <	Cu <	Pb <	Fe(III)	

column. Cadmium will replace all the metals shown except calcium. Through these stability trends, metals released to a water course may mobilize other metals bound to the sediments in the water.

The real environmental situation is often so complex that simple predictions based on stability constants may not be made, e.g. cadmium forms ternary complexes with pairs of ligands; mercury forms unknown complexes with the structurally undefined fulvic and humic acids; mercury metal is readily adsorbed on particulates in atmosphere and water. These heterogeneous natural systems are so complex that it is difficult to assess how much metal is transported through these processes and at what rate.

The role of microorganisms in the mobilization and conversion of metals is of great importance. This has been seen particularly in this review in terms of methylation and demethylation of metals. The other broad category of microbial transformation of metals is redox reactions, both reduction and oxidation. This topic has been reviewed recently and only broad outlines need be given here [1]. The microbial conversion of a toxic metal may be considered as an attempt by the organism to detoxify the metal by conversion to a less toxic form. However the new form may be even more toxic to other organisms, including man. Other microbial processes include hydrolysis or dehydration. Bacterial reductions are known for

As(V), Fe(III), Hg(I) and (II), Mn(IV), Se(IV) Te(IV);

oxidations are known for

As(III), Fe(O), Fe(II), Mn(II), Sb(III).

Molybdenum, copper and uranium are leached (oxidatively solubilized) by bacterial action [1].

The ability of the biota to accumulate and concentrate heavy metals is well known. The metal is usually in inorganic form, complexed to organic ligands in the biota, but mercury and lead have been analysed in fish in the organic form (as alkyl-metal compounds). Some Concentration Factors from sea water are given in Table 8. Concentration Factors are defined as concentration in the fresh organism divided by concentration in sea water for each element.

Table 8. Concentration factors from sea water. [Adapted from: Bowen, H.J.M.: Trace elements in biochemistry; Academic Press, London 1966]

Element	Plankton	Brown algae
Al	25,000	1,550
Cd	910	890
Co	4,600	650
Cr	17,000	6,500
Cu	17,000	920
Fe	87,000	17,000
I	1,200	6,200
Mg	0.59	0.96
Mn	9,400	6,500
Mo	25	11
N	19,000	7,500
Na	0.14	0.78
Ni	1,700	140
P	15,000	10,000
Pb	41,000	70,000
S	1.7	3.4
Si	17,000	120
Sn	2,900	92
V	620	250
Zn	65,000	3,400

The question of bioconcentration is discussed in detail in [207] but some general points might be made here. Nearly all of the heavy metal elements seem to be concentrated from sea water, and in general higher valent metals have a greater affinity for biota than lower valent metals. The order of Concentration Factors varies for each species and is not very readily correlated with orders of stability of organic complexes of metals, though heavier elements tend to be taken up more readily than light ones. Concentration data from freshwaters and soil solution are also given in this work.

In a similar manner attempts have been made to correlate toxicities with orders of stabilities of heavy metal complexes. These have been only partially successful if only because no one order of toxicity holds for biological organisms. In a very general sense only there is correlation e.g. the order of stability of metal chelates is

$$Hg > Cu > Ni > Pb > Co \sim Zn > Cd > Fe > Mn > Mg > Ca.$$

This correlates with our intuitive ideas of a rough order of toxicity.

Tables 9 and 10 give an indication of toxic properties of the heavy metal elements. The tables are compiled from source data in [207]. The most important mechanism for toxicity is the poisoning of enzymes – retention time in the organism is also contributory to this.

The precise environmental behavior and role of an element will be determined then by the interplay for all these factors. Reviews (often brief) exist in

Table 9. Toxicities of elements to fungi and angiosperms. (Data obtained from: Bowen, H.J.M.: Trace elements in biochemistry; Academic Press, London 1966)

Very toxic[a]		Moderately toxic[b]		Scarcely toxic[c]	
Ag		Al		Ca	
Be		Ba		Cs	
Cu		Bi		K	
Hg		Cd		Li	
Sn		Cr		Mg	
		Fe		Rb	
		Mn		Sr	
Co	Borderline	Tl			
Ni	Borderline	Zn		S^{VI}	Oxyanions
Pb	Borderline			Si	Oxyanions
Croxyanions	Borderline			Ti	Oxyanions
		As	Oxyanions		
		Mn	Oxyanions		
		Mo	Oxyanions		
		Sb	Oxyanions		
		Se	Oxyanions		
		Te	Oxyanions		
		V	Oxyanions		
		W	Oxyanions		

[a] Very toxic: Effects seen at concentrations below 1 $\mu g g^{-1}$ in a nutrient solution
[b] Moderately toxic: Effects seen between 1 and 100 $\mu g g^{-1}$
[c] Scarcely toxic: Effects rarely appear

Table 10. Oral toxicities to small mammals. (Data obtained from: Bowen, H.J.M.: Trace elements in biochemistry, Academic Press, London 1966)

Highly toxic[a]	Moderately toxic[b]	Slightly toxic[c]	Relatively harmless[d]
As	Cd	Al	Cs
Pu	Cu	Mo	Na
Se	Hg	Ta	I
Te	Pb	W	Rb
Tl	Sb	Zn	Ca
	V	Zr	K

[a] Highly toxic: LD_{50} between 1 and 10 mg Kg^{-1} bodyweight
[b] Moderately toxic: LD_{50} between 10 and 100 mg Kg^{-1}
[c] Slightly toxic: LD_{50} between 100 and 1,000 mg Kg^{-1}
[d] Relatively harmless: $LD_{50} > 1{,}000$ mg Kg^{-1}

some cases for other metallic elements, in many cases fluxes, residence times and chemical speciation are not well known.

Reference will be given to these reviews and the main findings briefly summarized (1,237). The elements cadmium and manganese have been subject to more detailed analysis than most.

Cadmium

Cadmium has a strong affinity for reduced sulfur (sulfide, cysteine) and should be at low aqueous concentration in anaerobic environments where sulfate is reduced to sulfide. Under mildly reducing conditions precipitation as the carbonate is possible. In the sea chloride species will be important. The carbonate is the stable solid for normal atmospheric concentrations of carbon dioxide. Cadmium particulates in water should exist as oxides, carbonate, solid solution of carbonate in calcium carbonate, adsorbed on aluminium, iron, silicon or manganese oxides, adsorped to clays, bound to organic particles, or as unsoluble sulfide. An equilibrium model system for cadmium in freshwater with a variety of other metals and ligands has been devised [238] and has been discussed above.

The main solubilizing agents for adsorbed cadmium are organic ligands and (in the sea) chloride ions. In the presence of oxygen, citrate and nitrolotriacetate are important ligands for binding cadmium. Under reducing conditions the sulfide and carbonate are predicted to predominate and the amount of cadmium in solution is small. Solubilization data (e.g. on dilution) are not clear – data is lacking on mobilization of cadmium from solid phases.

Evidence for the mobilization of cadmium by biomethylation will be discussed in the following Section. At this time it appears likely that cadmium will bioaccumulate through transportation through the zinc and manganese transport routes rather than through biomethylation. Data is lacking still on detailed cadmium emission points and rates. The mobilization rate from the sulfide phase in sediments is not well understood for cadmium. The accumulation rates in many environmental reservoirs are also poorly understood. As for other heavy metals the main residence zone for cadmium in aqueous systems will be the suspended and bottom sediments and little work so far has been carried out on the mobilization of the metal from such sediments.

Manganese

Essential differences between the manganese and mercury cycles lie in the non volatility of the former and the lack of any evidence for biogeneration of volatile methyl manganese species. In pre-man times the main flux between surface and atmosphere was transport of dust by winds. Today man generated manganese particles add to this flux – at about an equal flux rate.

Undersea volcanic activity also adds importantly to manganese flux (about 20% of the present day river flux). The river flux today appears to be three times that of the natural cycle – the increase is mainly to the suspended burden of the rivers and to ferric oxide coatings of suspended particles. The role of manganese nodule mining under the sea may be considered in more detail in the future in the context of man made heavy metal cycling.

Pristine and present day manganese cycles are shown in Fig. 10. The manganese cycle has recently been discussed in more detail [1, 226, 237, 238].

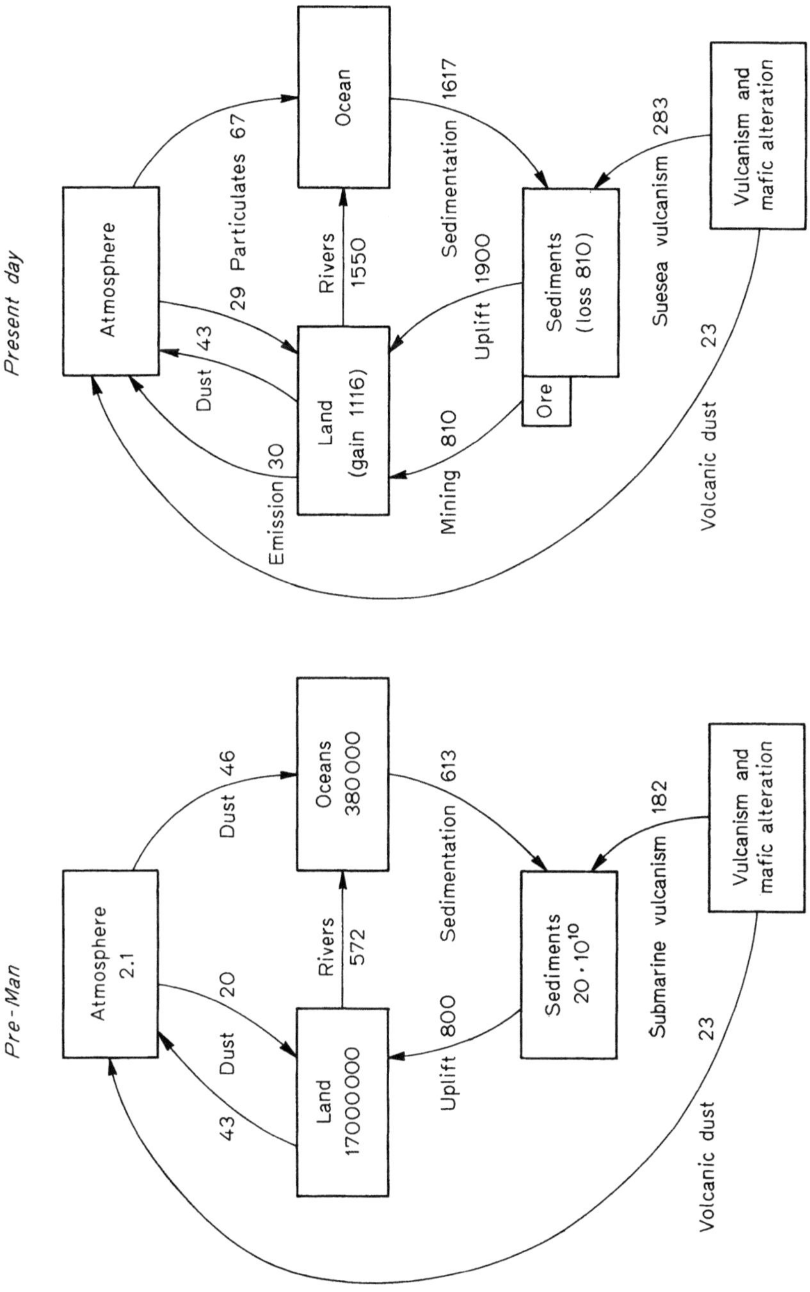

Fig. 10. Pre-man and present day manganese cycles [83]. (After: Garrels, R.M., MacKenzie, F.T., Hunt, C.: Chemical cycles and the global environment. W. Kaufman, Los Altos 1973, p. 132.) Fluxes in units of 10^{10} g pa. Reservoir masses in units of 10^{10} g

Other Elements

Few other toxic metal elements have been analyzed in any detail. Where they have health and toxicological activities have often taken priority over a study of their biogeochemical cycles. The US National Academy of Sciences has published a report on a Workshop on Geochemistry in the Environment in which geological and man derived occurrence of fluorine, iodine, chromium, lithium, cadmium, zinc, lead, selenium, tellurium, copper, molybdenum, beryllium, magnesium, manganese, nickel, silicon, strontium, tin, and vanadium are discussed [215]. Although each of these surveys is brief and toxic effects are emphasized, natural occurrence and industrial additions are mentioned and hence a useful overall impression is obtained. Cycling aspects are not covered though. The selenium cycle has been dealt with recently [239]. Marine pollution caused by mercury, cadmium, lead, copper, zinc, selenium, arsenic, vanadium, manganese, and iron has also been briefly discussed [240]. The proceedings of a conference on environmental arsenic held in Fort Lauderdale, Florida, USA have recently been published [239, 241]. Models of sedimentary cycles for mercury, manganese, silicon, iron, titanium, aluminum, mangesium, potassium sodium, chlorine, sulfur, calcium and carbon have recently been discussed in a recent review (mercury in some detail) [242]. Much detail on element concentrations, residence times, bioconcentrations of and pollution prospects exist in the book by Bowen [207].

Theoretical Treatments of Cycling Processes

These are covered mathematically in the review by MacKenzie and Wollast [242] and from the point of view of stability constant and solubility constant data in the US NRC report, Fates of Pollutants. [243] Studies on chemical cycles and the global environment are presented in a form suitable for use in undergraduate courses in the text by Garrels et al. [83].

Biomethylation of Other Elements

Heavy Metals

There are still few reports of biomethylation of other heavy metals. Those studies which have been reported are mainly abiotic model experiments using CH_3CoB_{12} and a metal salt substrate in aqueous or other solvent. These have demonstrated the feasibility of transfer of a methyl group to a metal but the metal-methyl grouping is usually highly unstable under the experimental conditions. In many cases then the environmental and biogeochemical significance of these experiments is not clear.

Huey [244, 245] et al have shown that a mercury-tolerant strain of *pseudomonas* in an agar culture medium containing glucose, yeast extract, hydrolyzed carein and various salts could volatilize cadmium(II) salts in the presence of cyanocobalamin ($CNCoB_{12}$). The implication was that dimethylcadmium was being produced though this was not definitely confirmed. This experiment, if confirmed under more general conditions, could prove to be of vital significance in the environmental area. Cadmium methyl compounds under aqueous abiotic model conditions are unstable and it has usually been assumed that an environmental biomethylation leading to accumulation of methyl cadmium could not occur. Cadmium(II) does not react with CH_3CoB_{12} in abiotic aqueous conditions [246].

An incubation of thallium(I) – as acetate – with anaerobic sediment bacteria from a lake under conditions of light and air exclusion yielded small amounts of $(CH_3)_2Tl^+$ – a thallium(III) species [247]. Monomethylthallium species are unstable and $(CH_3)_3Tl$ decomposes in water to $(CH_3)_2Tl^+$ and methane. $(CH_3)_2Tl^+$ is sufficiently water stable. Electrophilic attack by a methyl carbonium ion or a free radical mechanism could be responsible for the oxidation of thallium. The presumed initial CH_3Tl^{2+} may disproporationate or be further methylated to $(CH_3)_2Tl^+$. No mechanistic studies were carried out and no environmental work with real sediments has been reported. It is not known whether any cobalamin species are involved. It is of course conceivable that the methyl group could have arisn from the acetate moiety but this seems unlikely under the experimental conditions. It is possible that the reductive cleavage route for methyl release from $CHCoB_{12}$ may be operable in this case. Thallium(I) is isoelectronic with lead(II). There is therefore no firm evidence yet for *environmental* thallium methylation.

A number of in vitro model experiments concerning methylation of metals have been carried out. CH_3CoB_{12} had been demethylated by thallium(III) but not thallium(I) in vitro [248, 249]. Methyl cobaloxime, a model compound for CH_3CoB_{12}, will transfer a methyl group to thallium(III) to form CH_3Tl^{2+} (250). None of the methyl thallium species were stable under the experimental conditions.

The reaction of CH_3CoB_{12} with gold and platinum requires the simultaneous presence of higher and lower oxidation states for each metal [viz. Au(III)/Au(I) and Pt(IV)/Pt(II)] [248, 249]. A redox switch mechanism was proposed:

$$CH_3CoB_{12} + Pt^{II} \longrightarrow CH_3CoB_{12}\,Pt^{II}$$

$$CH_3CoB_{12}\,Pt^{II} + {}^*Pt^{IV} \longrightarrow H_2O \longrightarrow H_2OCoB_{12}^{+} + [CH_3Pt^{IV}]^{\dagger} + {}^*Pt^{II} + Cl^{-}.$$

More recent work with the $PtCl_6^{2-}$ and $PtCl_4^{2-}$ system has demonstrated the prior formation of a 1:1 CH_3CoB_{12}:$PtCl_4^{2-}$ complex with methyl carbanion transfer to the coordinated platinum with simultaneous oxidation by the $PtCl_6^{2-}$ species also present [251]. The initial 1:1 CH_3CoB_{12}:$PtCl_4^{2-}$ complex labilizes the cobalt carbon bond to electrophilic attack and intramolecular transfer to the platinum(II) species with simultaneous oxidation by the platinum(IV) species also present (i.e. redox switch). An alternative mechanism was methyl group transfer through direct electrophilic attach by $Pt^{IV}Cl_6^{2-}$ on the cobalt carbon σ bond labilized by complexation to $Pt^{II}Cl_4^{2-}$:

† Transient, the methyl group is observed as methyl chloride

$$CH_3CoB_{12}Pt^{II}Cl_4^{2-} + {}^*Pt^{IV}Cl_6^{2-} \longrightarrow H_2O \longrightarrow [CH_3^*Pt^{IV}Cl_5^{2-}]^\dagger + Pt^{II}Cl_4^{2-} + Cl^- + H_2OCoB_{12}^+.$$

The kinetics did not allow a choice between the two mechanisms. Recent work however suggests the occurrence of a redox switch mechanism. Reaction of CH_3CoB_{12} with a mixture of $PtCl_6^{2-}$ and $Pt(CN)_4^{2-}$ produces a methyl-platinum-cyanide complex via the redox switch route [252]:

$$CH_3CoB_{12} + Pt^{II}(CN)_4^{2-} \longrightarrow CH_3CoB_{12} \cdot Pt^{II}(CN)_4^{2-}$$

$$CH_3CoB_{12} \cdot Pt^{II}(CN)_4^{2-} + Pt^{IV}Cl_6^{2-} \longrightarrow H_2O \longrightarrow [CH_3Pt^{IV}(CN)_4Cl^{2-}]^\dagger + H_2OCoB_{12}^+ + Pt^{II}Cl_4^{2-} + Cl^-.$$

The reaction occurs faster with $PtCl_4^{2-}$ than $Pt(CN)_4^{2-}$. The methyl platinum species were identified by the methyl proton NMR peak. Arsenic(V) and selenium(VI) species will also replace platinum(IV) in this reaction. A 2:1 complex between CH_3CoB_{12} and $Pt(SCN)_4^{2-}$ has recently bccn crystallized. A similar route appears to operate for the gold(I) and -(III) system [252]. The nature of the methyl metal products of these reactions has been little investigated to date.

Taylor and Hanna however did succeed in isolating a water stable methyl platinum species from the reaction of CH_3CoB_{12} with $PtCl_6^{2-}$ and $PtCl_4^{2-}$ in aqueous solution at pH 2.0. Use of $^{14}CH_3CoB_{12}$ demonstrated attachment of a 14C methyl group to platinum in ratio 0.9–1.2 and an NMR absorption at 6.956 τ. CH_3CoB_{12} demethylation is stoichiometric with the amount of $PtCl_6^{2-}$ added. Catalytic quantities of $PtCl_4^{2-}$ accelerated the reaction rate but $PtCl_4^{2-}$ alone is unreactive. The final platinum product seems to differ according to whether the reaction is electrophilic (in darkness) or free radical (in the presence of light). Again the mechanism suggested was the two electron redox switch route discussed previously [203, 253, 254].

$$CH_3CoB_{12} + Pt^{II} \longrightarrow CH_3CoB_{12}\,Pt^{II} \text{ (1:1 complex)}$$

$$CH_3CoB_{12}\,Pt^{II} + {}^*Pt^{IV} \longrightarrow CH_3Pt^{IV} + {}^*Pt^{II} + H_2OCoB_{12}.$$

The nature of the NMR peak and coupling constants demonstrate the presence of methyl platinum groupings. However the position of the peak at 6.956 τ is much lower than that normally observed for methyl platinum complexes ($\sim 9.1 \pm 0.5$), i.e. the final product does not appear to be a simple $[CH_3PtCl_5]$ species; however condensation to the well known neutral "Cubane" system for platinum, e.g. $((CH_3)_3PtCl)_4$, or the analogous $(CH_3PtI_3)_4$ complex seems possible. A cube with 4 platinums, 4 methyl groups and 4 bridging atoms at opposite comers explains the Pt:C atomic ration. However the NMR data still reveals a lower τ value than expected for such a structure.

While the identity of the water stable methyl platinum system is as yet unclear this work suggests that biomethylation of platinum may occur under environmental conditions. Obviously platinum is a rare element but it is used as an industrial and automotive (pollution abatement) catalyst and is therefore present in the environment in some circumstances.

† Initial product

CH_3CoB_{12} is demethylated to aquocobala in the presence of $Ir^{IV}Cl_6^{2-}$ ions. However the product is methyl chloride and present views are that the mechanism does not involve even the transient intermediacy of methyl iridium species. A reaction with $Fe(CN)_6^{3-}$ also seems not to involve any methyl iron species [249] but also operates by a one electron oxidation by iron, generation of a methyl radical and final production of methyl chloride [255].

A kinetic study of the reaction of CH_3CoB_{12} with a palladium species, $PdCl_4^{2-}$ suggested a similar methylation mechanism as for the reaction with mercury, i.e. SE_2 attack and transfer of a methyl carbanion to palladium [256]. The methyl palladium species was not stable in water and was observed as methyl chloride.

Zinc(II) does not appear to react with CH_3CoB_{12} in aqueous solution. Neither does indium(III) [248, 249]. Aluminum(III) does not appear to demethylate the CH_3CoB_{12} model compound methyl cobaloxime [255].

The case of biological methylation of cobalt to produce methyl cobalamin in nature, although formally a biomethylation of a heavy metal, is considered to be outside the scope of this review.

Metalloids

The chief elements of concern here are arsenic, selenium, sulphur and tellurium. Several recent reviews of this area have appeared, e.g. [257, 258], one by Professor Challenger the pioneering worker in the biomethylation field [257], and the present review will give an account of the current situation in the area of biomethylation of these elements.

Work with *arsenic* originated with a search for the identity of toxic "Gosio Gas" evolved from arsenical wallpaper containing the mineral pigments Scheele's Green and Paris green. Later Gosio found that pure cultures of the mould *Penicillium brevicaulis* (now designated *Scopulariopsis brevicaulis*) containing arsenious oxide also evolved this vapour [259]. Challenger et al identified Gosio gas as trimethyl arsine, $(CH_3)_3As$, and hence methylation from inorganic arsenic salts had occurred [260–262]. Sodium methylarsenate ($CH_3AsO_3Na_2$) and sodium cacodylate ($(CH_3)_2AsO_2Na$) when added to a bread culture of this mould also evolved $(CH_3)_3As$. Higher alkyl analogues of the above salts produced $RAs(CH_3)_2$ demonstrating that methylation had not proceeded through arsenious oxide, As_2O_3. The methylation is believed to occur by (in effect) methyl carbonium ion transfer from S-adenosylmethionine [briefly $RS^+(CH_3)CH_2CH_2CH(NH_2)COO^-$] [261–263] following nucleophillic attack on the methyl group attached to positively charged sulphur by a trivalent arsenic grouping containing a lone electron pair. There is no separate existence of a methyl carbonium ion at any stage; the reaction is concerted. This implies that the reaction takes place in a reducing environment. All of the biomethylated arsines discussed by Challenger [257] were produced in well aerated mould cultures which nevertheless exhibited a strong reducing action. Strong evidence for methionine involvement arises from the production of 14 C labelled $(CH_3)_3As$ when $^{14}CH_3SCH_2CH_2CH(NH_2)COOH$

was added to the culture medium [260]. Soil organisms also were able to produce $(CH_3)_3As$ [264].

More recently McBride and Wolfe [265] have shown that arsenate is reduced and methylated under anaerobic conditions (hydrogen atmosphere) by *methanobacterium* (strain MoH) in the presence of CH_3CoB_{12} as methyl doner. Gaseous dimethylarsine $(CH_3)_2AsH$ was evolved and methyl arsonic acid $(CH_3AsO_3H_2)$ was also detected. Dimethyl arsonic acid [$(CH_3)_2AsO_2H$ – cacodylic acid] was detected under other conditions and both arsonic acids above could produce $(CH_3)_2AsH$ in cell extracts of the bacterium.

McBride et al. have questioned the involvement of CH_3CoB_{12} in arsenic methylation, particularly in view of the discovery of a methyl doner cofactor (CoM) found in methanogenic bacteria. CoM is 2,2′dithiodiethanesulphonic acid and it may be methylated chemically or biologically to form CH_3CoM, 2-(methylthio)ethanesulphonic acid [266]. It is not clear yet whether there are two independent methylation pathways for arsenic in anaerobic systems (one using CH_3CoB_{12}) or whether or not cobalamin and CoM are involved together with methyl donation from CH_3CoB_{12} to CoM and from CH_3CoM to arsenic.

$[SCH_2CH_2SO_3^-]_2$	$CH_3SCH_2CH_2SO_3^-$
CoM	CH_3CoM

In addition to these laboratory demonstrations of arsenic methylation methyl arsenic derivatives have been detected in natural waters, seashells, human urine [267], fish [268], mussels and lobster [269].

Recently there has been a report of arsenic methylation by pure bacterial cultures and by lake sediments. The methylation did not always proceed all the way to the volatile arsine derivatives; non volatile methyl arsenic derivatives (arsonic acids) sometimes remained in the aqueous layer. In this layer fish can not only absorb but also methylate arsenic compounds [270]. However volatile methyl arsine compounds were evolved into the gaseous phase in most experiments.

These experiments are significant in view of the wide distribution and use of arsenic and its compounds in the environment, and its toxicity. However only small amounts of arsenic are taken up by plants even from highly arsenical soils so it is hard to assess the true importance of arsenic methylation in environmental terms. A biological cycle for arsenic can be proposed:

Methylation to the methyl arsine stage has not been established to occur in the human body. If it does occur it could cease at the cacodylic acid stage or the arsines could be oxidized in the body.

With the same mold, S. brevicaulis, selenium and tellurium compounds may be biomethylated. Various oxyacids of selenium and tellurium evolve malodorous $(CH_3)_2Se$ and $(CH_3)_2Te$ [271, 272]. Other moulds or organisms [273–275] were also found capable of this operation [276]. A similar methyl carbonium transfer by attack of nucleophilic metalloidal species containing lone pair electrons seems to occur. Using labelled methionine Challenger and coworkers showed that over 90% of the methyl groups of $(CH_3)_2Se$ evolved from a culture of *A. niger* and selenate came from methionine [277, 278].

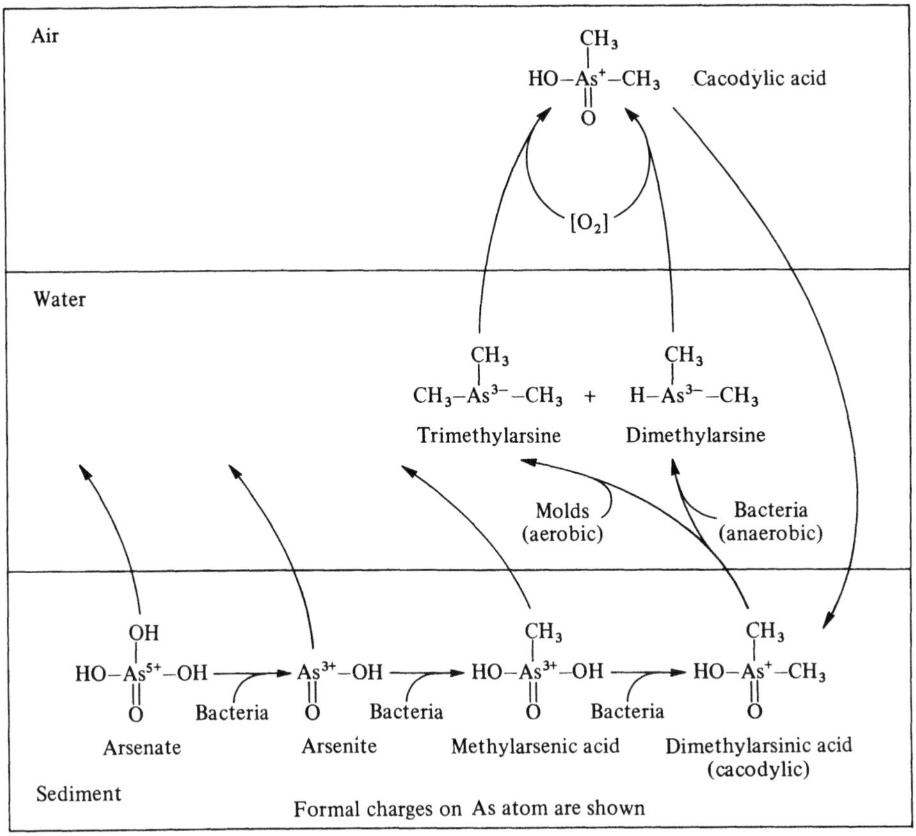

Fig. 11. The biological cycle for arsenic [10] (After: Wood, J.M., Science, 1974, *183*, 1049. Copyright 1974 by the American Association for the Advancement of Science)

It is well known that administration of selenite and tellurite to animals leads to exhalation of $(CH_3)_2Se$ [279] and $(CH_3)_2Te$ [280, 281] from the animals [282]. The $CH_3)_3Se^+$ ion has also been found in the urine of such animals (in this case, rats) [283–285].

More recently conversion of inorganic and organic selenium compounds to volatile compounds [$(CH_3)_2Se$, $(CH_3Se)_2$ and an unknown compound] by microorganisms present in lake sediments has occured [286]. The conversion could also be effected by bacteria and fungi. Sodium selenite on mixing with municipal sewage produces $(CH_3)_2Se$ [287]. Conversion was observed to arise from selenite, selenate, selenocystine, selenourea and seleno-D,L-methionine. Several bacteria were isolated and were shown to methylate selenite (*Aeromonas sp., Flavobacterium sp,* and a *Pseudomonas sp.*). Some higher plants methylate selenium [288, 289].

The known reduction in toxicity of methyl mercury in the presence of selenium salts appears to be caused by transfer of methyl groups from mercury to selenium and subsequent ventilation of $(CH_3)_2Se$ from the lungs of the animal [14]. Methyl transfer involved CH_3CoB_{12} for selenite and arises from $(CH_3)_2Hg$ for the anhydride form of selenium [14].

Strictly speaking methylation of sulfur lies outside the scope of this chapter. However in terms of biomethylation generally the topic is best dealt with here. Although cases of fission of dialkyl sulfides to RSH, and conversion of alkyl cysteines to alkyl thiol and $RSCH_3$ are known, only one microbial conversion of simple inorganic sulfur compounds to methyl sulfur derivatives is known. This is carried out by *Schizophyllum commune* and operates on sulfate [290].

The production of CH_3HgSCH_3 has been observed in shellfish. The CH_3S grouping is likely to derive from methane thiol generated by putrefaction [14]. Similarly the biological generation of alkyl sulphur compounds such as methionine ($CH_3S(CH_2)_2CH(NH_2)COOH$) falls outside the scope of this review, although formally a methylation of a sulphur atom must occur.

The mechanism of abiotic transfer to thiols has been studied and involved thiol radicals as follows [291]:

$$RS^0 + CH_3CoB_{12} \longrightarrow CoB_{12}^+ + RS^- + CH_3^0$$

$$RS^- + CoB_{12}^+ \longrightarrow RS^0 + CoB_{12}^0 \text{ [Cob[II] alamin]}$$

$$RS^0 + CH_3^0 \longrightarrow RSCH_3$$

These authors considered biological methyl transfer may be a one-electron reaction coupled with scavenging by thiolate [291]:

$$RS^- + CH_3CoB_{12} - e \longrightarrow CoB_{12}^* + CH_3SR$$

$$CoB_{12}^0 + e \longrightarrow CoB_{12}^-$$

$$CoB_{12}^- + \boxed{CH_3 - R} \longrightarrow CH_3CoB_{12}$$

methyl source

$* \; CoB_{12}^0$ = Cob (II) alamin – See above.

There seems little evidence for direct thiolate attack on CH_3CoB_{12} in the absence of an electron source [292]. Base attack on CH_3B_{12} fails even for cyanide [291].

The CoM enzyme isolated recently and referred to above may act as an intermediate methyl acceptor from CH_3CoB_{12}. The demethylation seems best explained in terms of thiol promoted homolytic cleavage of CH_3CoB_{12} to give cob(II) alamin and the methyl thioether:

$$CH_3CoB_{12} + [^-So_3CH_2CH_2S^0]^\dagger \longrightarrow CoB_{12}^* + \bar{S}O_3CH_2CH_2SCH_3$$

Although the elements around antimony in the Periodic Table are subject to biological methylation, the experiments carried out so far have not been able to demonstrate its occurence for this element.

† From CoM

A pathway for the biomethylation of iodine in sea water has been discussed. The route suggested is an electrophilic attack on the methylcobalt linkage of CH_3B_{12} by a polarized iodine molecule [14].

Acknowledgments

I am glad to acknowledge receipt of funding from the Northwest Area Foundation, Minnesota, USA and from Leicester Polytechnic, Leicester UK, which, together with the award of Senior Visiting Fellowships from NATO Brussels and from the Science Research Council, London, made possible the period of leave during which this Chapter was written. I am particularly glad to acknowledge the help of Prof. J.M. Wood, Director of the Gray Freshwater Biological Institute, University of Minnesota, for the provision of space in his laboratory and for many interesting and useful discussions on biomethylation during my period of Sabbatical leave at GFWBI.

References

1. Summers, A.O., Silver, S.: Ann. Rev. Microbiol. *32*, 637 (1978)
2. Geochemistry and the Environment, The Relation of Selected Trace Elements to Health and Disease; Nat. Acad. Sci. Washington (1977)
3. Prasad, A.S., Oberleas, D. (eds.): Trace Elements in Human Health and Disease; Academic Press, New York (1976)
4. Westoo, G.: Acta Chem. Scand. *20*, 2131 (1966)
5. Jensen, S., Jernelov, A.: Nature *223*, 753 (1969)
6. Wood, J.M., Kennedy, F.S., Rosen, C.G.: Nature *220*, 173 (1968)
7. Imura, W. et al., in: New Methods Environ. Chem. Toxicol.; Collect. Papers Res. Conf. New Methods Ecol. Chem. (Coulston, F., ed.). Int. Acad. Print. Co. Ltd, Tokyo, Japan 1973, p. 211
8. Wood, J.M., Fanchiang, Y.-T.: Proc. 3rd. European Symp. Vitamin B_{12} and Intrinsic Factor, Zürich 1979. de Gruyter Inc., Berlin and New York (in press)
9. Ridley, W.P., Dizikes, L.J., Wood, J.M.: Science *197*, 329 (1977)
10. Wood, J.M.: Science *183*, 1049 (1974)
11. Ridley, W.P., et al.: Environ. Health Perspectives *19*, 43 (1977)
12. Wood, J.M., et al.: Fed. Proc. *37*, 16 (1978)
13. Wood, J.M., et al.: Proc. Int. Symp. Bioinorg. Chem., Vancouver 1976 (Dolphin, D., ed.); p. 261
14. Wood, J.M., et al.: Proc. Int. Conf. Metals Environ. Toronto 1975 (Hutchinson, T.C., ed.); pub Plenum Press, p. 49
15. Wood, J.M.: Naturwissenschaften *62*, 1 (1975)
16. Wood, J.M.: La Recherche *70*, 711 (1976)
17. Wood, J.M., Fanchiang, Y.-T., Craig, P.J.: Some Bioinorganic Chemical Reactions of Environmental Significance; Raven Press; New York 1979 (in press)
18. Bertilsson, L., Neujahr, H.Y.: Biochemistry *10*, 2805 (1971)
19. DeSimone, R.E., et al.: Biochim. Biophys. Acta *304*, 851 (1973)
20. Yamamoto, H. et al.: Bull. Chem. Soc. Japan *48*, 844 (1975)
21. Chu, V.C.W., Gruenwedel, D.W.: Z. Naturforsch. C*31*, 753 (1976)
22. Craig, P.J., Morton, S.F.: J. Organometallic Chem. *145*, 79 (1978)
23. Robinson, G.C., Nome, F., Fendler, J.H.: J. Amer. Chem. Soc. *99*, 4969 (1977)
24. Bartlett, P.D., Craig, P.J., Morton, S.F.: Nature *267*, 606 (1977)

25. Huey, C. et al.: Proc. Int. Conf. Transport Persistant Chemicals Aquat. Ecosystem 1974, p. II, 73 (LeHam, Q.N., ed.); Nat. Res. Council Canada, Ottawa
26. Silver, S., Schottel, J., Weiss, A.: Proc. 3rd. Int. Biodegradation Symp. 1976, p. 899 (Sharpley, J.M., Kaplan, A.M., eds.). Applied Sci. Publ., London
27. Vonk, J.W., Sijpesteijn, A.K., van Leeuwenhoek, A.: J. Microbiol. *39*, 505 (1973)
28. Landner, L.: Nature *230*, 452 (1971)
29. Agaki, H., Takabatake, E.: Chemosphere *3*, 131 (1973)
30. Agaki, H., Fujita, Y., Takabatake, E.: Chemistry Lett. 171 (1975)
31. Hayashi, K. et al.: Chem. Commun. 158 (1977) (and references therein)
32. McBride, B.C., Edwards, T.L.: Biological Implications of Metals in the Environment, ERDA Symp. Ser. No. 42 (1977)
33. Cullen, W.R. et al.: J. Organometallic Chem. *139*, 61 (1977)
34. McBride, B.C. et al. in: Organometals and Organometalloids, ACS Symp. Series No. 82 (Brinckman, F.E., Bellama, J.M., eds.) 1978, p. 94
35. Craig, P.J., Bartlett, P.D.: Nature *275*, 635 (1978)
36. DeSimone, R.E.: Chem. Commun. *780* (1972)
37. Spangler, W.J. et al.: Science *180*, 192 (1973)
38. Bartlett, P.D., Craig, P.J., Morton, S.F.: Sci. Total Env. *10*, 245 (1978)
39. An Assessment of Mercury in the Environment. Rep. prepared by the Panel on Mercury; Nat. Acad. Sci., Washington, D.C. 1978
40. Korringa, P., Hagel, P. in: Proc. Intern. Symp. Problems Contam. Man Environ., 1973. Luxembourg, Comm. Euro. Communities 1974
41. Wollast, R., Billen, G., MacKenzie, F.T.: Proc. NATO Science Ctte. Conf. Ecol. Toxicol. Research (McIntyre, A.D., Mills, C.F., eds.). Plenum Press, New York and London 1975, p. 145
42. Abramovskiy, B.P. et al.: Second Joint US – USSR Symp. Comp. Analysis Environ., Honolulu 1975. U.S. Environ. Prot. Agency 1975, p. 14
43. U.S. Environ. Prot. Agency Final Rep. Mercury Balance prepared by URS Research Co., EPA – 560/3–75–007, Washington, D.C. 1975
44. This handbook (following volume)
45. Gavis, J., Ferguson, J.F.: Water Research *6*, 989 (1972)
46. Environm. Mercury and Man. U.K. Dept. Environ. Central Pollution Paper No. 1, Unit on Environ. Pollution; London, H.M.S.O. 1976
47. MacKenzie, F.T., Wollast, R. in: The Sea (Goldberg, E.D., et al. eds.). Wiley Interscience, New York 1977, Vol. 6, p. 739
48. Kothny, E.L. in: Trace Elements in the Environment (Gould, R.T., ed.); Adv. Chem. Series No. 123, Amer. Chem. Soc., Washington, D.C. 1973, p. 48
49. McCarthy, J.H. Jr. et al.: Circular 609, U.S. Geol. Survey, Washington, D.C. 1969
50. Johnson, D.L., Braman, R.S.: Environ. Sci. Technol. *8*, 1003 (1974)
51. Ref. [39], p. 34
52. Williston, S.H.: J. Geophys. Res. *73*, 7051 (1968)
53. Ref. [39], p. 2
54. Ref. [39], p. 16
55. Weiss, H.J. et al.: Geochim. Cosmochim. Acta *39*, 1 (1975)
56. Ref. [39], p. 32
57. Ref. [39], p. 35
58. WHO. Environm. Health Criteria 1, Mercury. Rep. a Meeting, Geneva 1975. WHO Geneva 1976
59. Shacklette, H.T., Boerngen, J., Turner, R.L.: Mercury in the Environ. – Surficial Materials of the Coterminous U.S., USGS Circular, Washington, D.C., U.S. Geological Survey 1971
60. Magos, L., Tuffery, A.A., Clarkson, T.W.: Brit. J. Ind. Medicine *21*, 294 (1964)
61. Furukawa, K., Suzuki, T., Tonamura, K.: Agric. Biol. Chem. *33*, 128 (1969)
62. Beckert, W.F. et al.: Nature *249*, 674 (1974)
63. Olsen, B.H., Cooper, R.C.: Nature *252*, 682 (1974)
64. Kudo, A. et al.: Nature *270*, 419 (1977)
65. Ericksson, E.: Tellus *12*, 63 (1960)
66. Anfalt, D. et al.: Svensk. Kemisk. Tidskrift *80*, 340 (1968)

67. Dyrssen, D., Wedborg, M. in: Marine Chemistry (Goldberg, E.D., ed.), Vol. 5, p. 181. John Wiley, New York 1974
68. Hem, J.D. in: Mercury in the Environ., U.S. Geol. Survey Prof. Paper 713; U.S. Govt. Printing Office, Washington, D.C. 1970, p. 19
69. Jewett, K.L., Brinckman, F.E., Bellama, J.M. in: Marine Chemistry in the Coastal Environm. ACS Symp. Series No. 18 (Church, T.C., ed.) 1975, p. 304
70. Hung, T.-C., Lin, T.-T.: Acta Oceanog. Taiwan. No. *6*, 30 (1976)
71. Chau, Y.K., Saitoh, H.: Intern. J. Environ. Anal. Chem. *3*, 133 (1973)
72. Andren, A.W., Harriss, R.C.: Geochim. Cosmochim. Acta *39*, 1253 (1975)
73. Gardner, D.: Nature *272*, 49 (1978)
74. Fitzgerald, W.F., Lyons, W.B.: Nature *242*, 452 (1973)
75. Lindberg, S.E., Harriss, R.C.: Environ. Sci. Technol. *8*, 459 (1973)
76. Lindberg, S.E., Andren, A.W., Harriss, R.C. in: Chemistry, Biology and the Estuarine System, Vol. I (Cronin, E.L., ed.); Academic Press, New York 1975, p. 64
77. Cranston, R.E., Buckley, D.E.: Environ. Sci. Technol. *6*, 274 (1972)
78. Thomas, R.L., Jaquet, J.-M.: J. Fish. Res. Board Can. *33*, 404 (1976) (and ref. therein)
79. Carr, R.A., Hoover, J.B., Wilkness, P.E.: Deep Sea Res. *19*, 747 (1972)
80. Ref. [39], p. 39
81. Mercury and the Environment. Studies of Mercury Use, Emission, Biological Impact and Control; Organization for Economic Cooperation and Development, Paris 1974
82. Shimp, N.F., Leland, H.V., White, W.A.: Environ. Geology Notes No. 32, Illinois State Geolog. Surv. 1970
83. Ref. [47], p. 762
84. Impact of Metals in the Biosphere, Dahlem Konferenzen. In: Global Chemical Cycles and their Alterations by Man (Stumm, W., Pytkowicz, R.M., eds.). Berlin, West Germany 1976
85. Ref. [47], p. 763
86. Kemp, A.L.W., Thomas, R.L.: Water, Air, Soil Pollut. *5*, 469 (1976) (and references therein)
87. Kudo, A. et al.: Prog. Water Technol. 1978, *10*, 329 (1978) (and ref. therein)
88. Craig, P.J., Morton, S.F.: Nature *261*, 125 (1976)
89. Hamdy, M.K., Noyes, O.E.: Appl. Microbiol. *30*, 424 (1975)
90. Olson, B.H., Cooper, R.C.: Water Res. *10*, 113 (1976)
91. Spangler, W.J. et al.: Appl. Microbiol. *25*, 488 (1973)
92. Ref. [37], p. 193
93. Tonamura, K., Furukawa, K., Yamada, M.: Environ. Toxicol. Pesticides (Matsumura, F., Bousland, G.M., Misato, T., eds.). Academic Press, N.Y. 1972, p. 115
94. Yamada, M., Tonamura, K.: J. Ferment. Technol. *50*, 159 (1972)
95. Ibid. *50*, 893 (1972)
96. Ibid. *50*, 901 (1972)
97. Brosset, C., Svedung, I.: Swed. Water and Air Pollution Res. Labor. Rep. No. B378, Gothenburg, Sweden 1977
98. Rowland, I.R., Davies, M., Grasso, P.: Arch. Environ. Health *32*, 24 (1977)
99. Lindberg, S.E., Harriss, R.C.: J. Water Pollut. Control Fed. *49*, 2479 (1977)
100. Ref. [27], p. 505
101. Hamdy, M.K., Noyes, O.R.: Applied Microbiol. *30*, 424 (1975)
102. Barber, R.T., Vijayakumar, A., Cross, F.A.: Science *178*, 636 (1972)
103. Miller, G.E. et al.: Science *175*, 1121 (1972)
104. Lakowicz, J.R., Anderson, C.J.: 1979 (submitted for publication)
105. Ref. [12], p. 18
106. Edwards, T., McBride, B.C.: Nature *253*, 462 (1975)
107. Rowland, I.R., Grasso, P., Davies, M.J.: Experimentia *31*, 1064 (1977)
108. Chau, Y.K., Saitoh, H.: Intern. J. Environ. Anal. Chem. *3*, 133 (1973)
109. Lofroth, G.: Ecol. Res. Bull. No. 4, Swed. Nat. Sci. Res. Council 1969
110. Kitamura, S.: Jumamoto Igk *37*, 494 (1963)
111. Ganther, H.E., Sande, M. L.: J. Food Sci. *39*, 1 (1974)
112. Ganther, H.E. et al.: Science *175*, 1122 (1972)
113. Rowland, J.R., Davies, M.J., Grasso, P.: Nature *265*, 718 (1977)
114. Craig, P.J., Bartlett, P.D.: Nature *275*, 635 (1978)

115. Krenkel, P., Reimers, R.S., Burrows, W.D.: Techn. Rep. No. 31. Vanderbilt University, Tennessee 1973 (and ref. therein)
116. Ref. [97], p. 10
117. Rogers, R.D.: J. Environ. Qual. *6*, 463 (1977)
118. Rogers, R.D.: U.S. Environ. Protection Agency Rep. No. EPA–600/3–77/007 1977 (Nat. Techn. Inform. Service, Springfield, Virginia 22161 USA)
119. Gilfillan, S.C.: J. Occup. Med. *7*, 53 (1965)
120. Waldron, H.A., Stoffen, D.: Sub-clinical Lead Poisoning. Academic Press, New York 1974
121. Patterson, C.C.: Arch. Environ. Health *11*, 344 (1965)
122. Nriagu, J.R. (ed.).: Biogeochemistry of Lead in the Environment. Elsevier-North Holland, Amsterdam 1978 (Vols. 1A and 1B).
123. Junge, C.E.: Air Chemistry and Radioactivity. Academic Press, N.Y. 1963
124. Patterson, C.C., in: Ref. [121]
125. Cawse, P.A.: Survey Atmosph. Trace Elements in the U.K.; A.E.R.E. Rep. No. R–7669 HMSO, London 1974
126. Kerin, Z.: Arch. Environ. Health *26*, 256 (1973)
127. Murozumi, M., Chow, T.J., Patterson, C.C.: Geochim. Cosmochim. Acta *33*, 1247 (1969)
128. Chow, T.J., Patterson, C.C.: Geochim. Cosmochim. Acta *26*, 263 (1962)
129. Poldervaart, A.: Chemistry of the Earth's Crust., Geol. Soc. Amer. Spec. Paper 1955, *62*, p. 119
130. Chow, T.J., Patterson, C.C.: Earth Planet., Sci. Lett. *1*, 397 (1966)
131. Patterson, C.C.: Marine Chem. *2*, 69 (1974)
132. Goldberg, E.D., Arrhenius, G.O.S.: Geochim. Cosmochim. Acta *13*, 153 (1958)
133. Durum, W.H., Haffty, J.: Geochim. Cosmochim. Acta *27*, 1 (1963)
134. Ref. [122], p. 285
135. Ref. [122], p. 204
136. Ref. [122], p. 205
137. Batley, G.E., Florence, T.M.: Marine Chem. *4*, 347 (1976)
138. Stumm, W., Bilinski, H. in: Adv. Water Pollut. Res., Proc. 6th. Intern. Conf. (Jenkins, S.H., ed.) 1972. Pergamon Press, New York, p. 39
139. Sillen, L.G. in: Oceanography (Sears, M., ed.). Amer. Assoc. Adv. Sci. Publ. 67, Washington 1961, p. 549
140. Goldberg, E.D. in: The Sea (Hill, M.N., ed.). Interscience, New York 1963, p. 3
141. Ref. [122], p. 264
142. Stumm, W., Morgan, J.J.: Aquatic Chemistry, Wiley Interscience, N.Y. 1970
143. Bilinksi, H., Stumm, W.: Swiss Fed. Inst. Technol., EAWAG News 1973, No. 1 (Jan).
144. Hem, J.D.: Geochim. Cosmochim. Acta *40*, 599 (1976)
145. Ramamoorthy, S., Kushner, D.J.: Nature *256*, 399 (1975)
146. Zirino, A., Yamamoto, S.: Limnol. Oceanogr. *17*, 661 (1972)
147. Dyrsson, D., Wedborg, M. in: Ref. [47] 1975, Vol. 5, p. 181
148. Ref. [122], p. 213
149. Ref. [122], p. 273, 274
150. Andren, A.W., Elzerman, A.W., Armstrong, D.E.: J. Great Lakes Res. *2* (suppl. 1), 101 (1976)
151. Perhae, R.M., Whelan, C.J.: J. Geochem. Explor. *1*, 47 (1972)
152. Leland, H.V., McNurney, J.M.: Ref. [25], p. II, p. 17
153. Ref. [122], p. 62
154. Wei, L.S.: Thesis, Agronomy Dept., Univ. Illinois, Urbana 1959
155. Ref. [122], p. 273
156. Gardner, L. R.: Geochim. Cosmochim. Acta *38*, 1297 (1974)
157. Ref. [122], p. 137
158. Robinson, E., Robbins, R.C.: Stanford Res. Inst. Rep. 1971, SCC-8507
159. Environment Canada Report 1973. Air Pollut. Control Directorate, Ottawa. Rep. No. APCD 73–7
160. Ref. [122], p. 138
161. Ref. [122], p. 141
162. Harrison, R.M., Perry, R., Slater, D.H.: Atmos. Environ. *8*, 1187 (1974)

163. Daines, R.H., Motto, H., Chilko, D.M.: Environ. Sci. Technol. *4*, 318 (1970)
164. Ref. [122], p. 254
165. Ter Haar, G.L., Bayard, M.A.: Nature *232*, 553 (1971)
166. Pierrard, J.M.: Environ. Sci. Technol. *3*, 48 (1969)
167. Laveskog, A. in: Proc. 2nd Intern. Clean Air Congress (Englund. H.M., Beery, W.T., eds.) Academic Press, New York 1971, p. 549
168. Harrison, R.M., Laxon, D.P.H.: Nature *275*, 735 (1978)
169. Ref. [122], p. 163
170. Ref. [122], p. 174
171. Harrison, R.M.: Sci. Total. Environ. *11*, 89 (1979)
172. Bryce-Smith, D., Mathews, J., Stephens, R.: Ambio *7*, 192 (1978)
173. Chow, T.J., Earl, J.L.: Science *169*, 577 (1970)
174. Livingstone, D.A. in: Data of Geochemistry 1963, 6th edn. (Fleischer, M., ed.), U.S. Geol. Survey Prof. Paper No. 440–G
175. Fleischer, M. in: Proc. Environ. Res. Conf. (Curry, M.G., Gigliotti, G.M., eds.) Nat. Environ. Res. Center, Cincinnati 1973, p. 3
176. Warren, H.V. et al.: Trace Substances Environ. Health *4*, 94 (1971)
177. Roberts, T.M., Goodman, G.T.: Trace Substances Environ. Health *7*, 117 (1973)
178. Lockertz, W.: Water, Air, Soil Pollut. *3*, 179 (1974)
179. Block, C., Dams, R.: Water, Air, Soil Pollut. *5*, 207 (1975)
180. Lindberg, S.E. et al.: Environ. Health Perspect. *12*, 9 (1975)
181. Page, A.L.: Rep. No. EPA 670/2–74–005 1974 Environ. Protn. Agency, Cincinnati
182. Ref. [122], p. 45
183. Swaine, D.J.: Tech. Comm. No. 48, Commonwealth Bureau Soil Sci., York, U.K. 1955
184. Van Loon, J.C. et al.: J. Water, Air, Soil Pollut. *2*, 473 (1973)
185. Bertinuson, J.R., Clark, C.S.: Interface *6*, 1073 (1973)
186. Bruland, K.W. et al.: Environ. Sci. Technol. *8*, 425 (1974)
187. Burrows, K.C., Hulbert, M.H.: Ref. [69], p. 382
188. Ref. [122], p. 58
189. Ref. [122], p. 60
190. Ref. [122], p. 63
191. Craig, P.J., Wood, J.M.: Biological Methylation of Lead. In: Proc. 2nd. Intern. Symp. Environ. Lead Res., Cincinnati 1978. Academic Press, New York 1979 (in press)
192. Wong, P.T.S., Chau, Y.K., Luxon, P.L.: Nature *253*, 263 (1975)
193. Chau, Y.K., Wong, P.T.S.: Ref. [34], p. 39
194. Jarvie, A.W.P., Markall, R.N., Potter, H.R.: Nature *255*, 217 (1975)
195. Schmidt, U., Huber, F.: Nature, *259*, 157 (1976)
196. Huber, F., Schmidt, U., Kirchmann, H.: Ref. [34], p. 65
197. Akagi, H., Fujita, Y., Takabatake, E.: J. Chem. Soc. Japan 1180 (1974)
198. Dumas, J.P. et al.: Proc. 12th Canadian Symp. Water Pollution Res. 1977 (in French)
199. Honeycutt, J.B. Jr., Riddle, J.M.: J. Amer. Chem. Soc. *82*, 3051 (1960)
200. Honeycutt, J.B. Jr., Riddle, J.M.: J. Amer. Chem. Soc. *83*, 369 (1961)
201. Riddle, J.M.: US Pat. 2,950,301 (1960)
202. Riddle, J.M.: US Pat. 2,950,302 (1960)
203. Taylor, R.T., Hanna, M.L.: J. Environ. Sci. Health Environ. Sci. Eng. *3*, 201 (1976)
204. Sirota, G.R., Uthe, J.F.: Annal. Chem. *49*, 823 (1977)
205. Agnes, G. et al.: Chem. Commun. 850 (1971)
206. Lewis, J., Prince, R.H., Stotter, D.A.: J. Inorg. Nucl. Chem. *35*, 341 (1973)
207. Bowen, H.J.M.: Trace Elements in Biochemistry. Academic Press, London 1966
208. Gordon, M.: Chem. Abstr. *47*, 4533 (1953)
209. Sainsbury, C.L., Hamilton, J.C., Huffman, C. Jr.: U.S. Geol. Survey Bull. 1968, No. 1242–F U.S. Govt. Printing Office, Washington, D.C.
210. Morik, J., Morlin, Z.: Chem. Abstr. *57*, 8840 (1962)
211. Schwarz, K., Milne, D.B., Vinyard, E.: Biochem. Biophys. Res. Commun. *40*, 22 (1970)
212. Monier-Williams, G.W.: Trace Elements in Food. Chapman-Hall, London 1949
213. De Groot, A.P., Feron, V.J., Til, H.P.: Food Cosmet. Toxicol. *11*, 19 (1973)
214. Zuckerman, J.J. et al.: Ref. [34], p. 411

215. Ref. [2], p. 88
216. Sawyer, A.K.: Organotin Compounds; Marcel Dekker, New York 1971 (3 volumes)
217. Ref. [34], p. 359 and p. 388
218. In: Ref. [84]
219. Brinckman, F.E., Iverson, W.P. in: Ref. [69], p. 319
220. Huey, C. et al. in: Ref. [25], p. II, 77
221. Dizikes, L.J., Ridley, W.P., Wood, J.M.: J. Amer. Chem. Soc. *100*, 1010 (1978)
222. Braman, R.S., Tomkins, M.A.: Anal. Chem. *51*, 12 (1979)
223. Jernelöv, A. in: Ref. [47] 1974, Vol. 5, p. 799
224. Encyclop. Brittanica, Macropaedia 1976, 15th edn., *6*, 702
225. MacKenzie, F.T., Wollast, R.: Ref. [47], 1977, Vol. 6, p. 764
226. Garrels, R.M., MacKenzie, F.T., Hunt, C.: Chemical Cycles and the Global Environment. William Kaufmann, Los Altos, Ca. 1973, p. 114
227. Pearson, R.G.: Chem. in Brit. *3*, 103 (1967)
228. Pearson, R.G.: J. Amer. Chem. Soc. *85*, 3533 (1963)
229. Pearson, R.G.: Science *151*, 172 (1966)
230. Pearson, R.G. (ed.): Hard and Soft Acids and Bases; Benchmark Papers in Inorg. Chem. Academic Press, New York 1973
231. Fates of Pollutants; Development Needs. US Nat. Res. Council, Washington 1977, p. 64
232. Ref. [231], p. 65
233. Baes, C.F. Jr. in: Cadmium – The Dissipated Element (Fulkerson, W., Goeller, H.E., eds.); Oak Ridge Nat. Lab., Tennessee 1973, p. 29
234. Ref. [231], p. 66
235. Jellinek, H.H.G., Sangal, S.P.: Water Res. *6*, 305 (1972)
236. Schindler, P.W.: Proc. 1st. Special Symp. Atmos. Contrib. Chem. Lake Waters, Geneva Park, Ontario 1975. In: J. Great Lakes Res. *2*, (supplement), 132 (1976)
237. MacKenzie, F.T., Wollast, R.: Ref. [47], 1977, Vol. 6, p. 777
238. Ref. [231], p. 74
239. Environ. Health Perspectives; U.S. Dept. HEW 1977, *19*, p. 1
240. Jernelöv, A.: Ref. [47] 1974, Vol. 5, p. 806
241. Ref. [239], especially p. 11
242. MacKenzie, F.T., Wollast, R.: Ref. [47], 1977, Vol. 6, p. 746
243. Ref. [231], p. 60
244. Huey, C.W. et al.: Intern. Conf. Heavy Metals Environ., Toronto 1975, Abstr. p. C214 (Hutchinson, T.C.; Programme: Coordinator)
245. Huey, C.W. et al.: Prog. Water Technol. (Krenkel, P.A., ed.) 1975, Vol. 7
246. In: Ref. [205]
247. Huber, F., Kirchmann, H.: Inorg. Chim. Acta *29*, L249 (1978)
248. In: Ref. [205]
249. Hill, H.A.O. et al.: Chem. Commun. 341 (1970),
250. Abley, P., Dockal, E.R., Halpern, J.: J. Amer. Chem. Soc. *95*, 3166 (1973)
251. Fanchiang, Y.-T., Ridley, W.P., Wood, J.M.: J. Amer. Chem. Soc. *101*, 1442 (1979)
252. Fanchiang. Y.-T. Wood, J.M.: Personal commun. 1979
253. Taylor, R.T., Hanna, M.L.: Bioinorgan. Chem. *6*, 281 (1976)
254. Taylor, R.T., Happe, J.A., Wu, R.: J. Environ. Sci. Health A13(9), 707 (1978)
255. Ref. [206], p. 344
256. Scovell, W.M.: J. Amer. Chem. Soc. *96*, 3451 (1974)
257. Challenger, F.: Ref. [34], p. 1
258. McBride et al.: Ref. [34], p. 94
259. Gosio, B.: Arch. Ital. Biol. *35*, 201 (1901)
260. Challenger, F.: Chem. Rev. *36*, 315 (1945)
261. Challenger, F.: Quart. Rev., Chem. Soc. *9*, 255 (1955)
262. Challenger, F.: Aspects of the Organic Chemistry of Sulfur. Butterworth, London 1959
263. Cantoni, G.L.: J. Amer. Chem. Soc. *74*, 2942 (1952)
264. Cox, D.P., Alexander, M.: Applied Microbiol. *25*, 408 (1973)
265. McBride, B.C., Wolfe, R.S.: Biochem. *10*, 4312 (1971)
266. Wolin, M.J., Wolin, E.A., Wolfe, R.S.: Biochem. Biophys. Res. Commun. *12*, 465 (1963)

267. Braman, R.S., Foreback, C.C.: Science *182*, 1247 (1973)
268. Johanson, D.L., Braman, R.S.: Deep Sea Res. *22*, 503 (1975)
269. Edmonds, J.S., Francesconi, K.A.: Nature *265*, 436 (1977)
270. Wong, P.T.S. et al.: Proc. 11th Ann. Conf. Trace Substances Environ. Health 1977 (June), Columbia, Missouri, USA
271. Challenger, F., North, H.E.: J. Chem. Soc. 68 (1934)
272. Bird, M.L.Challenger, F.: J. Chem. Soc. 163 (1939)
273. Flemming, R.W., Alexander, M.: Applied Microbiol. *24*, 424 (1972)
274. Burkes, L., Flemming, R.W.: Bull. Environ. Contam. Toxicol. *12*, 308 (1974)
275. Francis, A.J., Duxburg, J.M., Alexander, M.: Appl. Microbiol. *28*, 248 (1974)
276. Bird, M.L. et al.: Biochem. J. *43*, 78 (1948)
277. Dransfield, P.B., Challenger, F.: J. Chem. Soc. 1153 (1955)
278. Challenger, F., Lisle, D.B., Dransfield, P.B.: J. Chem. Soc. 1760 (1954)
279. McConnell, K.P., Portman, O.W.: J. Biol. Chem. *195*, 277 (1952)
280. Challenger, F.: Ref. [262]
281. Challenger, F.: Ref. [261]
282. Klug, H.L., Froom, J.D.: Proc. S.D. Acad. Sci. *64*, 247 (1965)
283. Byard, J.L.: Arch. Biochem. Biophys. *130*, 556 (1969)
284. Palmer, I.S. et al.: Biochem. Biophys. Acta *177*, 336 (1969)
285. Palmer, I.S. et al.: Biochem. Biophys. Acta *208*, 260 (1970)
286. Chau, Y.-K. et al.: Science *192*, 1130 (1976)
287. Alexander, M.: Cornell Univ. Water Resource Mar. Sci. Cent. Tech. Rep. 1972, No. 47
288. Lewis, B., Johnson, C.M., Delwiche, C.C.: J. Agric. Food Chem. *14*, 638 (1966)
289. Evans, C.S., Asher, C.J., Johnson, C.M.: Austr. J. Biol. Sci. *21*, 13 (1968)
290. Birkinshaw, J.H., Findlay, W.P.K., Webb, R.A.: Biochem. J. *36*, 526 (1942)
291. Agnes, G. et al.: Biochim. Biophys. Acta *252*, 207 (1971)
292. Schrauzer, G.N.: Acc. Chem. Res. *1*, 97 (1968)
293. Bernhauer, K., Irion, E.: Biochem. Z. *339*, 521 (1964)

Natural Organohalogen Compounds

D. J. Faulkner
Scripps Institution of Oceanography
La Jolla, CA 92093, USA

Introduction

Over the past decade, marine natural products chemists have demonstrated that marine organisms produce a wide variety of halogenated natural products. The halogenated natural products range in size from simple halomethanes to relatively large and complex toxins. The principal sources of halogenated natural products are marine bacteria, marine algae and sponges, with smaller contributions from other marine invertebrates. Despite their importance in medicine, the halogenated antibiotics such as aureomycin and griseofulvin, which are manufactured by fermentation processes involving terrestrial microorganisms, have not been included as *natural* organohalogen compounds.

This review of halogenated marine natural products will emphasize the major structural types of chemicals encountered, their chemotaxonomic relationships and, wherever possible, their potential environmental impact. Although there has been little research in these areas, the biosynthesis of halogenated organic compounds in marine organisms and the fate of halogenated marine natural products in seawater will be discussed briefly. Since it would be unrealistic to review every known halogenated marine natural product, I have selected representative examples from among groups of chemically similar compounds.

Marine Bacteria

Several species or strains of marine bacteria have been shown to contain brominated pyrrole derivatives. In 1966, Burkholder et al. [1] showed that the antimicrobial agent produced by the marine bacterium *Pseudomonas bromoutilis* was a highly brominated pyrrole derivative *1*. The same pyrrole *1* was subsequently found in four other antibiotic-producing strains of marine bacteria, including *Chromobacterium* sp. [2], which also contained tetrabromopyrrole *2* and hexabromobipyrrole *3*. It is important to note that these halogenated bacterial metabolites were isolated because they possessed antimicrobial properties and not as a result of a systematic chemical investigation

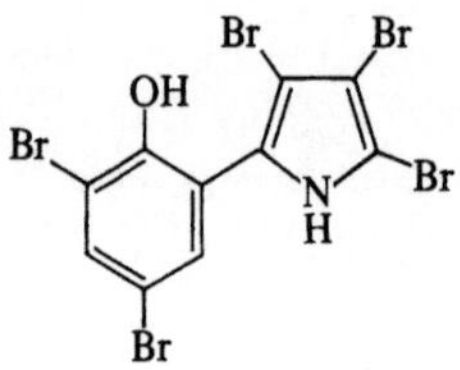

1

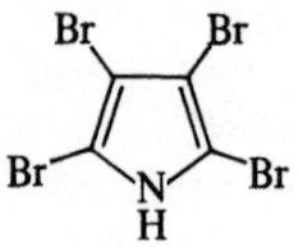

2

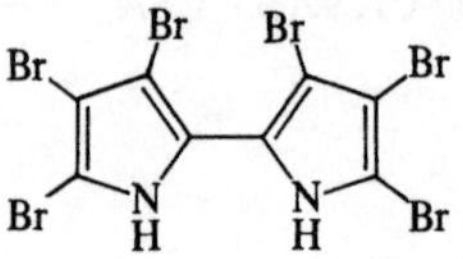

3

of marine bacteria. The production of antimicrobial compounds by marine bacteria is thought to provide a competitive advantage to the producing organism. However, when *Chromobacterium* sp. was grown in an enriched culture medium, sufficient amounts of the antibiotic pyrrole *1* were released into the medium to cause autotoxicity. This provides one simple illustration of the fact that the enriched culture medium required to grow a sufficient quantity of bacterial cells for chemical studies differs so considerably from the natural conditions found in the open ocean, on the surface of particles, or in the sediments, that we may never be able to determine the importance of marine bacteria in the production of halogenated natural products.

Marine Algae

Considering the importance of the phytoplankton in the marine ecosystem, surprisingly little is known about the natural products chemistry of diatoms or dinoflagellates. There are no reports of the isolation of halogenated compounds from pure cultures of either diatoms or dinoflagellates.

Studies of the toxic or antibiotic constituents of blue-green algae (Cyanophyta) have resulted in the isolation of several halogenated metabolites. Fenical and Look [3] have isolated the chlorinated phenol *4* from a pure culture of the blue-green alga *Anacystis marina*. The phenol *4* inhibited the growth of test microorganisms and may also inhibit the growth of competing organisms in the natural ecosystem.

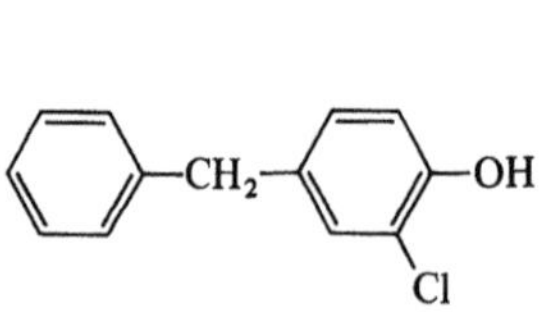

4

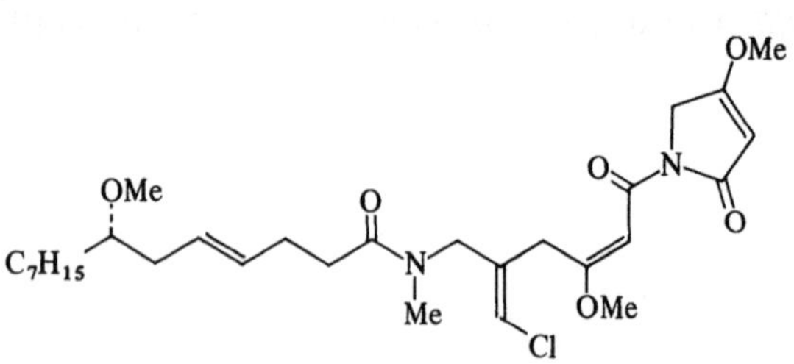

5

A close similarity between non-halogenated compounds found in the blue-green alga *Lyngbya majuscula* and the halogenated compounds aplysiatoxin *94* and stylocheilamide *95*, both found in the sea hare *Stylocheilus longicauda* (Sect. "Other Marine Invertebrates"), suggested that these halogenated metabolites could have been produced by blue-green algae. Subsequently, three bromine-containing toxins related to aplysiatoxin *94* were found as minor constituents of a mixture of *Oscillatoria nigroviridis* and *Schizothorix calcicola* [4]. The same mixture of blue-gren algae also contained *E*-l-chlorotridec-l-ene-6,8-diol as a major constituent [5]. Moore [6] has recently reported the presence of two chlorinated metabolites, malyngamide A *5* and malyngamide C, in samples of *Lyngbya majuscula*. Research by Moore's group suggests that not all samples of a particular species of blue-green alga contain the same compounds, but the reason for this chemical diversity is not yet known.

There have been no reports of halogenated metabolites from brown algae (Phaeophyta) although there has been some speculation that brown algae may be the source of iodinated metabolites (see Sect. "Other Marine Invertebrates"). The green algae (Chlorophyta) do not generally contain halogenated metabolites. The single exception was a calcareous green alga *Cymopolia barbata*, which contained five brominated metabolites [7]. The simplest compounds, cymopol *6* and its monomethyl ether, are prenylated hydroquinones in which the aromatic ring has been brominated. The more interesting metabolites are cyclocymopol *7* and the corresponding monomethyl ether, in which the isoprenoid chain has been cyclized with incorporation of a second bromine atom. The biosynthesis of cyclocymopol *7* must involve two separate halogenation mechanisms, both of which are more commonly found in red algae.

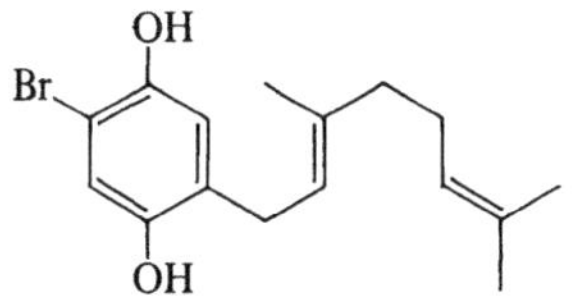

6

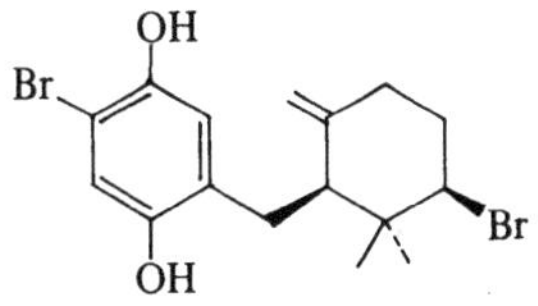

7

The greatest variety of halogenated natural products has been found in the red algae (Rhodophyta). The large number (>250) of halogenated metabolites from red algae precludes a detailed description of each compound. The compounds to be discussed include examples of all major structural types and illustrate the various halogenation pathways that can be encountered.

The simplest halogenated metabolites were found in the family Bonnemaisoniaceae, particularly in members of the genus *Asparagopsis* [8]. Both *A. taxiformis* from Mexico and Hawaii and *A. armata* from Spain contain very high concentrations (2%–3% dry weight) of bromoform. Among the minor constituents were methyl iodide, chloroform, carbon tetrachloride and carbon tetrabromide, all compounds that are normally considered to be synthetic molecules. Since these compounds are all relatively volatile and may be lost during work-up, the abundances of the halomethanes in *Asparagopsis* species could not be estimated with any accuracy. The presence of halogenated acetones and halogenated acetic acids in *Asparagopsis* suggested that the halomethanes and halogenated acetic acids may be formed by the haloform reaction of halogenated acetones. The presence of halogenated acrylic acids in *Asparagopsis* has been rationalized by suggesting that they were formed by a Favorsky rearrangement of the halogenated acetones. The most complex molecules isolated from *Asparagopsis* were the halogenated butenones of which 1,1,4,4-tetrabromo-3-buten-2-one is a good example. Combined gas chromatography-mass spectrometry (GC-MS) allowed the detection of over 100 different halogenated C_1–C_4 compounds containing various combinations of bromine, chlorine and iodine.

Members of the genus *Bonnemaisonia* all contain halogenated compounds in the C_7–C_9 range. The major metabolite (0.01% wet weight) of *B. hamifera* was 1,1,3,3,-tetrabromo-2-heptanone [9] while *B. nootkana* contained mainly *trans*-1,3,3-tribromo-l-heptene oxide *8* [10]. *B. asparagoides* was shown to

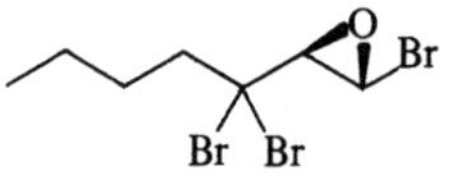

8

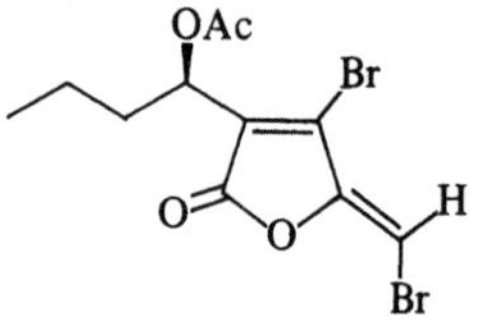

9

contain five halogenated 1-octen-3-ones with *E*-1-bromo-1,2,4-trichloro-1-octen-3-one as the most abundant constituent [11]. It is interesting to note that *Bonnemaisonia asparagoides* contains compounds with both bromine and chlorine substituents while only brominated metabolites were found in *B. nootkana* and, with the exception of a trace of a compound containing both iodine and bromine, in *B. hamifera*.

Both *Delisea fimbriata* [12] and *Ptilonia australasica* [13] have been shown to contain 1,1,2,4-tetrabromo-1-octen-3-one and other related metabolites. *Delisea fimbriata* also contained a number of halogenated C_9-γ-lactones, of which the acetate *9* was the most carefully studied [13, 14].

Any attempt to assess the importance of the halogenated compounds from the Bonnemaisoniaceae must contend with many unknowns. The natural products chemists have described a large number of compounds, which have been reviewed in great detail by Moore [15], but no reliable quantitative data exist. Hager and coworkers [16] found that *Asparagopsis taxiformis* had a total halogen content of 0.22% of dry weight with a bromine/total halogen ratio of 55%. For *Bonnemaisonia hamifera* the respective figures were 0.12% total halogen containing 88% bromine. Other researchers have reported that there is a great deal of variability between samples of the same species (8a). Under these circumstances, quantitative data for individual compounds seem unreliable and a global estimate of algal production of halogenated compounds seems impossible.

Two other algae that were shown [16] to have high total halogen contents were *Laurencia pacifica* (0.29% dry weight) and *Plocamium pacificum [≡ Plocamium cartilagineum?]* (0.64% dry weight). These data could have been predicted from the studies of the natural products from *Plocamium* and *Laurencia* species. *Plocamium* species have been shown to produce halogenated monoterpenes while *Laurencia* species produced a very wide range of both terpenoid and nonterpenoid halogenated metabolites.

It is rather unfortunate that the first halogenated monoterpene *10* was initially isolated from the digestive gland of the sea hare *Aplysia californica* rather than from its true algal source, *Plocamium cartilagineum* [17]. Subsequent examination [18] of *P. cartilagineum* from La Jolla, California revealed the presence of eleven other monoterpenes, four of which (two diastereoisomeric pairs) were derived by loss of HCl or BrCl from monoterpene *10,* while four others (again two diastereoisomeric pairs) could be represented by structure *11.* Furthermore, examination of individual plants showed that there were at least three different "chemical types" of *P. cartilagineum* in the La Jolla area, all of which contained appreciable quantities of halogenated acyclic monoterpenes but with different arrays of compounds [19]. It is therefore no surprise that the halogenated monoterpenes from *Plocamium cartilagineum* vary considerably according to collecting locations [20]. The discovery of halogenated cyclic monoterpenes from *P. cartilagineum* collected in Britain [21] and Spain [22] led researchers to abandon the use of cyclic *vs.* acyclic halogenated monoterpenes as a chemotaxonomic marker for distinguishing *Plocamium* species; in La Jolla, however, cyclic halogenated monoterpenes were found to be characteristic of *P. violaceum.*

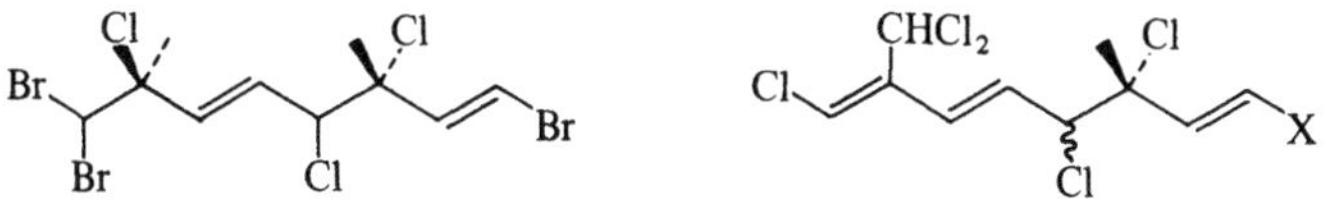

10   11 X = H or Br

Plocamium violaceum from La Jolla contained halogenated cyclic monoterpenes of two structural types. Violacene *12* [23] can be considered as a cyclized monoterpene, while violacene-2 *13* [24] is a non-isoprenoid and must be considered as a rearranged monocyclic monoterpene. The structural assignment of violacene *12,* initially incorrect [25], illustrated the difficulty in assigning the relative locations of bromine and chlorine in polyhalogenated molecules. Violacene-2 *13* undergoes an autocatalytic dehydrohalogenation to give the chlorinated styrene *14,* one of the few chlorinated aromatic

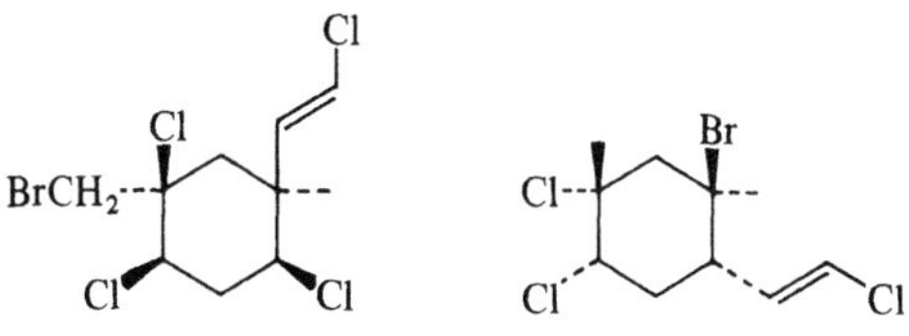

12   13

compounds easily derived from algal metabolites. Some samples of *P. violaceum* from Monterey County, California, contain halogenated acyclic monoterpenes, such as *15*, which could be the immediate precursors of the cyclic derivatives [26].

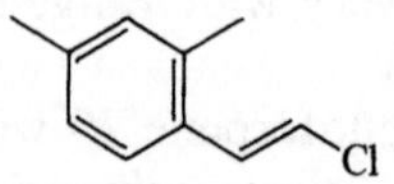

14

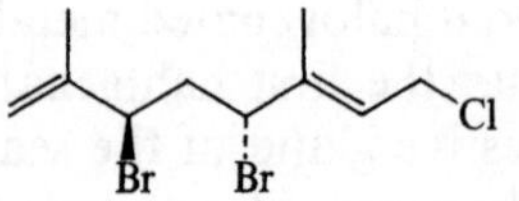

15

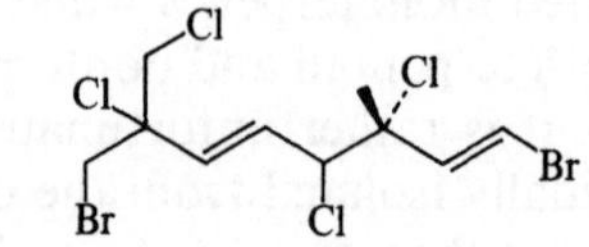

16

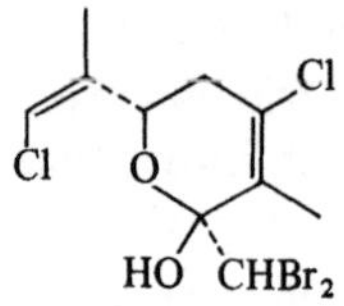

17

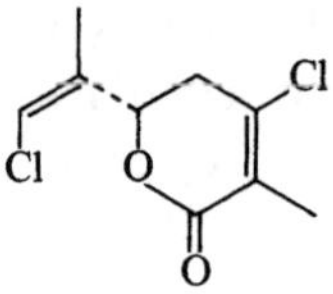

18

Plocamium oregonum [27] was shown to contain halogenated acyclic monoterpenes, such as oregonene A *16*, which were reminiscent of *P. cartilagineum* metabolites. Although *Plocamium costatum* [28] contains a cyclic ether *17* and a lactone *18* together with acyclic compounds, these are all considered acyclic compounds. The transformation from *17* to *18* is presumed to involve the evolution of dibromomethane. *Plocamium mentensii* contained a cyclic halogenated monoterpene similar to violacene *12*.

Numerous cyclic and acyclic halogenated monoterpenes have been isolated from Hawaiian and Japanese *Chondrococcus hornemanni* [29]. The acyclic compounds are based on the myrcene skeleton and vary in complexity from 2-chloromyrcene *19* to the highly halogenated derivative *20*. The cyclic halogenated monoterpenes such as chondrocole A *21* are based on an unusual carbon skeleton. *Ochtodes secundiramea* from Belize contains mainly cyclic halogenated monoterpenes such as ochtodene *22* [30].

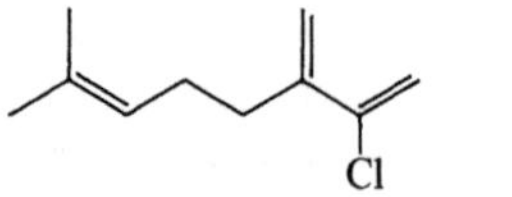

19

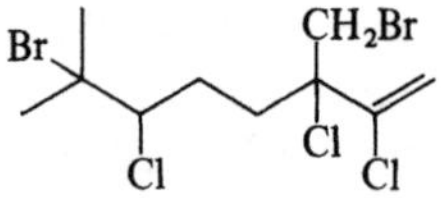

20

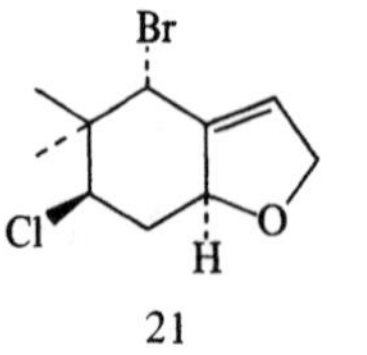

21

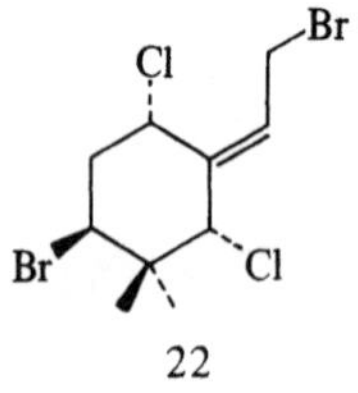

22

The importance of the halogenated monoterpenes can be judged from the following data: *Plocamium* species contain very high total halogen values; the halogenated monoterpenes, though often unstable to work-up and chromatography conditions, can still be isolated in up to 0.5% dry weight of the algae; the algae that produce halogenated monoterpenes have been found in all the world's oceans, in tropical, temperate and polar regions; along the Pacific coast of North America, *Plocamium* species occur from the intertidal zone to over 35 m depth.

Red algae of the genus *Laurencia* produce a wide variety of halogenated metabolites which, with a few exceptions, fall into three structural types: the sesquiterpenes, the diterpenes and the "linear" acetylenes. Each of these structural types can be further subdivided. The sesquiterpenes are represented by halogenated bisabolanes, cuperanes, chamigranes, selinanes and several new carbon skeletons [31]. The chemical diversity is quite remarkable.

Laurinterol *23,* isolated from *Laurencia intermedia* [32], was among the first halogenated terpenes to be described. Laurinterol and several other related *para*-bromophenols, which differ in the arrangement of groups around the five-membered ring, have been identified as metabolites of other *Laurencia* species [33] and are responsible for the antimicrobial properties of members of this genus [34]. Two brominated cuperanes *24* with bromine substitution in the five-membered ring were found in *L. glandulifera* and *L. nipponica* [35].

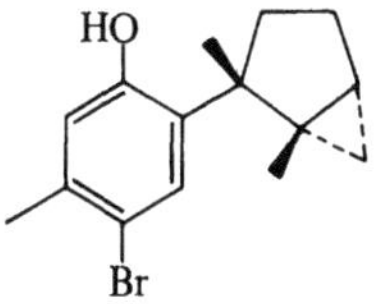

23

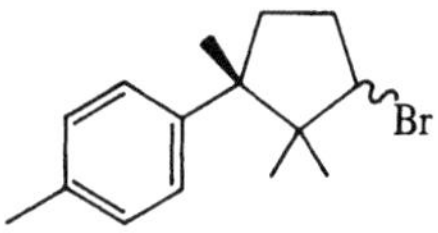

24

Bisabolane derived metabolites have been found in *L. intricata* and *L. caespitosa.* Preintricatol *25* [36] was isolated from *L. intricata* while the ethers caespitol *26* [37] and isocaespitol *27* [38] could be obtained from *L. caespitosa.* An interesting feature of isocaespitol *27* is the unusual diaxial chlorine and bromine substitution pattern: isocaespitol *27* rearranged to caespitol on melting. Compounds in the monocyclofarnesane series have been found in *L. synderae* and *L. obtusa.* The major sesquiterpenoid in *L. obtusa* from England is the ether *28* [39]. *β*-Synderol *29* was isolated from *L. synderae* [40]. Either of these monocarbocyclic skeletons, bisabolane or monocyclofarnesane, could be the biosynthetic precursor of the halogenated chamigrenes.

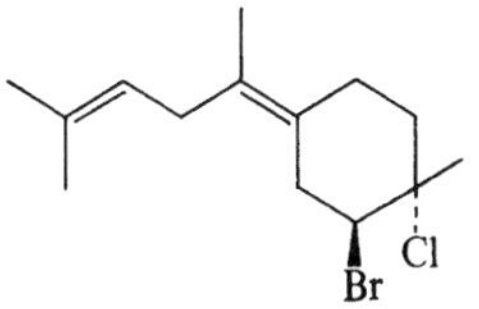

25

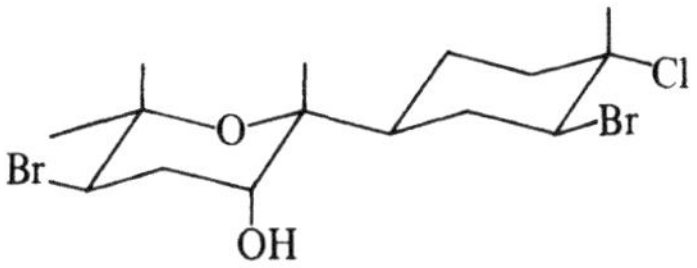

26

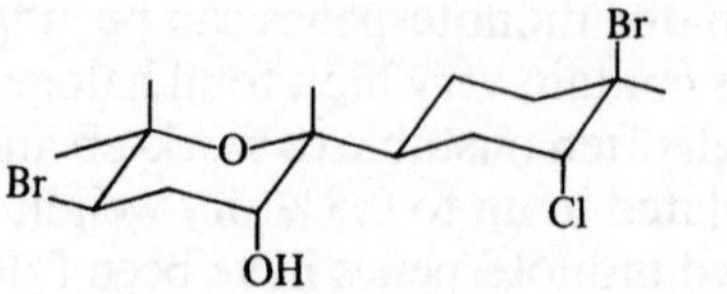
27

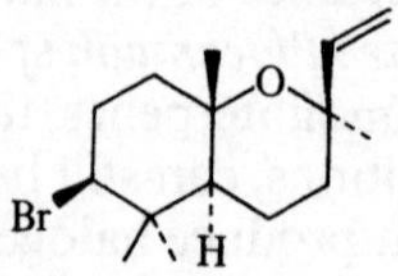
28

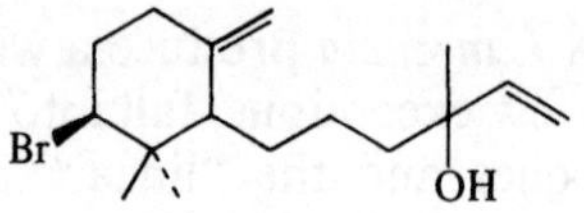
29

Pacifenol *30*, a halogenated chamigrene, was the first natural product reported to contain both chlorine and bromine [41]. Both pacifenol *30* from *L. pacifica* and johnstonol *31* from *L. johnstonii* [42] are probably artifacts of isolation resulting from acid-catalyzed rearrangements of prepacifenol *32* [43] and prepacifenol epoxide *33* [44]. There are many halogenated chamigrenes [45] derived from the relatively simple 10-bromo-α-chamigrene *34* found in *L. pacifica* [46]. Both the epoxide of bromo-chamigrene *35* and an isomer resulting from ring contraction, spirolaurenone *36*, have been found in *L. glandulifera* [47]. Rearrangement of the chamigrene skeleton can result in the novel sesquiterpenes perforatone *37* and the α-chloroketone *38*, which were both found in *L. perforata* [48].

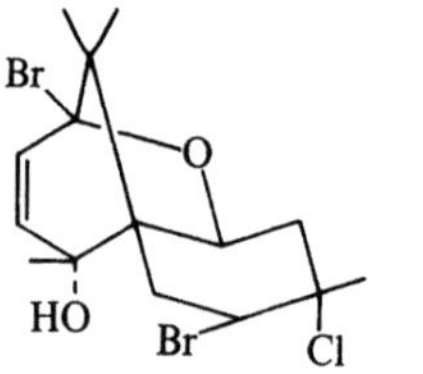
30

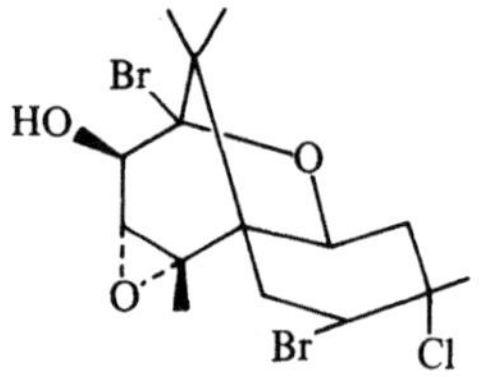
31

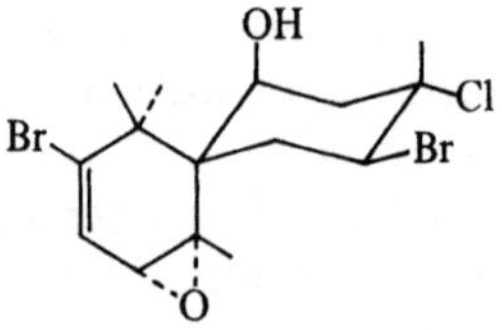
32

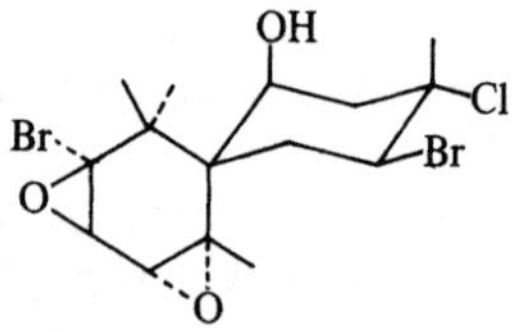
33

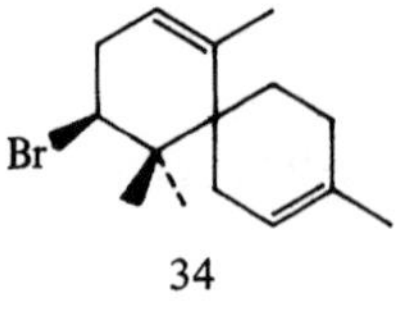
34

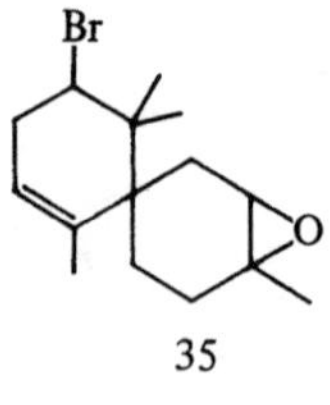
35

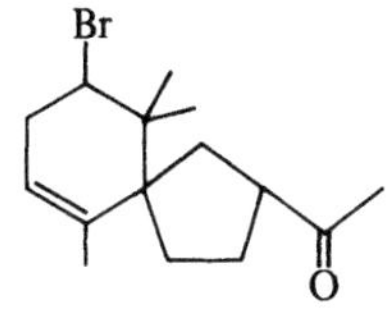

36

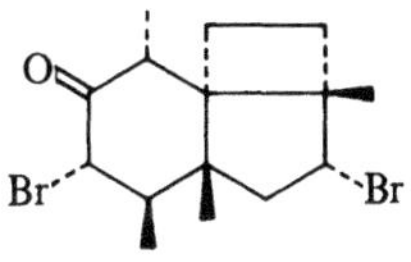

37

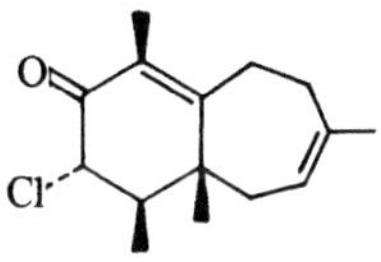

38

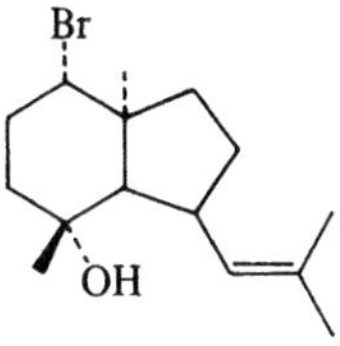

39

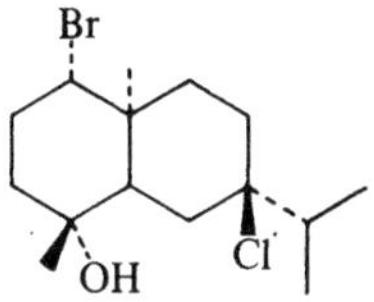

40

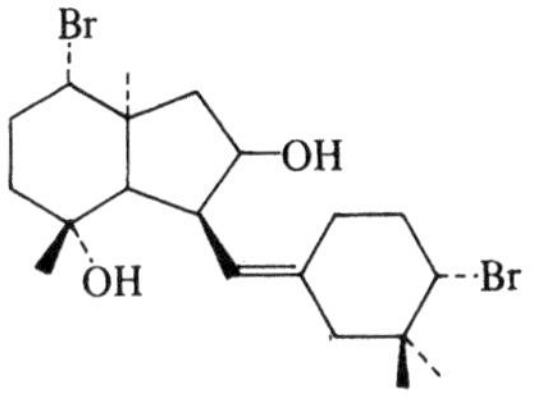

41

The unusual sesquiterpene oppositol *39*, obtained from *L. subopposita* [49], has a carbon skeleton which has not been found in terrestrial sources. Its closest relative is the halogenated selinene *40*, found in an unidentified *Laurencia* species [50].

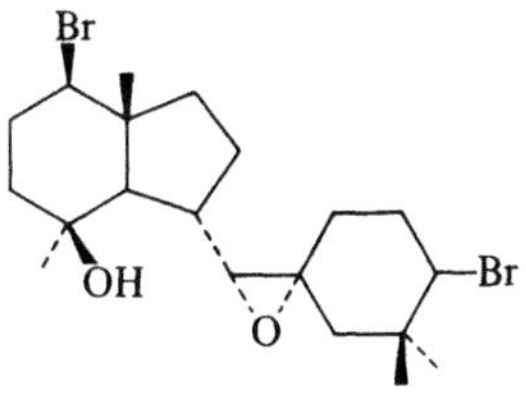

42

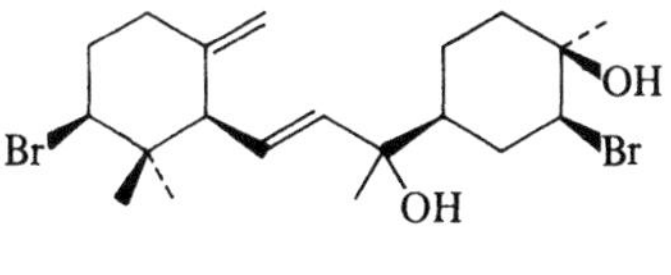

43

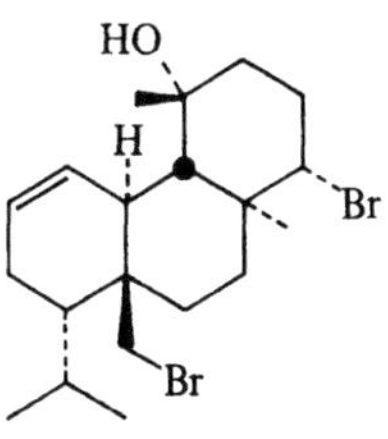

44

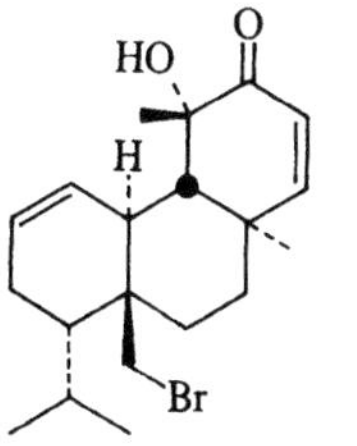

45

Laurencia ireii contains diterpenes that incorporate the oppositol sesquiterpenoid ring skeleton. The structures of iriediol *41* and irieol A *42*, which were determined by X-ray crystallography [51], both contain two bromine atoms at opposite ends of the molecules. Their biosyntheses almost certainly include two very different bromination steps. A series of seven related "irieols" has recently been described [52].

The structures of obtusadiol *43* from *L. obtusa* [53] and bromosphaerol *44* from *Sphaerococcus coronopifolius* [54] also contain two well-separated bromine atoms, while sphaerococcenol A *45* [55], containing only one bromine atom, has presumably lost the second bromine atom by elimination of HBr from a *β*-bromoketone. The remaining brominated diterpene from a red alga is concinndiol *46* from *Laurencia concinna* [56].

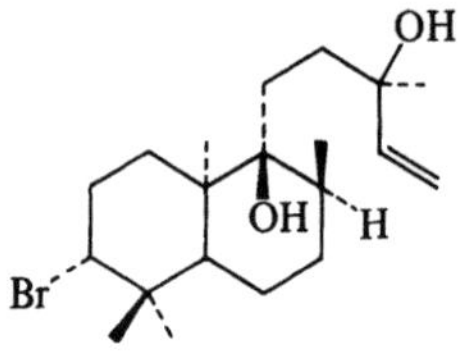

46

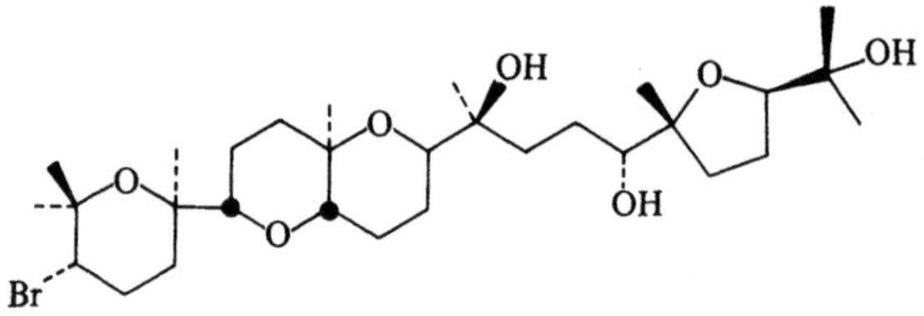

47

One of the most unusual halogenated terpenes to be described is thysiferol *47*, a metabolite of *Laurencia thysifera* that can be derived from the squalene carbon skeleton [57].

Halogenated acetylenes have been found in many species of *Laurencia*. Despite the fact that they invariably have a fifteen-carbon molecular formula, these compounds are *not* sesquiterpenoids but are derived from the carboxylic acid biosynthetic pathway. The simplest of this group of compounds is laurencin *48*, isolated from *L. glandulifera* [58], that contains an α, β-unsaturated acetylene attached to a brominated 8-membered ether. The unsaturated acetylene functionality, either *cis* or *trans*, and the brominated cyclic ether rings are the characteristic features of this class of compounds. Other examples are laureatin *49*, isolaureatin *50* and laurefucin *51* [59]. Some of the acetylenes from *Laurencia* species contain both chlorine and bromine atoms.

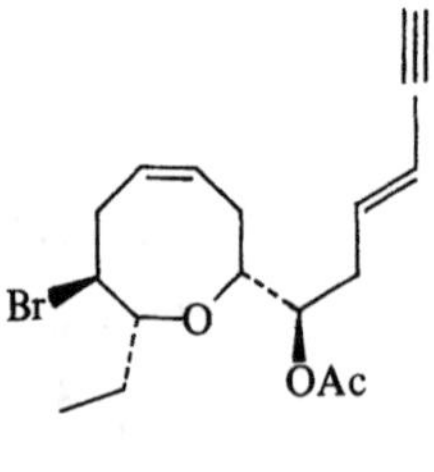

48

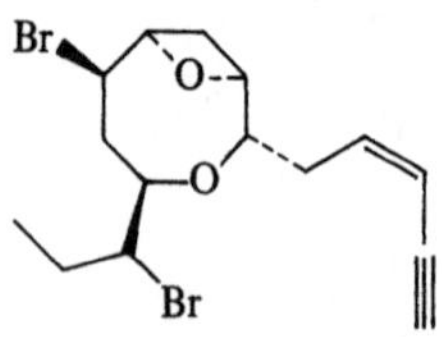

49

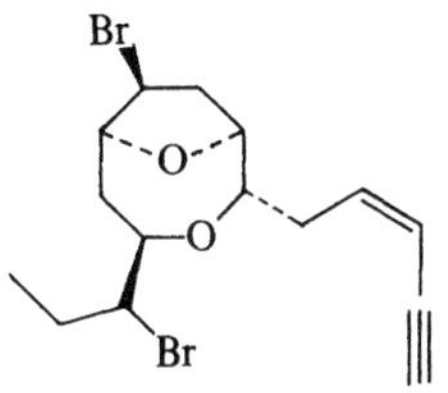

50

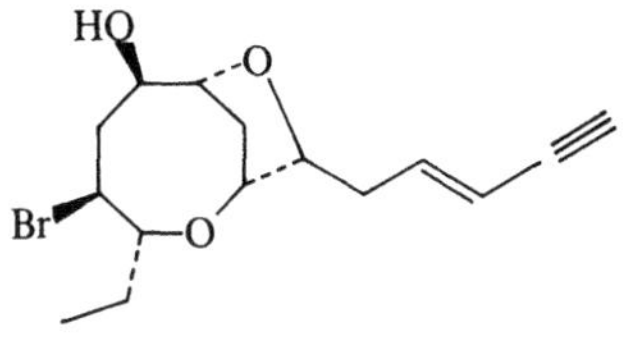

51

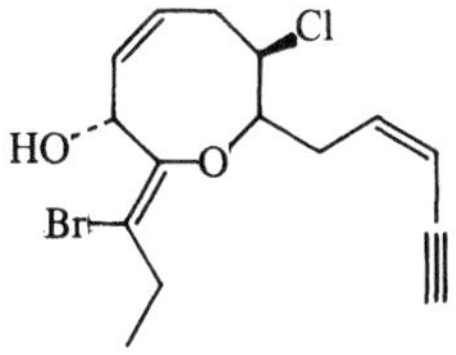

52

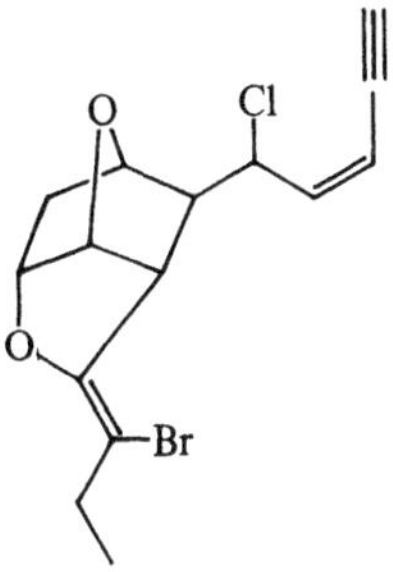

53

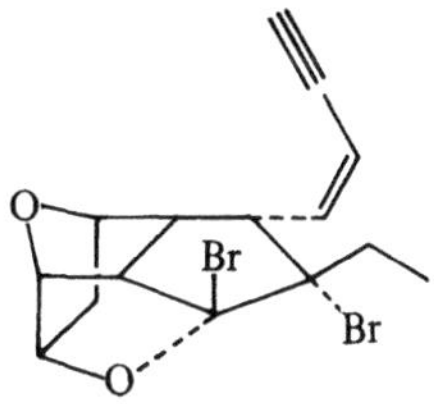

54

Two examples are chondriol *52* from *Laurencia yamada* (syn. *Chondria oppositiclada*) [60], and maneonene A *53* from a variety of *Laurencia nidifica* [61]. Isomanoenone A *54*, also from *L. nidifica,* contained an additional carbocylic ring [62].

The halogenated acetylenes have been found only in *Laurencia* species and in organisms that feed on *Laurencia.* They are fairly unstable, particularly when exposed to an acidic medium, but can often be examined by gas chromatography. They have been reported present in *Laurencia* species throughout a wide geographical area.

Many red algae, especially members of the Rhodomelaceae, contain brominated phenols [63]. Lanasol *55* is the most abundant and widely distributed of the brominated phenols [64] having been obtained both as the free phenol and as a di-potassium salt of a disulfate *56* [65]. The first reports of brominated phenols in marine algae date from around 1950, with 5-bromo-3,4-dihydroxybenzaldehyde *57* from *Polysiphonia morrowii* being the first compound to be identified [66]. Since that time, over twenty brominated phenols have been identified, mostly from algae obtained from the coast of Europe. Two research groups have surveyed large numbers of red algae for the presence of simple brominated phenols [64, 67]. The brominated phenols were identified by combined gas chromatography-mass spectrometry of the per-trimethylsilyl ethers. The procedures used to isolate the brominated phenols were sufficiently severe that sulfates would have been hydrolyzed to the free phenols and some benzylic ethers could have been formed by reaction with solvents [68]. Although it would be unwise to consider that the gas chromatographic record gives an accurate picture of the compounds in the alga, the method can be used to define the basic brominated phenol entities in an alga. A second method of analysis favored by French workers involved permethylation of the brominated phenols. Using this method, they have isolated and identified several unusual brominated methyl phenyl ethers such as the enol ether *58* [69], presumably derived from the corresponding pyruvic acid, and the dimer *59* [70], derived from condensation of lanasol with 2,4-dibromo-1,3,5-trihydroxy-benzene *60*, which co-occurs with *59* in *Rytiphlea tinctoria*.

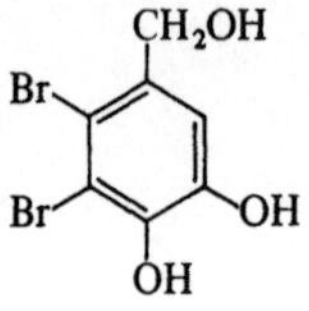

55

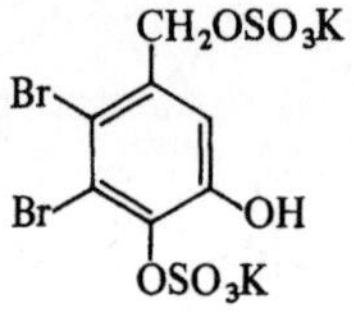

56

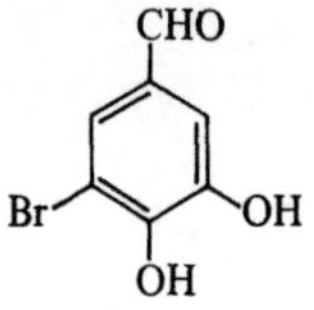

57

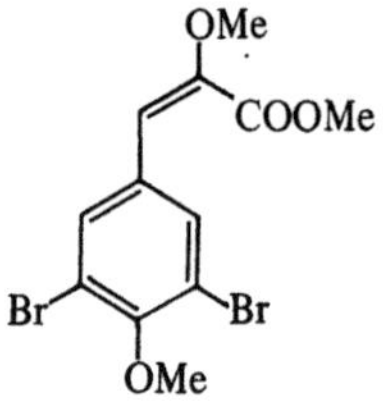

58

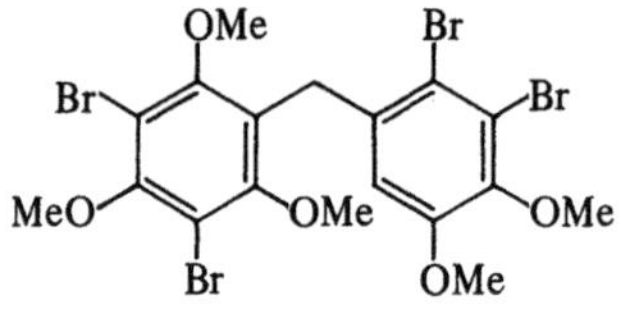

59

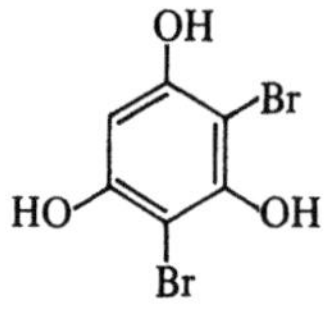

60

The brominated phenols are probably the most stable of the halogenated algal metabolites and are therefore most likely to appear as contaminants in other analyses. The red algae that produce the brominated phenols are often abundant along the coastline of industrial nations where analyses of seawater for synthetic halogenated hydrocarbons are often performed. Furthermore, some of the red algae that produce halogenated phenols are epiphytic on the large brown algae that might be selected for studies of bio-accumulation of industrial pollutants.

There have been three recent reports of halogenated nitrogenous compounds from marine algae. The simple amide, dichloroacetamide, was found in *Marginisporum aberrans* [71]. Polyhalogenated indoles containing both chlorine and bromine were isolated from *Rhodophyllis membranaceae* [72], while tri- and tetra-brominated indoles were found in *Laurencia brongniartii* [73].

Sponges

Since many of the antimicrobial metabolites of sponges (Porifera) were found to be halogenated compounds, a disproportionately large number of halogenated sponge metabolites have been described. Unlike algal metabolites that often contain both chlorine and bromine, sponge metabolites invariably contain either chlorine or bromine substituents.

The chlorinated sponge metabolites are relatively rare. A group of six carbonimidic dichlorides have been obtained from the Pacific sponge *Pseudaxinyssa pitys* [74]. No other naturally occurring carbonimidic dichlorides have been described. Three structural classes of sesquiterpene carbon skeletons were found among the carbonimidic dichlorides: acyclic *61*, monocyclic *62*, and bicyclic *63*. Despite the reactivity of the carbonimidic dichloride functionality toward nucleophiles, the compounds have no apparent physiological activity. These compounds would not be expected to persist in the marine environment since each of the chlorine atoms in *62*, for example, is relatively reactive toward hydrolysis.

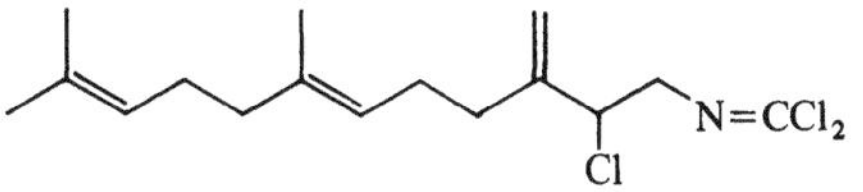

61

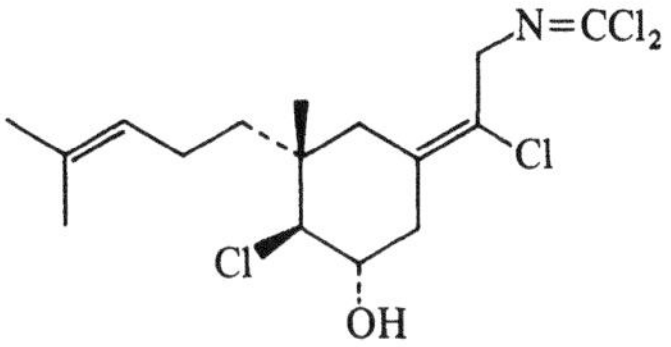

62

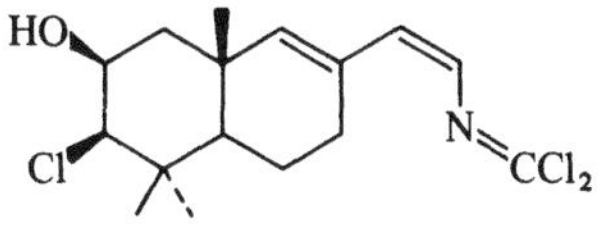

63

Four chlorine-containing metabolites have been isolated from samples of *Dysidea herbacea*. The metabolites dysidin *64*, dysidenin and isodysidenin *65* are all characterized by the presence of trichloromethyl groups. The structures of dysidin *64* [75] and isodysidenin *65* [76] were determined by X-ray crystallography. Dysidenin [77] was found to be an isomer of isodysidenin, probably at the C-5 carbon. A diketopiperazine *66* has also been isolated from an Australian sample of *D. herbacea* [78]. In each of these compounds, the biosynthesis is thought to involve the halogenated amino acid trichloroleucine.

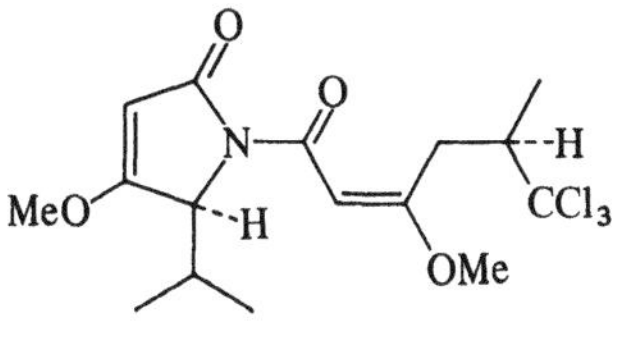

64

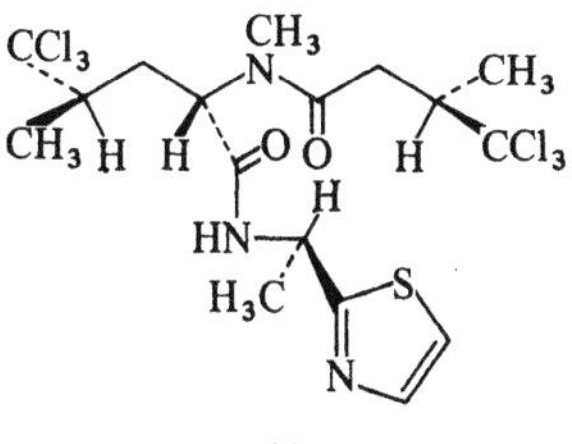

65

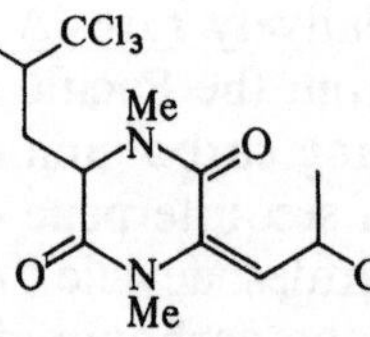

66

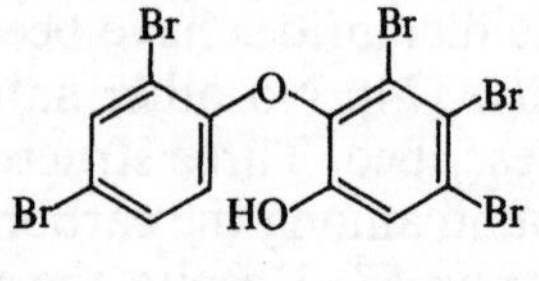

67

A sample of *Dysidea herbacea* from the Caroline Islands contained a series of polybrominated biphenyl ethers exemplified by the pentabrominated phenol *67* [79]. Since non-halogenated terpernoids have also been described from this source [80] there has been some speculation that all the compounds from *Dysidea herbacea* might be synthesized by symbiotic blue-green algae.

The greatest variety of brominated metabolites from sponges have been isolated from *Verongia* (≡ *Aplysina*) species. The dienone *68* and the dimethoxy ketal *69* were first isolated from *Verongia fistularis* and *V. cauliformis* [81]. Since the dimethoxy ketal *69* was isolated from a methanol extract, it was first thought that the dienone *68* was a true metabolite while the ketal *69* was an artifact of extraction. Subsequent isolation of the mixed ketal *70* from an ethanolic extract of a *Verongia* species has led to the hypothesis that all three metabolites were artifacts [82]. The true metabolite of *Verongia*, which gives artifacts on solvent extraction, has not yet been identified.

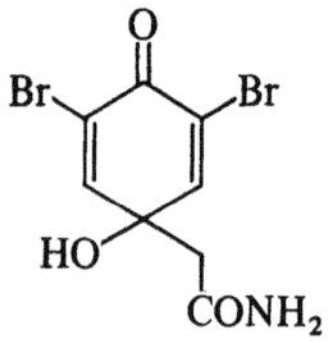

68

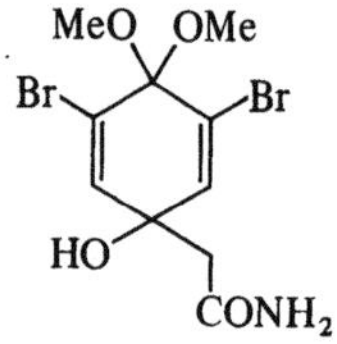

69

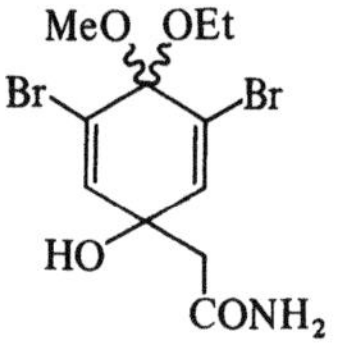

70 (2 diastereoisomers)

The nitrile, aeroplysinin-I *71* [83], and the lactone, aeroplysinin-II *72* [84], have both been isolated from *Verongia* species. Both the dibromophenol *73* [85] and the rearranged dibromohydroquinone *74* [86] have also been found as *Verongia* metabolites. The dimeric compounds aerothionin *75* and homoaerothionin *76* [87] both contain the carbon skeleton and the heteroatom substitution pattern of dibromotyrosine *77* from which all the *Verongia* metabolites are thought to be drived biosynthetically. The isoxizolidone *78* represents a slight change from the other metabolites of *Verongia* since it contains an additional three-carbon unit [88]. Other *Verongia* metabolites of higher molecular weight have recently been described [89]. Many of the metabolites of *Verongia* inhibit the growth of bacteria but none of the known compounds appear to be as active as the fresh extracts of *Verongia*. This suggests that additional halogenated metabolites can be expected from *Verongia* species.

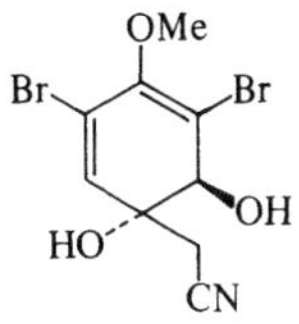

71

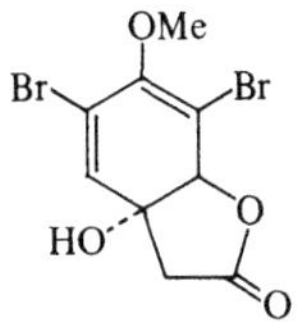

72

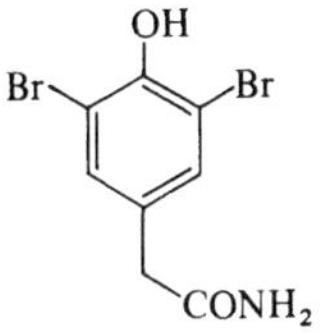

73

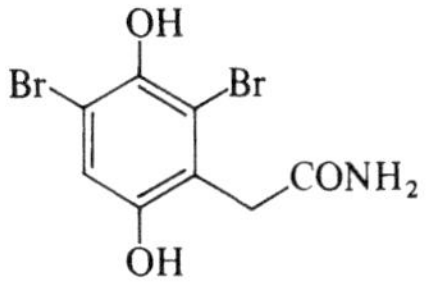

74

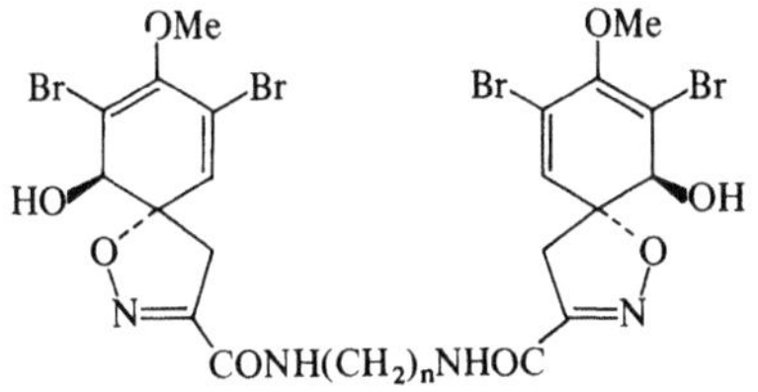

75 n = 4
76 n = 5

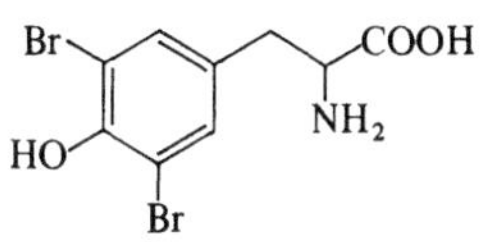

77

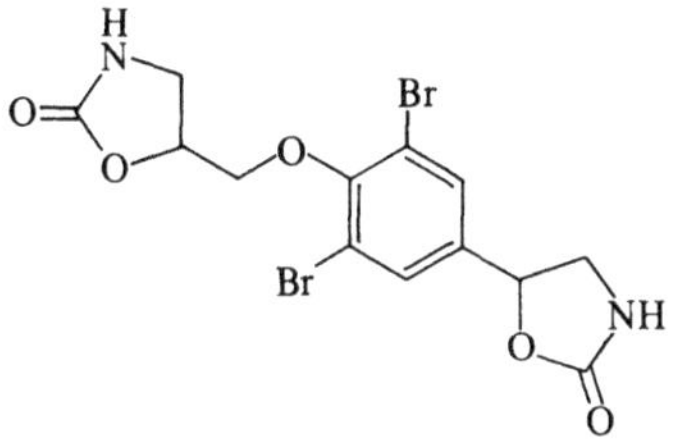

78

Sponges of the genus *Agelas* produce a variety of brominated pyrroles. A study of the metabolites of *Agelas oroides* [90] resulted in the isolation of 4,5-dibromopyrrole-2-carboxylic acid *79* together with the corresponding nitrile and primary amide. The water soluble material from this sponge contained oroidin *80*, together with other brominated metabolites which are currently under investigation. Other species of *Agelas* contain 4-bromopyrrole-2-carboxylic acid and its derivatives [91]. A related sponge, *Phakellia flabellata*, contained dibromophakellin *81*, a compound that can be regarded as having a "cyclized oroidin" skeleton [92].

Several halogenated indoles have been isolated from sponges. 5,6-dibromotryptamine and the corresponding N-methyl derivative were isolated from the Caribbean sponge *Polyfibrospongia maynardii* (≡ *Smenospongia* sp.?)

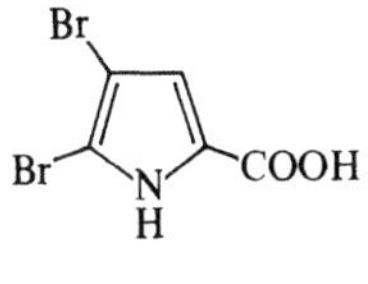

79

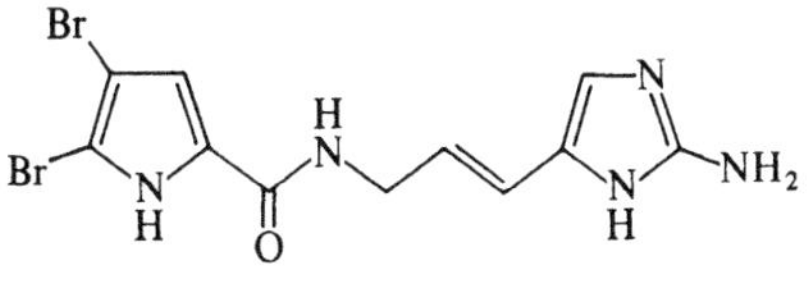

80

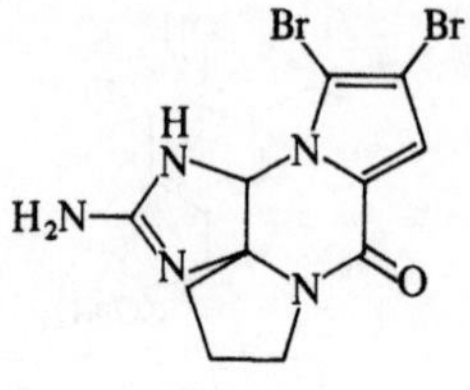

81

[93]. 5-Bromo-N,N-dimethyl-tryptamine and 5,6-dibromo-N,N-dimethyl-tryptamine were isolated from *Smenospongia aurea* and *Smenospongia echina* respectively [91]. The North Atlantic sponge *Pachymatisma johnstoni* contained 6-bromohypaphorine *82*, an amino-acid derivative which gave rise to an abnormal mass spectrum with ions having m/e greater than the molecular weight of the molecule [94]. The burrowing sponge *Cliona celata* from British Columbia waters contains an unstable antibacterial compound that was isolated as its peracetyl derivative *83*, which also gave an abnormal mass spectrum [95].

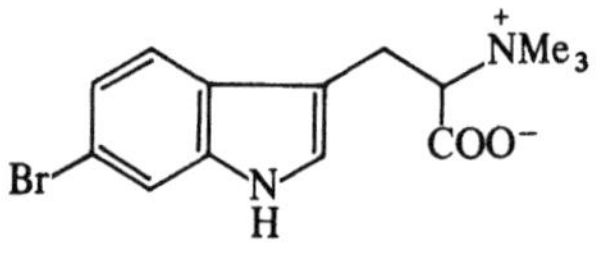

82

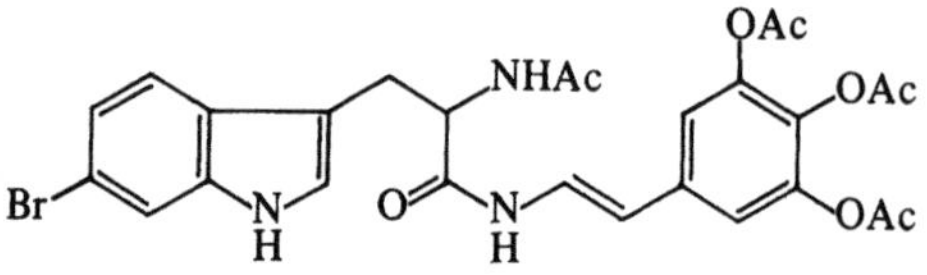

83

An interesting dibrominated acetylenic acid *84* was isolated from *Xestospongia muta*, one of the most abundant Caribbean sponges [96]. The Mediterranean sponge *Reniera fulva* contains a brominated acetylenic ketone *85* and the corresponding alcohol [97].

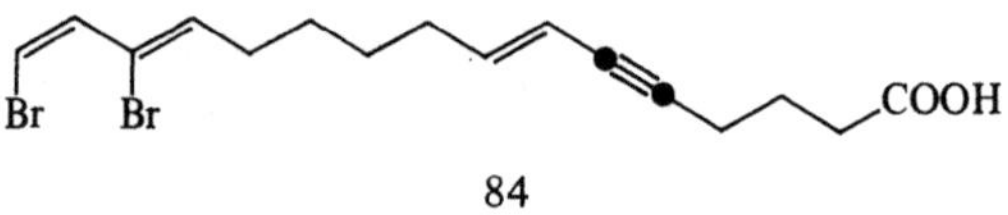

84

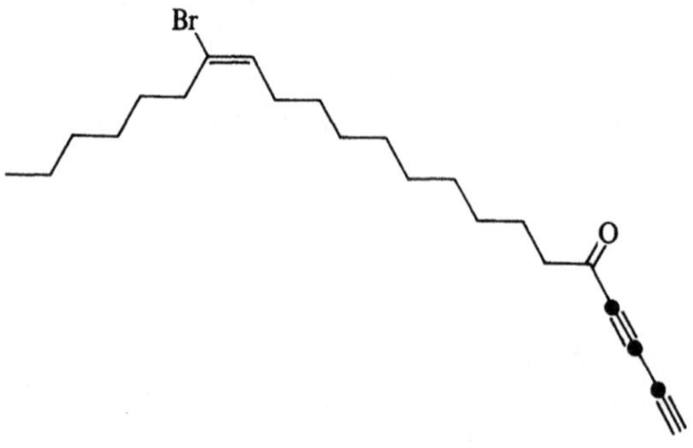

85

Other Marine Invertebrates

Halogenated natural products are rarely found in coelenterates. Chlorinated diterpenes have been isolated from one gorgonian (sea whip) and two sea pens. The Caribbean gorgonian *Briareum asbestinum* contains a series of closely related metabolites, one of which, briarein A *86,* has been identified by X-ray analysis [98]. A closely related compound, stylatulide *87* was isolated from the sea pen *Stylatula* sp. [99], while ptilosarcone and ptilosarcenone were found in the sea plume *Ptilosarcus gurneyi* [100].

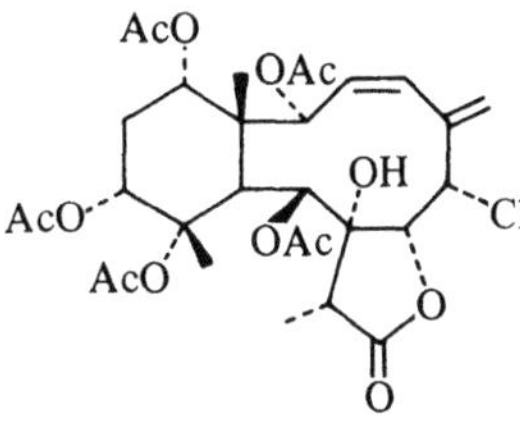

86

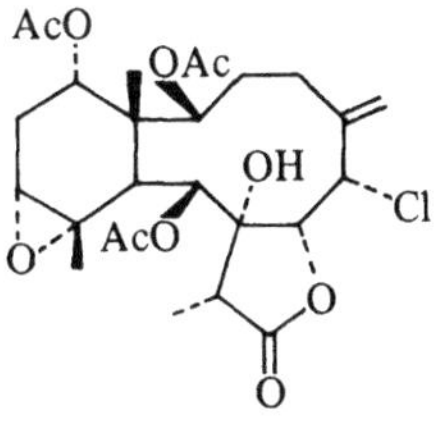

87

Whereas the algae, bacteria, sponges and coelenterates are assumed to produce halogenated metabolites, the opisthobranch molluscs (sea hares) are known to accumulate organohalogen compounds from dietary sources [101]. Studies of the contents of the digestive glands of *Aplysia californica* revealed the presence of metabolites from algae of the genera *Laurencia* and *Plocamium* [102]. Laurinterol *23,* aplysin *88,* pacifenol *30* and johnstonol *31,* all found in *Laurencia pacifica,* were isolated from *A. californica* along with polyhalogenated monoterpenes from *Plocamium* species. Aplysin *88* had first been isolated from *A. kurodai* by Yamamura and Hirata in 1963 and was one of the first halogenated natural products to be described [103]. The diterpene aplysin-20 *89,* also isolated from *A. kurodai* [104] is similar to concinndiol *46* from *Laurencia concinna,* though the two organisms were from different locations.

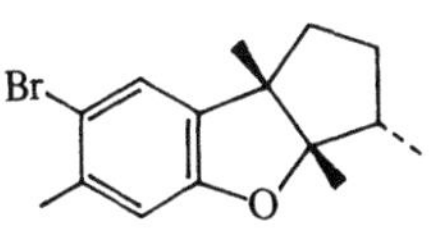

88

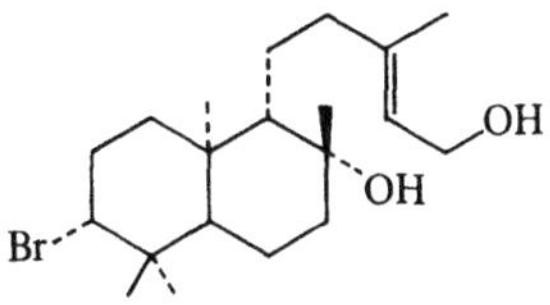

89

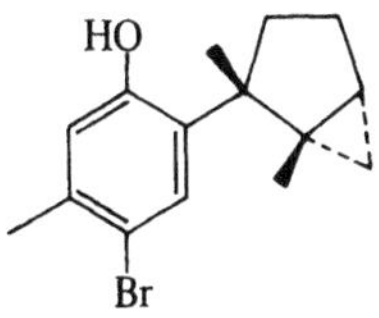

23

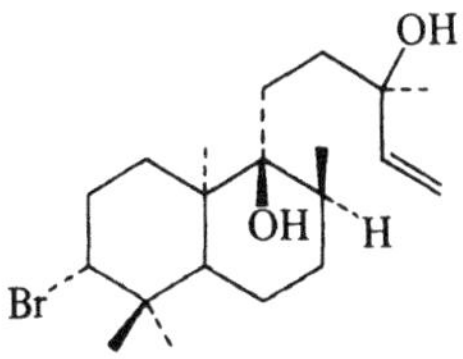

46

There are numerous examples of sea hare metabolites that resemble algal metabolites. Angasiol *90* [105] and aplysiastatin *91* [106], both from *Aplysia angesi*, resemble iriediol *41* and β-synderol *29* respectively. The acetylene dactylyne *92* from *A. dactylomela* [107] is reminiscent of the acetylenes *48–52*. The allene *93* isolated from *A. braziliana* [108] is probably derived from a *Laurencia* acetylene.

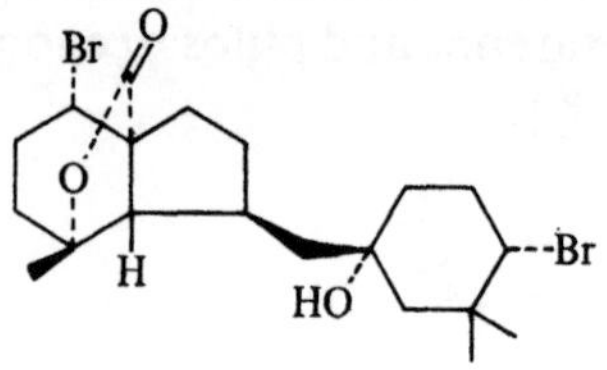

90

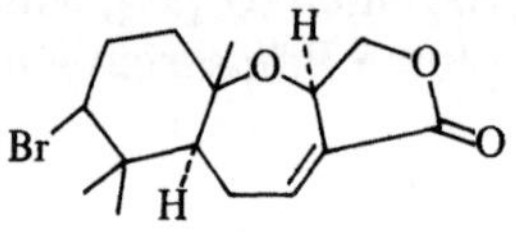

91

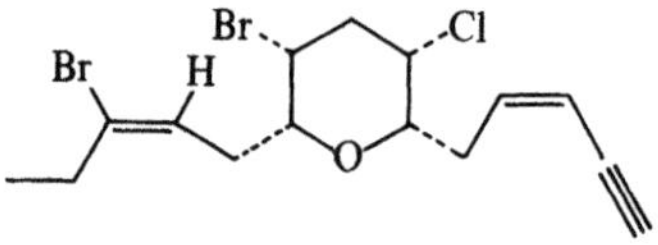

92

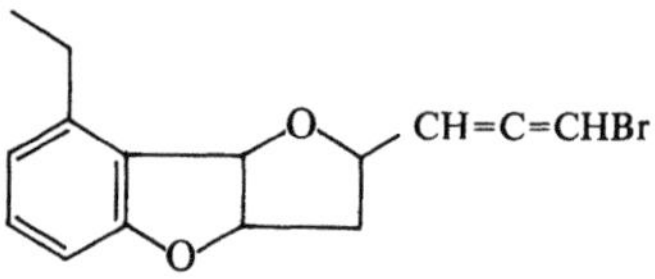

93

The toxic substance from the Hawaiian sea hare *Stylocheilus longicauda* was shown to be the complex bromophenol aplysiatoxin *94* [109]. Subsequent investigations of *S. longicanda* resulted in the isolation of stylocheilamide *95* [110]. Both compounds are undoubtedly metabolites of blue-green algae. One of the most toxic compounds from a mollusc is surugatoxin *96*, obtained from the carnivorous gastropod *Babylonia japonica* [111]. Since the gastropods were toxic only in Suruga Bay, there was good reason to suspect that the toxin had been concentrated by passage through a marine food chain. Recent research suggests that the toxin was produced by a marine bacterium that proliferated in polluted waters [112].

Some nudibranchs feed exclusively on marine sponges. Those feeding on sponges containing halogenated metabolites will contain the same or related metabolites. For example, *Tylodina fungina* concentrates the dienone *68* from *Verongia* species.

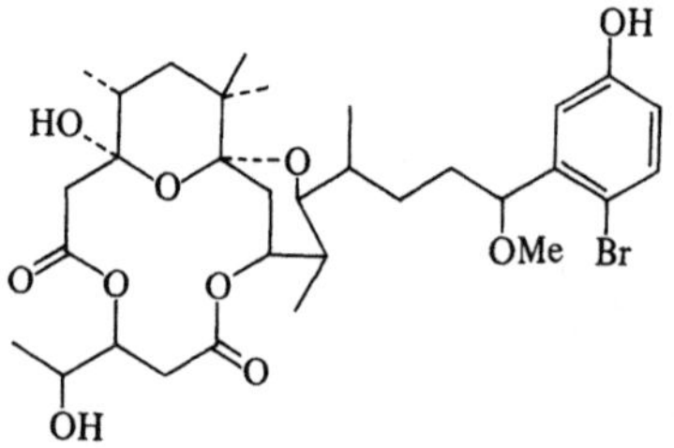

94

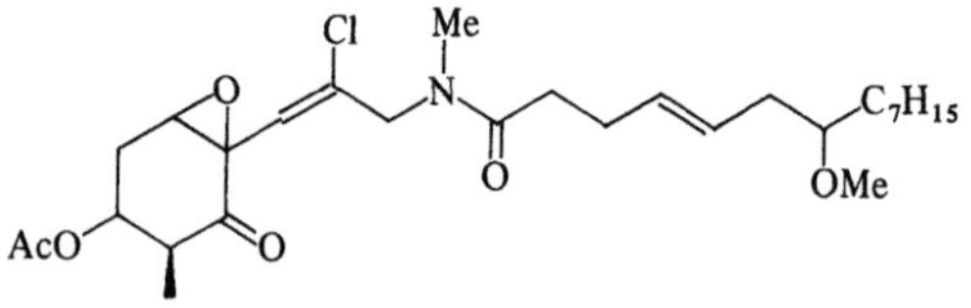

95

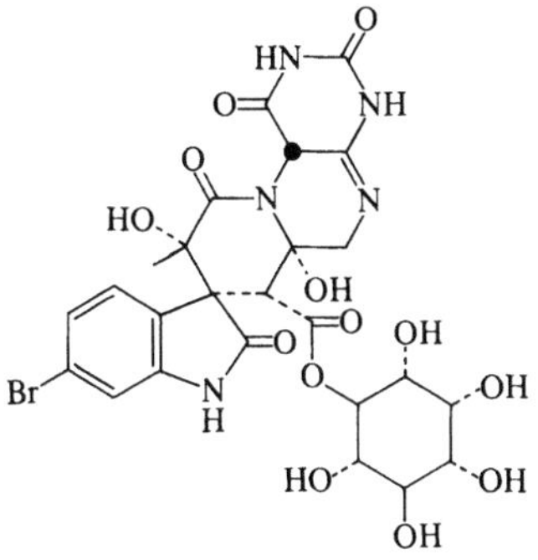

96

The most famous of marine natural products is the dye Tyrian purple *97*, a brominated analogue of indigotin produced by various Murex species. The history and the chemistry of production of Tyrian purple have been reviewed [113].

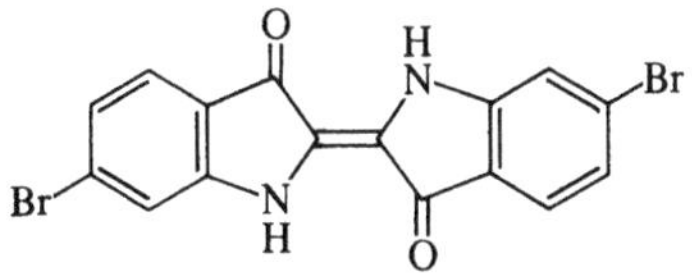

97

Many worms contain halogenated natural products but it is not known whether these halogenated metabolites are synthesized by the worms or derived from the bacteria or detritus on which the worms feed. An argument in favor of the idea that worms, like molluscs, contain metabolites from their dietary constituents is that worms of different phyla (Annelida, Sipunculida, Platyhelminthes, Nemertinea, and Enteropneusta) all contain very similar compounds. In addition, the chemical content of one worm has been shown to vary with the collecting location.

Acorn worms (Hemichordata) were once described as having an iodoform-like odor. The odor has since been attributed to at least two compounds but neither of these contain iodine. The major compound in *Balanoglossus biminiensis* was 2,6-dibromophenol [114]. *B. misakiensis* contained both 2,6-dibromophenol and 2,4,6-tribromophenol while *B. carnosus* has the same two compounds together with 2,4-dibromophenol [115]. Both 2,6-dibromophenol and 2,4,6-tribromophenol were also found in the tube-worm *Phoronopsis viridis* [116]. The annelid worm *Thelepus setosus* is the source of 3,5-dibromo-4-hydroxybenzyl alcohol, the corresponding aldehyde and the interesting antifungal compound thelepin *98* [117]. The acorn worm *Ptychodera flava laysanica* has shown considerable chemical variations, depending on the collection locality [118]. The major metabolites were tetrabromohydroquinone and tribromohydroquinone, depending on the sample, and some dimers and trimers were also isolated. The odor of *P. flava* was attributed to 3-chloroindole *99* [119]. Other halogenated indoles such as 3-bromoindole, 6-bromo-3-chloroindole, 5,7-dibromo-6-methoxy indole, 3,5,7-tribromoindole, and 6-methoxy-3,5,7-tribromoindole *100* have all been found in *P. flava* and *B. carnosus* [115]. Perhaps the most unusual finding was the discovery of Tyrian purple *97* and other halogenated indigotin dyes in a population of *P. flava*.

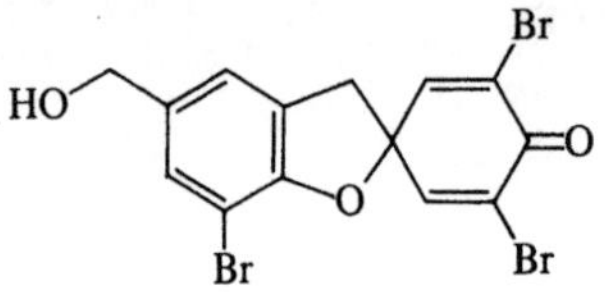

98

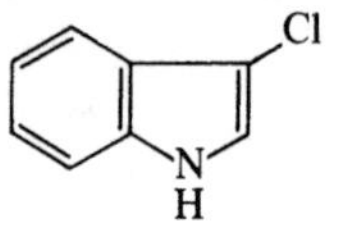

99

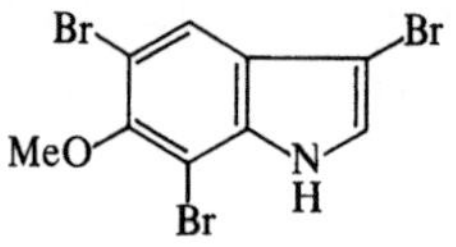

100

Some Unusual Metabolites from Terrestrial Organisms

Our knowledge of the organohalogen compounds from terrestrial organisms has increased little since the review by Siuda and DeBernardis in 1973 [120]. The majority of the terrestrial halogenated natural products contain chlorine, mainly as chlorinated aromatic compounds from microorganisms. The most interesting of the terrestrial products are the fluorinated carboxylic acids found in higher plants. The toxicity of *Dichapetalum cymosum* is due to the presence of fluoroacetic acid [121]. A number of fluorinated fatty acids such as ω-fluorooleic acid and ω-fluoropalmitic acid were isolated from *D. toxicarium* [122]. Since most halogenated materials from terrestrial organisms are very minor constituents or are manufactured products from miocroorganisms, they are not expected to make an important impact on environmental chemistry. They have been adequately reviewed elsewhere [120].

Biosynthesis of Halogenated Natural Products

An understanding of the biosynthesis of halogenated natural products is invaluable not only to determine the structures of new metabolites but also to differentiate between natural and synthetic molecules. The biosynthetic hypotheses to be discussed have been generated to explain the halogenated natural products already known. There is some support for these hypotheses from studies of biomimetic synthesis and from the single study of the enzymatic halogenation process.

An enzyme has been extracted from the red alga *Bonnemaisonia hamifera* that is capable of incorporating bromine into a number of organic substrates [123]. For example, incubation of the bromoperoxidase preparation with hydrogen peroxide, bromide ion, and 3-ketooctanoic acid gave dibromomethane, bromoform and *n*-pentyl bromide. The mechanism of bromination was assumed to involve an enzyme-bound bromonium ion which is often represented by "Enz-Br^+" or simply "Br^+".

The biomimetic syntheses of marine natural products have shown that it is possible to generate a "Br^+" species to initiate cyclization reactions [124]. An important consequence of this mechanism is that bromination of olefins in nature usually obeys the Markovnikoff Rule (bromine adds to an unsym-

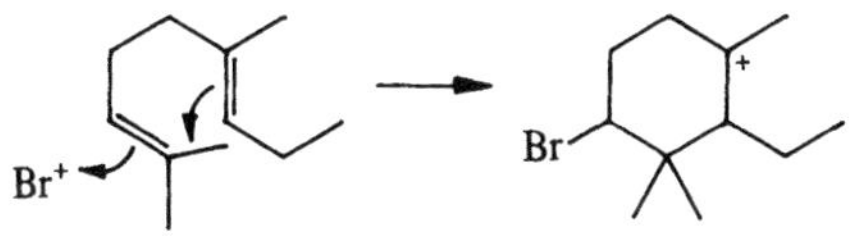

metrical olefin to give bromine at the less substituted carbon and generate a carbonium ion at the more substituted carbon). Thus, most compounds containing bromine and chlorine on adjacent carbon atoms (cf. *10, 12, 20, 25, 26, 30–33*) have bromine at the less substituted carbon and chlorine at the more substituted carbon because the mechanism involves "Br^+" and "Cl^-" [125]. Exceptions to this rule have been found (cf. *27*) and it is therefore particularly important to confirm the positions of the halogen atoms using ^{13}C NMR spectroscopy [126]. Bromophenols are usually brominated in the *ortho* and *para* positions on the aromatic ring; hence the characteristic 3,5-dibrominated compounds *68–78* derived from dibromotyrosine that are found in *Verongia* (≡ *Aplysina*) sponges. Formation of the acetylenes *48–51* requires the reaction of a bromonium species with an olefinic bond followed by reaction with an alcohol functionality to form the cyclic bromoether moieties.

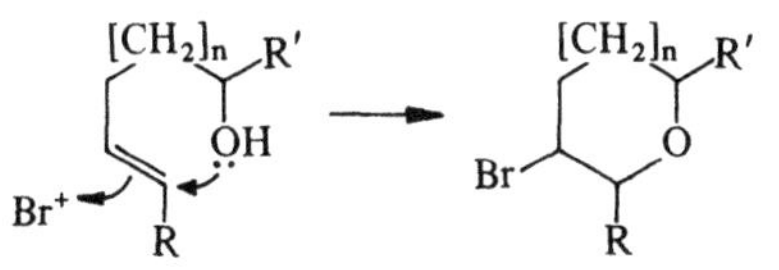

The biosynthesis of compounds containing only chlorine is less well understood. Compounds containing a chlorohydrin functionality probably result from addition of chloride ion to an epoxide and may even be artifacts. However, the chlorinated monoterpenes, the carbonimidic dichlorides *61–63* and compounds *64–66*, containing a trichloromethyl group, all appear to require a chloroperoxidase during biosynthesis.

The probable specificity of enzyme-mediated halogenation can be invoked to explain why brominated natural products are more common than chlorinated metabolites. If the formation of halogenated natural products involved halide ions, one would expect many more chlorinated metabolites than brominated metabolites, reflecting the relative abundance of halide ions in seawater. Since the biosynthesis of halogenated metabolites usually involves an oxidized halogen species, brominated and iodinated metabolites are favored because of the lower oxidation potentials involved. The combination of oxidation potential and relative abundance of halide ions in seawater has resulted in a greater number of brominated marine natural products than chlorinated or iodinated metabolites.

Halogenated Natural Products in Seawater

There is good evidence that marine organisms release halogenated metabolites into seawater. Although quantitative measurements of the concentrations of marine natural products in seawater are rare, indirect observations leave little room for doubt. Halogenated metabolites have been implicated in chemical defense mechanisms of marine organisms, which in turn implies that the organisms must release these compounds. Even if there were no active excretion of halogenated metabolites, the compounds would be released when the organisms were eaten or otherwise destroyed.

Measurements of methyl iodide in seawater indicated that the concentrations increased in the vicinity of *Laminaria digitata* beds [127] but there was no direct proof that the brown alga produced the methyl iodide. The brominated phenol lanasol *55* was detected by combined gas chromatography-mass spectrometry among the dissolved organic compounds in seawater samples collected in the vicinity of *Polysiphonia brodiaei,* a known source of lanosol [64]. Unidentified halogenated compounds could be detected by GC analysis using an electron capture detector in hexane extracts of water samples taken from a La Jolla tidepool [128]. As expected, the concentrations of halogenated metabolites increased during the period that the tidepool remained isolated.

Indirect evidence for the transport of halogenated metabolites through seawater is provided by reports of low levels of known algal metabolites in organisms which are totally unrelated to the normal algal source. For example, where *Microcladia coulteri* existed in the vicinity of *Plocamium* species, extracts of *M. coulteri* contained very small amounts of the halogenated monoterpenes normally found in *Plocamium* [129]. Since *M. coulteri* from an "isolated" habitat contained no traces of halogenated monoterpenes [130], it was assumed that halogenated monoterpenes were released from the nearby *Plocamium* plants and subsequently adsorbed onto the surface of *Microcladia.* Reports [131] of the occurrence of laurinterol *23,* a metabolite of various *Laurencia* species, in many unrelated algae and in sponges, echinoderms and bryazoans provide further support for the hypothesis that algal metabolites are released into seawater.

Although one must be cautious in interpreting the results of laboratory experiments with marine bacteria, the isolation of the bromophenol *1* from the culture medium in which *Chromobacter* had been grown suggested that this toxic compound was excreted into seawater as a potential inhibitor of other microorganisms. There is no evidence that halogenated metabolites from sponges or coelenterates are released into seawater. Studies of the sea hare *Aplysia californica* provided evidence that halogenated compounds were excreted through the skin. Radiolabelled laurinterol *23* was converted into aplysin *88* in the digestive gland but all radioactive compounds had been excreted within 90 days [132]. It was proposed that halogenated compounds were stored in the digestive gland, transported to the skin, and then released in the mucus secretion.

In contrast to synthetic halogenated aromatic compounds, halogenated natural products appear to be rapidly degraded by microorganisms. This observation is somewhat surprising since many halogenated phenols inhibit the growth of marine microorganisms. This has led to the suggestion that somewhat specialized bacteria are required for the degradation of these compounds.

Since halogenated natural products are produced by marine organisms and released into seawater, there exists the possibility that these natural materials can interfere with assays for chlorinated hydrocarbon pollutants. Due to the relatively rapid degradation of natural materials, they are likely to be detected only in the vicinity of beds of red algae. Using gas chromatography with an electron capture detector, operated under the prescribed conditions for chlorinated hydrocarbon pollutant analysis, many natural products have the same retention times as pollutants [128]. The simple solution to distinguishing between natural and synthetic materials is to use combined gas chromatography-mass spectrometry for the analysis. For examples of industrial chemicals such as bromoform, chloroform, and carbon tetrachloride that are also produced by marine algae, the relative contributions from natural and synthetic sources have not been determined. There have been no reliable estimates of the rates of production or degradation of halogenated natural products. There have been no attempts to determine the "standing crop" of halogenated natural products, but researchers in this field have suggested informally that natural production of organohalogen compounds might be comparable with industrial production.

References

1. Burkholder, P.R., Pfister, R.M., Leitz, F.M.: Appl. Microbiol. *14,* 649 (1966)
2. Andersen, R.J., Wolfe, M.S., Faulkner, D.J.: Marine Biol. *27,* 281 (1974)
3. Fenical, W., Look, S.A.: Personal communication
4. Mynderse, J.S., Moore, R.E.: J. Org. Chem. *43,* 2301 (1978)
5. Mynderse, J.S., Moore, R.E.: Phytochem. *17,* 1325 (1978)
6. Moore, R.E.: Personal communication
7. Högberg, M.-E., Thomson, R.H., King, T.J.: J. Chem. Soc, Perkin I, 1696 (1976)
8. a) McConnell, O.J.: Ph. D. Thesis, UC San Diego 1978
   b) Fenical, W.: Tetrahedron Lett. 4463 (1974)
   c) Burreson, B.J., Moore, R.E., Roller, P.P.: Tetrahedron 473 (1975). Burreson, B.J., Moore, R.E., Roller, P.P.: Agri. Food Chem. *24,* 856 (1976); Woolard, F.X., Moore, R.E., Roller, P.P.: Tetrahedron *32,* 2843 (1976)
   d) McConnell, O.J., Fenical, W.: Phytochem. *16,* 367 (1977)
9. Suida, J.F. et al.: J. Am. Chem. Soc. *97,* 937 (1975)
10. McConnell, O.J., Fenical, W.: Tetrahedron Lett. 4159 (1977)
11. McConnell, O.J., Fenical, W.: Tetrahedron Lett. 1851 (1977)
12. Rose, A.F., Pettus, J.A., Sims, J.J.: Tetrahedron Lett. 1847 (1977)
13. Kazlauskas, R. et al.: Tetrahedron Lett. 37 (1977)

14. Pettus, Jr., J.A., Wing, R.M., Sims, J.J.: Tetrahedron Lett. 41 (1977)
15. Moore, R.E. in: Marine Natural Products, Chemical and Biological Perspectives, (ed.) P.J. Scheuer, Vol. 1, Academic Press, New York 1978, p. 59
16. Theiler, R.F., Suida, J.F., Hager, L.P. in: Drugs and Food from the Sea (eds.) P.N. Kaul and C.J. Sindermann, Univ. Oklahoma: Norman 1978, p. 154
17. Faulkner, D.J. et al.: J. Am. Chem. Soc. *95,* 3413 (1973)
18. Mynderse, J.S., Faulkner, D.J.: Tetrahedron *31,* 1963 (1975)
19. Mynderse, J.S., Faulkner, D.J.: Phytochem. *17,* 237 (1978)
20. a) Crews, P., Kho, E.: J. Org. Chem. *39,* 3303 (1974)
    b) Norton, R.S., Warren, R.G., Wells, R.J.: Tetrahedron Lett., 3905, (1977) and [21] and [22]
21. Higgs, M.D., Vanderah, D.J., Faulkner, D.J.: Tetrahedron *33,* 2775 (1977)
22. González, A.G. et al.: Phytochem. *17,* 947 (1978)
23. Van Engen, D. et al.: Tetrahedron Lett. 29 (1978)
24. Mynderse, J.S. et al.: Tetrahedron Lett. 2175 (1975)
25. Mynderse, J.S., Faulkner, D.J.: J. Am. Chem. Soc. *96,* 6711 (1974)
26. Crews, P., Kho-Wiseman, E.: J. Org. Chem. *42,* 2812 (1977)
27. Crews, P.: J. Org. Chem. *42,* 2634 (1977)
28. Stierle, D.B., Wing, R.M., Sims, J.J.: Tetrahedron Lett. 4455 (1976). Kazlauskas, R. et al.: Tetrahedron Lett. 4451 (1976)
29. a) Ichikawa, N., Naya, Y., Enomoto, S.: Chem. Lett. 1333 (1974)
    b) Burreson, B.J., Woolard, F.X., Moore, R.E.: Tetrahedron Lett. 2155 (1975)
    c) Burreson, B.J., Woolard, F.X., Moore, R.E.: Chem. Lett. 1111 (1975)
    d) Woolard, F.X. et al.: Tetrahedron Lett. 2367 (1978)
30. McConnell, O.J., Fenical, W.: J. Org. Chem. *43,* 4238 (1978)
31. Martín, J.D., Darias, J. in: Marine Natural Products: Chemical and Biological Perspectives, (ed.), P.J. Scheuer, Vol. 1, Academic Press, New York 1978; p. 125
32. Irie, T. et al.: Tetrahedron *26,* 3271 (1970)
33. a) Kazlauskas, R. et al.: Aust. J. Chem. *29,* 2533 (1976)
    b) Caccamese, S., Rinehart, Jr., K.L. in: Drugs and Food from the Sea – Myth or Reality? (ed. P.N. Kaul and C.J. Sinderman), Univ. Oklahoma, Norman 1978; p. 187
34. Sims, J.J. et al.: Antimicrob. Agents and Chemother. *7,* 320 (1975)
35. Suzuki, T., Suzuki, M., Kurosawa, E.: Tetrahedron Lett. 3057 (1975)
36. White, R.H., Hager, L.P. in: The Nature of Seawater (ed. E.D. Goldberg), Dahlem Konferenzen, Berlin 1975, p. 633
37. González, A.G. et al.: Tetrahedron Lett. 1249 (1974)
38. González, A.G. et al.: Tetrahedron *31,* 2449 (1975)
39. Faulkner, D.J.: Phytochemistry *15,* 1992 (1976)
40. Howard, B.M., Fenical, W.: Tetrahedron Lett. 41 (1976)
41. Sims, J.J. et al.: J. Am. Chem. Soc. *93,* 3774 (1971)
42. Sims, J.J. et al.: Tetrahedron Lett. 195 (1972)
43. Sims, J.J. et al.: J. Am. Chem. Soc. *95,* 972 (1973)
44. Faulkner, D.J., Stallard, M.O., Ireland, C.: Tetrahedron Lett. 3571 (1974)
45. Sims, J.J., Lin, G.H.Y., Wing, R.M.: Tetrahedron Lett. 3487 (1974), González, A.G. et al.: Tetrahedron Lett. 3051 (1976). Waraszkiewicz, S.M., Erickson, K.L.: ibid. 2003 (1974). See also [36]
46. Howard, B.M., Fenical, W.: Tetrahedron Lett. 2519 (1976)
47. Suzuki, M., Kurosawa, E., Irie, T.: Tetrahedron Lett. 4995 (1970), Suzuki, M., Kurosawa, E., Irie, T.: Tetrahedron Lett. 821 (1974)
48. González, A.G. et al.: Tetrahedron Lett. 2499 (1975)
49. Wratten, S.J., Faulkner, D.J.: J. Org. Chem. *42,* 3343 (1977)
50. a) Rose, A.F., Sims, J.J.: Tetrahedron Lett. 2935 (1977)
    b) Howard, B.M., Fenical, W.: J. Org. Chem. *42,* 2518 (1977)
51. Fenical, W. et al.: Tetrahedron Lett. 3983 (1975)
52. Howard, B.M., Fenical, W.: J. Org. Chem. *43,* 4401 (1978)
53. Howard, B.M., Fenical, W.: Tetrahedron Lett. 2453 (1978)
54. Fattorusso, E. et al.: Gazz. Chim. Ital. *106,* 779 (1976)

55. Fenical, W., Finer, J., Clardy, J.: Tetrahedron Lett. 731 (1976)
56. Sims, J.J. et al.: Chem. Commun. 470 (1973)
57. Blunt, J.W. et al.: Tetrahedron Lett. 69 (1978)
58. a) Irie, T., Suzuki, M., Masamune, T.: Tetrahedron *24*, 4193 (1968)
    b) Cameron, A.F. et al.: J. Chem. Soc. (B) 559 (1969)
59. a) Irie, T., Izawa, M., Kurosawa, E.: Tetrahedron *26*, 851 (1970)
    b) Kurosawa, E. et al.: Tetrahedron Lett. 3857 (1973)
    c) Furusaki, A. et al.: Tetrahedron Lett. 4579 (1973)
    d) Kurosawa, E., Fukuzawa, A., Irie, T.: Tetrahedron Lett. 4135 (1973)
60. Fenical, W., Gifkins, K.B., Clardy, J.: Tetrahedron Lett. 1507 (1974)
61. Waraszkiewicz, S.M., Sun, H.H., Erickson, K.L.: Tetrahedron Lett. 3021 (1976)
62. Sun, H.H., Waraszkiewicz, S.M., Erickson, K.L.: Tetrahedron Lett. 4227 (1976)
63. For a more complete review see: Faulkner, D.J. in: Top. Antibiotic Chem. (ed. P.G. Sammes), Vol. 2, Ellis Horwood, Chichester, U.K. 1978, p. 39
64. Pedersen, M., Saenger, P., Fries, L.: Phytochem. *13*, 2273 (1974)
65. Glombitza, K.-W., Stoffelen, H.: Planta Med. *22*, 391 (1972)
66. Saito, T., Ando, Y.: Nippon Kaguka Zasshi *76*, 478 (1955)
67. Glombitza, K.-W. et al.: Planta Med. *25*, 105 (1974)
68. Stoffelen, H. et al.: Planta Med. *22*, 396 (1972)
69. Chautraine, J.-M., Combaut, G., Teste, J.: Phytochem. *12*, 1793 (1973)
70. Chevolot-Magneur, A.-M. et al.: Phytochem. *15*, 767 (1976)
71. Ohta, K., Takagi, M.: Phytochem. *16*, 1085 (1977)
72. Brennan, M.R., Erickson, K.L.: Tetrahedron Lett. 1637 (1978)
73. Carter, G.T. et al.: Tetrahedron Lett. 4479 (1978)
74. a) Wratten, S.J., Faulkner, D.J.: J. Amer. Chem. Soc. *99*, 7367 (1977)
    b) Wratten, S.J. et al.: Tetrahedron Lett. 1391 (1978)
    c) Wratten, S.J., Faulkner, D.J.: Tetrahedron Lett. 1395 (1978)
75. Hofheinz, W., Oberhänsli, W.E.: Helv. Chim. Acta. *60*, 660 (1977)
76. Charles, C. et al.: Tetrahedron Lett. 1519 (1978)
77. Kazlauskas, R. et al.: Tetrahedron Lett. 3183 (1977)
78. Kazlauskas, R., Murphy, P.T., Wells, R.J.: Tetrahedron Lett. 4945 (1978)
79. Sharma, G.M., Vig, B.: Tetrahedron Lett. 1715 (1972)
80. Kazlauskas, R., Murphy, P.T., Wells, R.J.: Tetrahedron Lett. 4949 (1978)
81. a) Sharma, G.M., Burkholder, P.R.: Tetrahedron Lett. 4147 (1967)
    b) Sharma, G.M., Vig, B., Burkholder, P.R.: J. Org. Chem. *35*, 2823 (1970)
82. Andersen, R.J., Faulkner, D.J.: Tetrahedron Lett. 1175 (1973)
83. Fattorusso, E., Minale, L., Sodano, G.: J. Chem. Soc. Perkin I, 16 (1972). Fulmore, W. et al.: Tetrahedron Lett. 4551 (1970)
84. Minale, L. et al.: Chem. Commun. 674 (1972)
85. Stempien, Jr. et al. in: Food-Drugs from the Sea Proc., Mar. Techn. Soc. 1972; p. 105
86. Krejcarek, G.E. et al.: Tetrahedron Lett. 507 (1975)
87. Moody, K. et al.: J. Chem. Soc. Perkin *I*, 18 (1972). (Structure confirmed by X-ray: J. Clardy, personal communication)
88. Borders, D.B., Morton, G.O., Wetzel, E.R.: Tetrahedron Lett. 2709 (1974)
89. Schmitz, F.J.: IUPAC Symp. Marine Nat. Prod. Sorrento 1978
90. a) Forenza, S. et al.: Chem. Commun. 1129 (1971)
    b) Garcia, E.E., Benjamin, L.E., Fryer, R.I.: Chem. Commun. 78 (1973). (The structure of oroidin has been confirmed by X-ray analysis: J. Clardy, personal communication)
91. Unpublished research from this laboratory
92. Sharma, G.M., Burkholder, P.R.: Chem. Commun. 151 (1971)
93. Van Lear, G.E., Morton, G.O., Fulmor, W.: Tetrahedron Lett. 299 (1973)
94. Raverty, W.D., Thomson, R.H., King, T.J.: J. Chem. Soc. Perkin *I*, 1204 (1977)
95. Andersen, R.J.: Tetrahedron Lett. 2541 (1978)
96. Schmitz, F.J., Gopichand, Y.: Tetrahedron Lett. 3637 (1978)
97. Cimino, G., De Stefano, S.: Tetrahedron Lett. 1325 (1977)
98. Burks, J.E. et al.: Acta Cryst. *33*, 704 (1977)
99. Wratten, S.J. et al.: J. Amer. Chem. Soc. *99*, 2824 (1977)

100. Wratten, S.J. et al.: Tetrahedron Lett. 1559 (1977)
101. Faulkner, D.J., Ireland, C. in: Marine Natural Products Chemistry (eds. D.J. Faulkner and W. Fenical) Plenum, New York 1977, p. 23
102. Stallard, M.O., Faulkner, D.J.: Comp. Biochem. Physiol. *49B*, 25 (1974)
103. Yamamura, S., Hirata, Y.: Tetrahedron *19*, 1485 (1963)
104. Matsuda, H. et al.: Chem. Commun. 898 (1967)
105. Pettit, G.R. et al.: J. Org. Chem. *43*, 4685 (1978)
106. Pettit, G.R. et al.: J. Amer. Chem. Soc. *99*, 262 (1977)
107. McDonald, F.J. et al.: J. Org. Chem. *40*, 665 (1975)
108. Kinnel, R. et al.: Tetrahedron Lett. 3913 (1977)
109. Kato, Y., Scheuer, P.J.: Pure Appl. Chem. *41*, 1 (1975)
110. Rose, A.F. et al.: J. Amer. Chem. Soc. *100*, 7665 (1978)
111. a) Hashimoto, Y. et al.: Bull. Jap. Soc. Sci. Fish. *33*, 661 (1967)
b) Kosuge, T. et al.: Tetrahedron Lett. 2545 (1972)
112. Personal communication from Y. Shimizu, who has translated Japanese reports
113. Baker, J.T.: Pure Appl. Chem. *48*, 35 (1976) (review)
114. Ashworth, R.B., Cormier, M.J.: Science *155*, 1588 (1967)
115. Higa, T.: Amer. Chem. Soc. Meeting, Honolulu 1979
116. Sheikh, Y.M., Djerassi, C.: Experientia *31*, 265 (1975)
117. Higa, T., Scheuer, P.J.: Tetrahedron *31*, 2379 (1975)
118. Higa, T., Scheuer, P.J. in: NATO Conf. Marine Nat. Prod. (eds. D.J. Faulkner and W. H. Fenical). Plenum, New York 1977, p. 35
119. Higa, T., Scheuer, P.J.: Naturwissenschaften *62*, 395 (1975)
120. Siuda, J.F., DeBernardis, J.F.: Lloydia *36*, 107 (1973)
121. Marius, J.S.C.: Onderstepoort J. Vet. Sci. *20*, 67 (1944)
122. Ward, P.F.V., Hall, R.J., Peters, R.A.: Nature *201*, 611 (1964)
123. Thieler, R. et al.: Science *202*, 1094 (1978)
124. Wolinsky, L.E., Faulkner, D.J.: J. Org. Chem. *41*, 597 (1976)
125. Faulkner, D.J., Stallard, M.O.: Tetrahedron Lett. 1171 (1973)
126. Sims, J.J., Rose, A.F., Izac, R.R. in: Marine Natural Products, Chemical and Biological Perspectives (ed. P.J. Scheuer) Vol. 2, Academic Press, New York 1978, p. 297
127. Lovelock, J.E.: Nature *256*, 193 (1975)
128. Fenical, W.H.: Personal communication
129. Crews, P. et al.: Phytochemistry *15*, 1707 (1976)
130. Unpublished observations – this laboratory
131. a) Ohta, K., Takagi, M.: Phytochemistry *16*, 1062 (1977)
b) Rinehart, Jr. et al. in: Dahlem Conf. Rep. Nature of Sea Water, (ed. E.D. Goldberg) Abakon Verlags-Ges., Berlin 1975, p. 651
132. Stallard, M.O., Faulkner, D.J.: Comp. Biochem. Physiol. *49B*, 37 (1974)

Subject Index

GPSR Compliance

The European Union's (EU) General Product Safety Regulation (GPSR) is a set of rules that requires consumer products to be safe and our obligations to ensure this.

If you have any concerns about our products, you can contact us on ProductSafety@springernature.com

In case Publisher is established outside the EU, the EU authorized representative is:

Springer Nature Customer Service Center GmbH
Europaplatz 3
69115 Heidelberg, Germany

Batch number: 10838062

Printed by Printforce, the Netherlands